Springer Series in **Materials Science** 13

Edited by K. Alex Müller

Springer Series in **Materials Science**

Volume 1 **Chemical Processing with Lasers**
By D. Bäuerle

Volume 2 **Laser-Beam Interactions with Materials**
Physical Principles and Applications
By M. von Allmen

Volume 3 **Laser Processing of Thin Films and Microstructures**
Oxidation, Deposition and Etching of Insulators
By I. W. Boyd

Volume 4 **Microclusters**
Editors: S. Sugano, Y. Nishina, and S. Ohnishi

Volume 5 **Graphite Fibers and Filaments**
By M. S. Dresselhaus, G. Dresselhaus,
K. Sugihara, I. L. Spain, and H. A. Goldberg

Volume 6 **Elemental and Molecular Clusters**
Editors: G. Benedek, T. P. Martin, and G. Pacchioni

Volume 7 **Molecular Beam Epitaxy** Fundamentals and Current Status
By M. A. Herman and H. Sitter

Volume 8 **Physical Chemistry of, in and on Silicon**
By G. F. Cerofolini and L. Meda

Volume 9 **Tritium and Helium-3 in Metals**
By R. Lässer

Volume 10 **Computer Simulation of Ion – Solid Interactions**
By W. Eckstein

Volume 11 **Mechanisms of High Temperature Superconductivity**
Editors: H. Kamimura and A. Oshiyama

Volume 12 **Laser Technology in Microelectronics**
Editors: S. Metev and V. P. Veiko

Volume 13 **Semiconductor Silicon**
Materials Science and Technology
Editors: G. Harbeke and M. J. Schulz

G. Harbeke M. J. Schulz (Eds.)

Semiconductor Silicon

Materials Science and Technology

Proceedings of the Summer School
Erice, Trapani, Sicily, July 3–15, 1988

With 260 Figures

Springer-Verlag Berlin Heidelberg New York
London Paris Tokyo Hong Kong

Prof. Günther Harbeke
Paul-Scherrer-Institut, c/o Laboratories RCA Ltd., Badener Str. 569
CH-8048 Zürich, Switzerland

Prof. Max J. Schulz
Institut für Angewandte Physik, Universität Erlangen-Nürnberg, Glückstr. 9
D-8520 Erlangen, Fed. Rep. of Germany

ISBN-13: 978-3-642-74725-0 e-ISBN-13: 978-3-642-74723-6
DOI: 10.1007/978-3-642-74723-6

Softcover reprint of the hardcover 1st edition 1989

2154/3150-543210 – Printed on acid-free paper

Preface

The summer school on "Silicon: Materials Science and Technology" was held at the Ettore Majorana Centre for Scientific Culture in Erice, Sicily. It was the 16th course of a series in the International School of Materials Science and Technology. The course attracted participants from all over the world. 30 lectures were presented by internationally reputed experts in the field.

The proceedings include all the lecture topics of the course. Short notes on the scientific work of the attendees are integrated with the main papers.

The semiconductor material silicon is of paramount importance in microelectronics. The various aspects of the silicon materials science and its technology were critically reviewed and the forseeable future developments of devices and integrated circuits were discussed during the two weeks of the course. The topics ranged from the fundamental physics to the applied science of processing technology and device fabrication covering crystal growth, processing technology, defects, measurement methods, thin films, and device problems. The course brought together theoretical and experimental physicists and engineers from universities, research institutes and industry.

The unique setting of the Ettore Majorana Centre in the ancient town of Erice created a friendly atmosphere for fruitful scientific exchange. Long discussions continued over lunch and only ended late at night in the wine cellar of the school.

The editors wish to thank all the lecturers and participants for their stimulating contributions to the success of the course and to the present book. The organizers are especially grateful to Dr. J.L. Freeouf, IBM Watson Research Center, Yorktown Heights, USA, who stepped in for Prof. C. Calandra at very short notice to give the lectures on silicide films and their interface to silicon.

All the course participants express their appreciation of the fine facilities provided by the Ettore Majorana Centre and are indebted to its director, Prof. A. Zichichi, to Dr. A. Gabriele, Dr. Pinola Savalli, and the staff of the Centre. We are also grateful for generous support, in particular for student grants, given to us by the sponsoring institutions. The advice and recommendations given to us by the director of the school, Prof. M. Balkanski and the international advisory committee were very helpful in setting up this course and in editing the book.

Zurich, Erlangen
December 1988

G. Harbeke
M. Schulz

Directors and Sponsors

Directors of the Course:
G. Harbeke, PSI c/o RCA Laboratories, Zurich, CH
M. Schulz, University of Erlangen-Nürnberg, Erlangen, FRG

Director of the School:
M. Balkanski, University Pierre et Marie Curie, Paris, F

Director of the Ettore Majorana Centre for Scientific Culture:
A. Zichichi, University of Bologna, I

The conference was sponsored by:
European Physical Society (EPS)
Italian Ministry of Education
Italian Ministry of Scientific and Technological Research
Sicilian Regional Government

The conference was supported by:
AT&T Italia
European Physical Society (EPS)
European Research Office, London
Italian Ministry of Education
Italian Ministry of Scientific and Technological Research
Italian National Group for the Structure of Matter
Sicilian Regional Government
Siemens AG, Munich

Contents

Part IV Characterization Methods

Part V Insulating Films

Part VI Silicide Films

Part VII Devices

Part I

Crystal Growth

Czochralski Growth of Silicon

W. Zulehner

Wacker-Chemitronic GmbH, D-8263 Burghausen, Fed. Rep. of Germany

1. Introduction

Today's electronics is almost exclusively based on devices made from silicon - both in low power and in high power electronics. Low power electronics consumes the biggest portion of semiconductor silicon. Worldwide, about 5000 metric tons of dislocation free silicon single crystals will be produced in the current year 1988. About 80% of this basic material is grown by the Czochralski method, the rest mainly by the float-zone method. From the 5000 tons of single crystal silicon, an enormous number of electronic devices can be manufactured. For example, if one would only prepare devices or chips for pocket calculators, one could produce some 100 billion pocket calculators from this quantity of silicon.
The Czochralski-grown silicon is especially suited for low-power devices, particularly for highly integrated devices which are needed in logic and control electronics.

2. Czochralski Growth of Silicon

2.1 Fundamentals of CZ Si Growth

The basic principle of the pulling of single crystals by the Czochralski method is shown in Fig. 1. A melt is held in a crucible, a seed crystal is first dipped into the melt and is then slowly withdrawn vertically to the melt surface, whereby the liquid Si crystallizes at the seed. This method is named after J. Czochralski [17C] who established it in 1916 to determine the crystallization velocity of metals.

There are two other competing pulling methods for silicon single crystals, which need no crucible. Their principle arrangements are shown in Fig. 2. The first one is the float-zone pulling method, the only serious competitor to Czochralski crystal pulling. The second one is the pedestal pulling method, which is only used for the production of crystals with small diameters, for example "seed" crystals.

For the growth of silicon and germanium single crystals the Czochralski method was first applied in 1950 by Teal, Little and Bühler [50T, 52T]. The silicon single crystals of the early years were not dislocation-free. In 1958, Dash [59D]

Springer Series in Materials Science, Vol. 13
Semiconductor Silicon Editors: G. C. Harbeke and M. J. Schulz

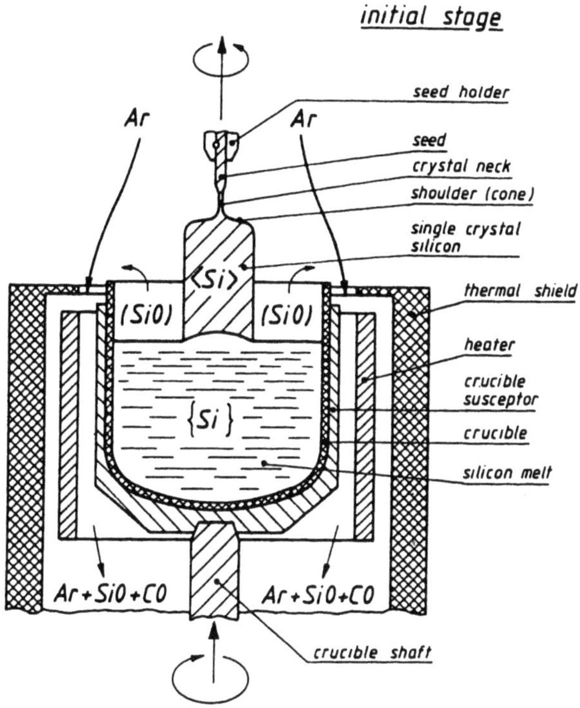

Fig.1: Czochralski single crystal pulling (crucible pulling)

developed a special pulling technique which enabled crystals to be grown dislocation-free. Today, only dislocation-free Si crystals are used in electronics.

2.2 Starting Materials

The silicon single crystals for the production of electronic devices must be of very high purity, particularly with respect to electrically active impurities. For this reason, the polycrystalline Si used for crystal pulling has to be of equal purity. The concentrations of boron and phosphorus, for example, have to be below $5x10^{12}/cm^3$ or 0,1 ppba.

Such pure poly-Si is obtained by distilling trichlorosilane ($SiHCl_3$), or similar silicon compounds, to very high purity and by reacting it with high-purity hydrogen according to the reaction given in Fig. 3. The poly silicon is rod-shaped and has in general a very small grain size. For crucible pulling, the poly silicon rods have to be crushed into chips or sawn into portions and then cleaned. Under clean-room conditions the poly silicon is filled into the crucibles. In most cases the dopant is filled into the crucible together with the poly silicon, either in a diluted form as silicon dopant-alloy or in elemental form. Volatile dopants, for example, antimony, are added later on directly to the melt.

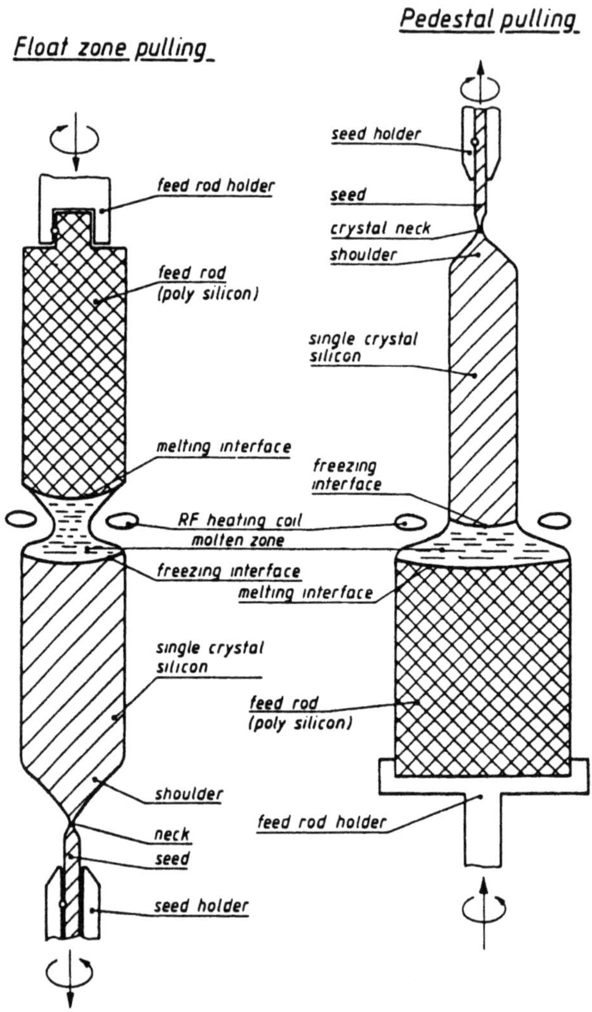

Fig.2: Crucible free pulling of silicon crystals

2.3 Crucible Materials

Molten silicon reacts to a large extent with **every** crucible material. This is due to its high binding forces for the light elements and to the high dissolving power of the melt for most elements. It is thus a question of finding a crucible material which reacts sufficiently slowly with the melt and whose component elements do not deteriorate the silicon crystal with respect to its electronic qualities.

Nearly all elemental crucible materials, especially metals and their alloys, are quickly dissolved by liquid silicon.

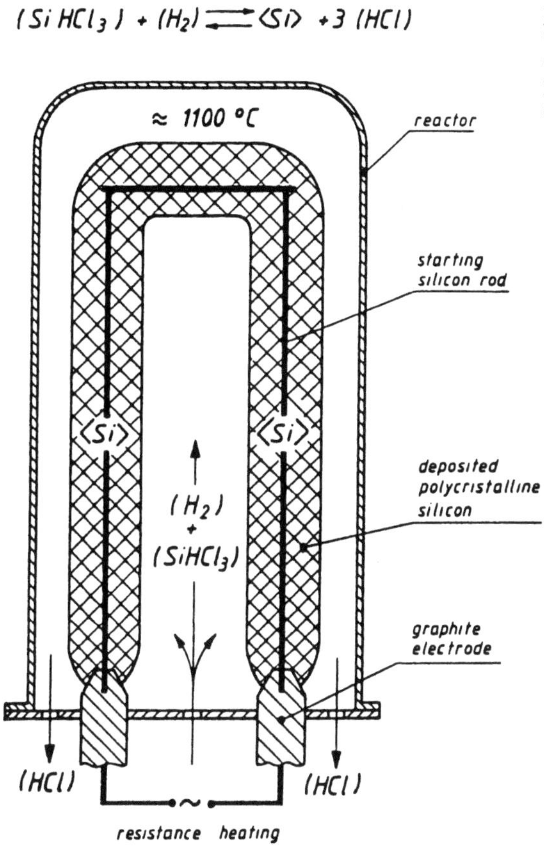

Fig.3: Chemical vapor deposition of electronic grade polycrystalline silicon: $SiHCl_3 + H_2 = Si + 3\ HCl$

Only dense carbon, particularly vitreous carbon, dissolves slowly, so that a silicon melt can be held in such a crucible over a longer period, for example to pull a polycrystalline rod. There are several compound materials - namely some carbides, nitrides and oxides - which also react sufficiently slowly. All these materials are dissolved by liquid silicon with an erosion rate of at least several microns per hour. For example, vitreous silica dissolves with an erosion rate of 7 µm/h near the melting point of silicon in an argon atmosphere of pressure 1 bar. Due to the continuous dissolution, the silicon melt is soon enriched to higher concentrations with the constituents of the crucible material. Therefore, no crucible materials are allowed which contain metals or dopants - for example, titanium carbide (TiC), boron nitride (BN) or aluminium oxide (Al_2O_3). Only the silicon compounds - silicon carbide, nitride or oxide - are considered as potential crucible materials.

Up to now, only vitreous silica has been used in the production of CZ Si crystals. The choice arises, because the product of the reaction between silicon melt and silica crucible - silicon monoxide - evaporates easily from the melt. Therefore, under usual pulling conditions, the melt never reaches the solubility limit for oxygen and no disturbance of crystal growth by oversaturation with oxygen occurs. Under usual pulling conditions less than 1 % of the reacting oxygen arrives in the crystal and more than 99 % of the oxygen evaporates out of the melt in the form of SiO. Vitreous silica is rather soft at temperatures around the melting point of silicon (1420°C). The silica crucible, therefore, is supported by a graphite susceptor as shown in Fig.1.

In contrast to silicon monoxide, with its relatively high vapour pressure of 12 mbar at the melting point of silicon, the vapour pressures of silicon carbide and silicon nitride are very low. Due to this fact, the silicon melt in a crucible of silicon carbide or nitride is unable to reduce its content of C or N by evaporation. Hence, the silicon melt is soon enriched with C or N up to the solubility limit, which is rather low in both cases, see Table 1. If the solubility limit is reached, precipitates of C or N form both in the crystal and in the melt and small carbide and nitride particles develop. At this moment, dislocations are generated in the silicon crystal. They multiply and spread over the whole cross-section of the crystal and backwards into the crystal over a greater distance driven by the cooling strain. The crystal grows, generating further dislocations and later on, it becomes polycrystalline. Such a crystal is useless for electronic applications.

Table 1: C, N and O in Si at the melting point of Si, [74N]

element	C	N	O	C	N	O
solubility in solid Si	3	0,045	31	6	0,09	63
solubility in liquid Si	40	60	22	81	121	44
dimension	$\times 10^{17}/cm^3$			ppma		
equilibrium distribution coefficient	C: k_o = 0,07 ± 0,01 N: k_o = 7 x 10^{-4} O: k_o = 1,4 ± 0,3					

In spite of these difficulties, some people are working to develop crucibles of silicon nitride for silicon crystal pulling. They aim to reduce the oxygen content, to combine the qualities of float-zone and crucible pulled crystals and to make use of some special features of nitrogen in silicon.

2.4 The Pulling Atmosphere

Due to the high vapour pressure of silicon monoxide, melting and pulling in high vacuum is not suitable, because the melt tends to boil by the strong evaporation of silicon monoxide, and this boiling disturbs the crystal pulling. Therefore, pulling below 5 mbar is not usual. At pressures above 5 mbar, no boiling in the melt occurs during pulling, but it can happen when the melt is overheated, e.g. during melting. To avoid condensation of silicon monoxide in the atmosphere above the melt and to prevent supersaturation of the melt with oxygen, the space above the melt has to be purged intensively with argon, see Fig.1. Argon flows of about 1 Nm^3/h, which are 100 m^3/h at 10 mbar, are typical in CZ Si crystal pulling. The usual pulling pressures lie between 5 and 1000 mbar; all pressures in this range are applicable. The pulling technique, furnace construction, the type of doping element and the desired oxygen content determine the pulling pressure to be used.

The second important purpose of the intensive argon flow is to carry away the carbon monoxide (CO) from the hot parts of the furnace - especially from the heater - to the vacuum pump, so that no backstreaming of carbon monoxide to the melt is possible. If carbon monoxide comes into contact with the silicon melt, it is immediately aborbed and the carbon content of the melt increases accordingly. Carbon monoxide is generated by the reduction of silicon monoxide at the hot graphite parts. In addition, the backstreaming of impurities, which evaporate out of the hot graphite parts, must be prevented by the gas flows.

2.5 Dislocation Free Growth

After melting the poly-Si and adjusting the melt temperature for crystal pulling, a seed crystal of proper orientation is dipped into the melt. Even if the seed crystal is dislocation free, at each dipping stage dislocations are generated in the seed by the thermal strain and the surface tension. These dislocations must subsequently be eliminated, otherwise a dislocated and thus useless crystal grows. The change to dislocation free growth is achieved with the so-called "Dash Technique" [59D], which comprises the following steps:

a) Gradual reduction of the crystal diameter to about 3 mm.
b) Simultaneous raising of the pulling rate to several mm/min.

Normally, after a few centimeters, the crystal growth becomes dislocation-free what is manifested by the development of "ridges" on the crystal surface. The dislocation-free state is rather stable and the diameter can be enlarged slowly or rapidly until it has reached the desired value. The region of small diameter is called "crystal neck", the procedure is called "crystal necking" and the transition from the neck to the cylindrical body is called "shoulder". A decrease in diameter is achieved by faster pulling and an increase in diameter by slower pulling compared to growth of constant diameter.

The elimination of dislocations results from four effects:
1) The dislocations grow laterally out of the crystal,
2) the cooling strains are too low for cross slip,
3) the dislocations move too slowly,
4) the axial dislocations mostly become oblique as a consequence of climbing.

Only dislocations, whose axis is parallel to the pulling direction, can grow throughout all the crystal down to its end. In the case of <110> crystals many 60° dislocations - the most frequent slip dislocation type - are lying parallel to the crystal axis. Therefore, <110> crystals are most difficult to grow dislocation free. Once a crystal is dislocation free, it is difficult to generate a first new dislocation in the perfect lattice, it needs a greater disturbance or activation energy. It is possible, therefore, to grow crystals of large diameters of up to 200 mm and more, despite the high cooling strains which exist in such large crystals. This is clearly observed if the dislocation free growth of a large crystal is disturbed, for instance, by a small solid particle which comes into contact with the edge of the crystal-melt interface. When Si solidifies around the particle, dislocations are generated.

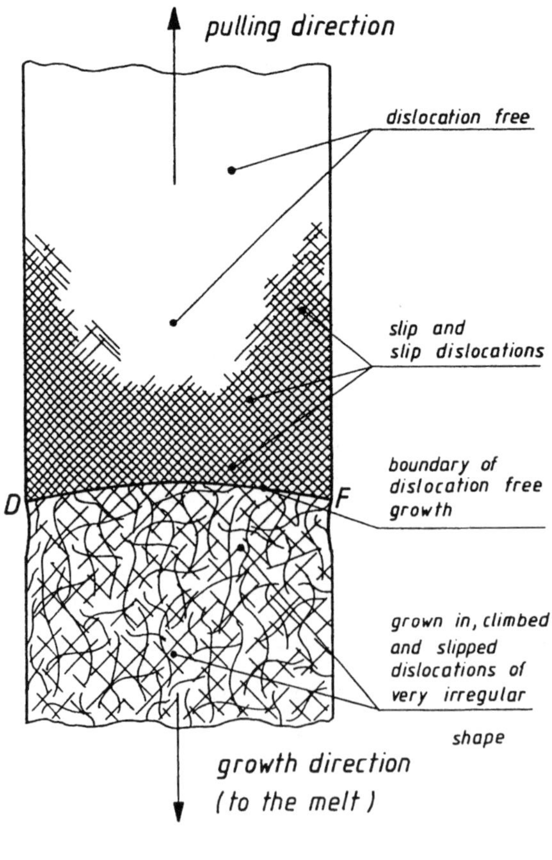

Fig. 4: Transition from dislocation free to dislocated growth.

Driven by the cooling strain, the dislocations move, multiply and spread over the whole cross-section of the formerly dislocation free crystal, as well as backwards into the crystal.

This situation is schematically drawn in Fig. 4, which shows an axial section through a crystal, with a change from dislocation free to dislocated growth. The slightly curved line DF represents the last solid-liquid interface of the formerly dislocation free part of the crystal. Although the crystal has been hotter in its centre, the dislocations do not reach as far upwards in its centre as in its outer region. This can be explained by the faster cooling of the outer parts, which causes shrinkage of the outer parts onto the the inner parts and with this, tensile strain in the outer parts parallel to the crystal surface. The distribution of slip dislocations in the part above line DF has been drawn according to real cases, Fig. 5. Figure 5 shows the slip dislocations in the dislocation free part of a crystal. This crystal was annealed at 1000°C for several hours, to produce layers of oxygen precipitates in the dislocation free part. In the slipped region, the dislo- cations have prevented the formation of these precipitates. There, the oxygen has precipitated at the dislocations.

To complete a crystal dislocation free, a second cone must be grown at the bottom end of the crystal, prior to removing the crystal from the melt. If the crystal diameter is reduced to nearly zero, then the crystal remains completely dislocation

Fig. 5: Axial cross section through a CZ-Si crystal with bended layers of oxygen precipitates and with slip dislocations (lower part)

free after separation from the melt. If the diameter of the solid-liquid interface is not nearly zero at sepa-ration, then dislocations are generated by cooling strains and freezing melt drops at the bottom end of the crystal. They spread backwards into the crystal - as described earlier - over a distance which is about equal to the crystal diameter.

2.6 Heat Flow Conditions

Fig. 6 shows schematically the heat flows for CZ crystal pulling for the initial and the final stage of crystal growth in axial sections through furnace, crucible, melt and crystal. The arrows indicate the approximate direction of the heat flows. The closed lines in the melt indicate the approximate direction of the melt flows. In order to have nearly pure thermal convection (the thermocapillary convection is neglected), the crucible and crystal rotations are taken to be nearly zero:
In the initial stage, the crucible is situated deep in the heater and the heat is transmitted to the melt mainly laterally. The coldest area in the melt is the crystal-melt interface, with the temperature of freezing silicon. At the crucible bottom, the crucible shaft carries away heat to the lower part of the furnace chamber. The hottest points in the melt are situated in the area marked with "M". As a consequence, the melt ascends near the crucible wall and

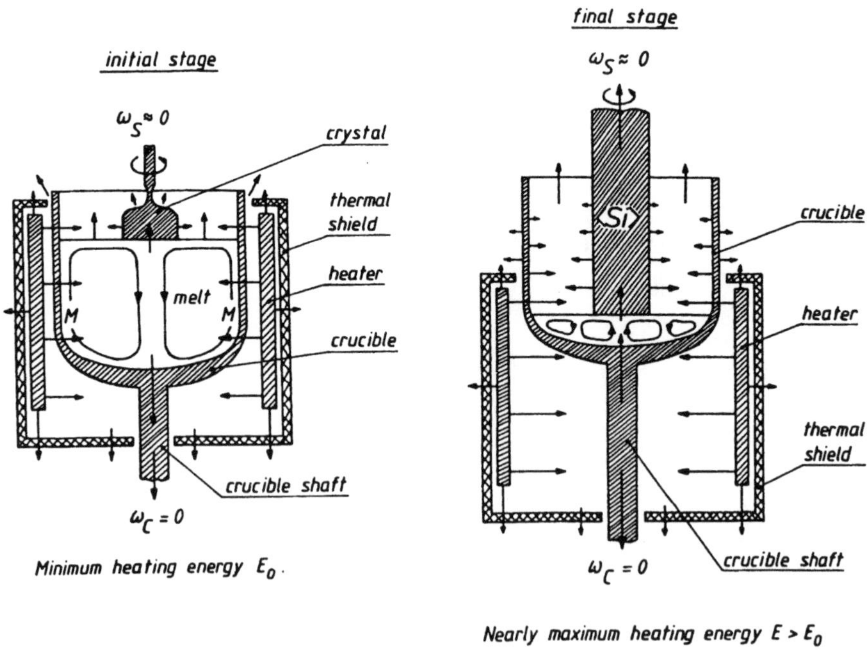

Fig. 6: Heat flow at CZ-crystal growth in the initial and in the final stage of the pulling process

descends in the middle of the crucible. The melt which flows immediately below the surface evaporates SiO to a great extent and, when arriving at the crystal, contains less oxygen than before, leading to a lower oxygen content in the crystal.

The heat flows change gradually during crystal pulling. The crucible moves slowly out of the heater and the thermal shield, whereby the heat emitting parts of the crucible above melt and thermal shield grow and the heat absorbing part inside the shield shrinks. Due to increasing irradiation, the crystal inside the crucible becomes hotter, the pulling speed has to be reduced, and the conditions for precipitation inside the crystal change. To compensate for the loss of energy by radiation, the heater temperature, and hence, the heating power must be raised gradually.

Depending on construction, the crucible shaft may carry heat up to the crucible at the end of pulling, whereas it had carried heat away at the beginning. As a consequence, the melt flow pattern changes strongly in the final stage of the pulling process. Now, the melt flows directly from the crucible bottom to the crystal without evaporation of SiO in between. A higher oxygen content and a better radial dopant distribution in the crystal should result.

2.7 Stirring of the Melt

In Fig.6, crystal and crucible rotation were zero or nearly zero, respectively. In real cases the crystal and the crucible have to be rotated to compensate for temperature

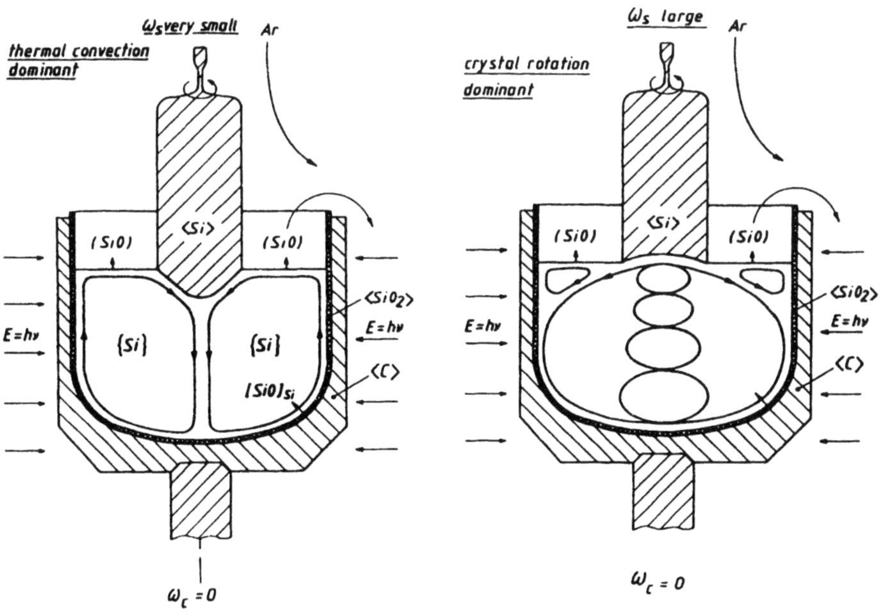

Fig. 7: Melt flow characteristics in CZ-Si melts, [82Z]

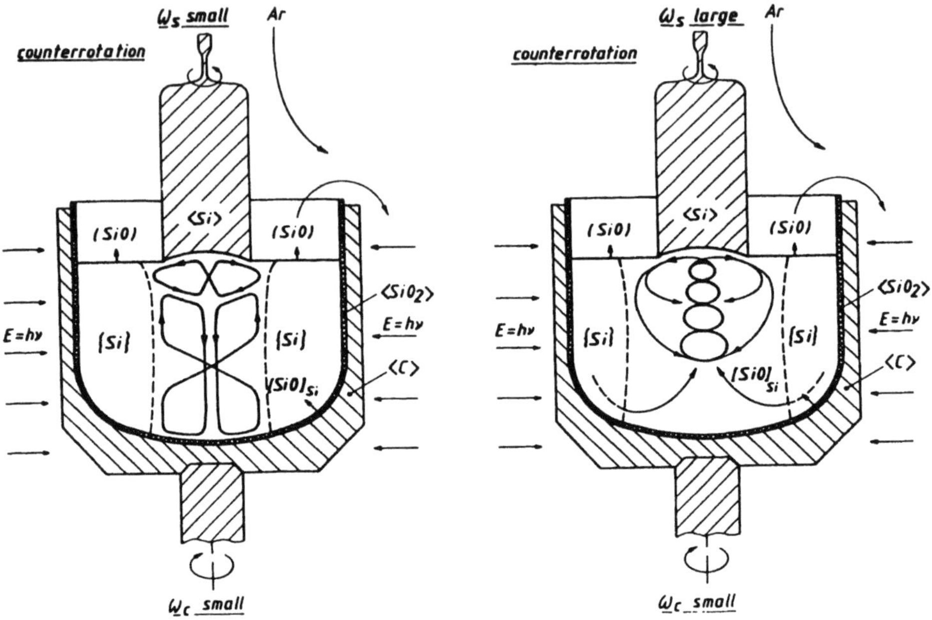

Fig. 8: Melt flow characteristics in CZ-Si melts, counterrotation of crystal and crucible, [82Z]

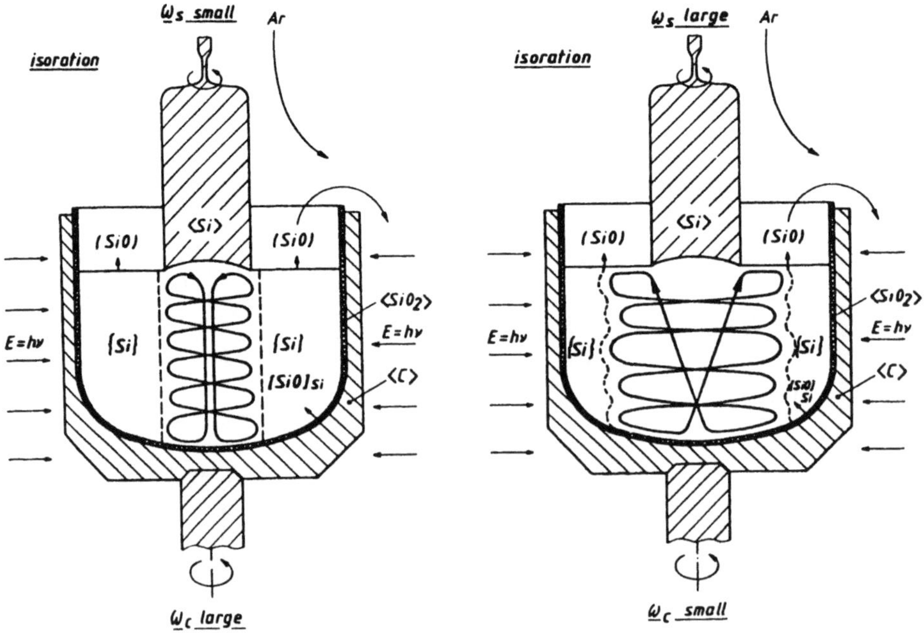

Fig. 9: Melt flow characteristics in CZ-Si melts, isorotation of crystal and crucible, [82Z]

inhomogeneities in furnace and melt, which would prevent uniform crystal growth. The flow characteristics change again with crystal and crucible rotation. In most cases, crystal and crucible are counter rotated and the oxygen content in the crystals is highest at the beginning and lowest at the end. That means, the flow characteristics must be very different from that of Fig.6, where thermal convection was dominant. The possibili- ties to counteract the thermal convection by crystal and crucible rotation depend strongly on the strength of the thermal convection. In the case of weak thermal convection the flow pattern can be totally changed by crystal and/or crucible rotation.

In the Figs. 7 to 9, melt flows according to Carruthers and Nassau [67C, 68C] are shown. In all cases, the thermocapillary convection - the Marangoni convection - is neglected. Nevertheless, they give a good approximation to the reality if the thermal convection is not too strong.

2.8 Oxygen Levels and Distributions

The flow characteristics of a melt is by far the most dominant factor for the resulting oxygen content and its distribution in the crystal. Figure 10 shows the axial oxygen distribution of 11 individual CZ crystals grown from silica crucibles, without application of a magnetic field. All crystals are of

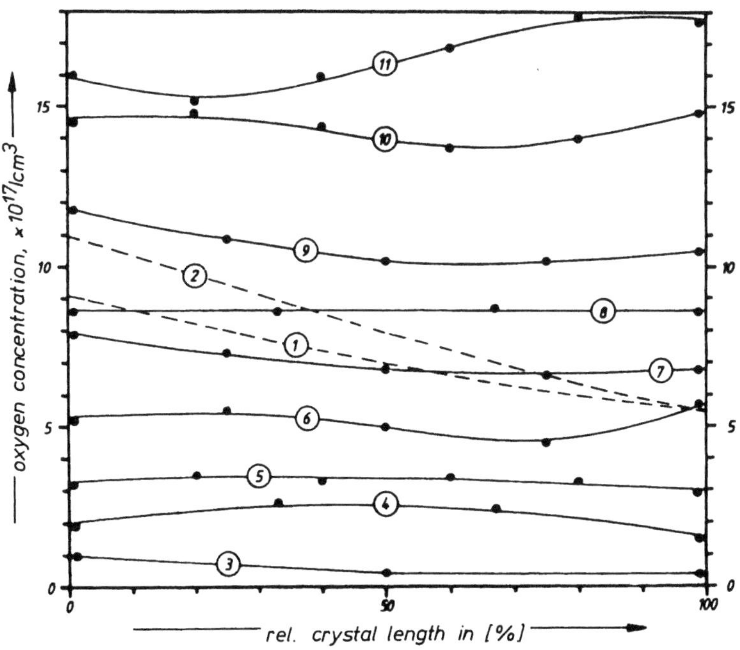

Fig. 10: Axial oxygen distribution in 11 CZ-Si crystals

the same length, diameter (2"Φ), orientation and doping (except crystal 3). Crystals 1 and 2 - given by the broken lines - were grown with strong thermal convection and represent an older pulling technique. For this technique, a steady axial decrease in oxygen content by a factor of about two is typical, beginning with about 11 x $10^{17}/cm^3$ at the seed end and ending with about 5,5 x $10^{17}/cm^3$ at the bottom end of the crystal. The other crystals 3 to 11 were grown with a newer technique which gives a weaker thermal convection. Their pulling differs only in crystal and crucible rotation, except crystal 3, which was pulled with another pressure and throughput of argon and which has another dopant (indium, instead of phosphorous). As can be seen from the curves 4 to 11 of the diagram, it is possible to realize oxygen contents in 2"Φ-CZ crystals which differ by a factor of about 10 only by variation of crystal and crucible rotation. All other pulling and growth parameters were held constant. Another feature of this newer pulling technique is the more uniform axial oxygen distribution.

The radial oxygen distribution in a CZ crystal is also strongly influenced by the melt flow characteristics. Two examples for this are shown in Fig. 11.

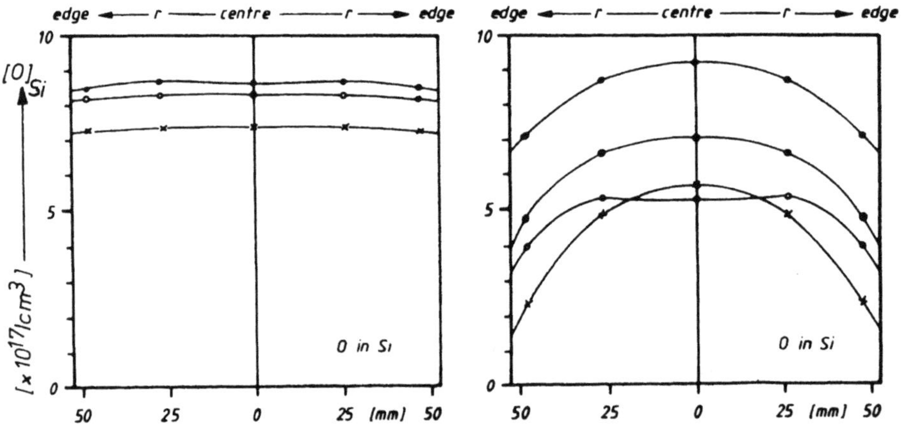

Fig. 11: Radial oxygen distribution in two CZ-Si single crystals at three resp. four different positions of the crystal length. In the crystal of the left diagram the oxygen distribution was more homogeneous both in radial and axial direction whereas in the right the oxygen distribution is very non uniform. In both crystals, the higher oxygen values were at the seed end and the lower values at the bottom end

2.9 Marangoni Convection

The thermocapillary or "Marangoni" convection [1871M] plays an important role in the radial distribution of all impurities. The thermocapillary convection is driven by differences in the surface tension. In the case of CZ growth, differences in surface tension originate from the temperature differences at the surface of the melt, especially near the edge of the crystal. The surface tension drives the surface-near melt from the hotter to the colder points, that means, from the free melt surface towards the crystal edge. This is shown in Fig. 12. The influence of the Marangoni convection on impurity incorporation at crystal growth depends on the strength of mechanical and thermal convection, on crystal diameter and on crystal rotation rate. In very small crystals, the Marangoni convection is mostly dominant. The larger the crystal diameter the smaller is its influence in total. Especially with the crystal rotation, the penetration depth of the Marangoni convection can be reduced.

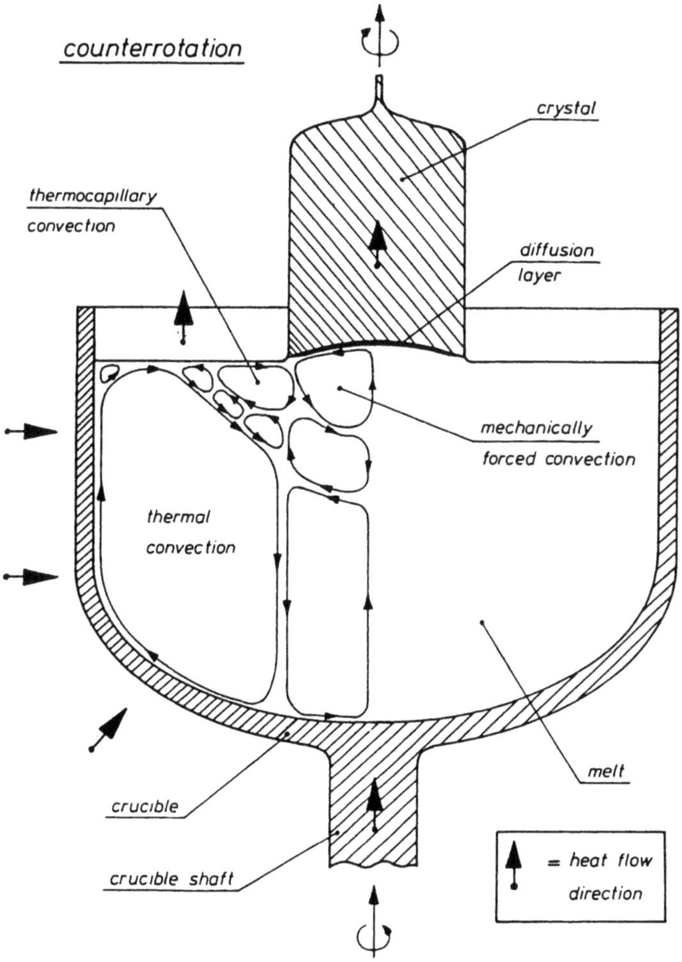

Fig. 12: Melt flow characteristics in a CZ-Si crucible including thermal convection, mechanically forced convection and thermocapillary (= Marangoni) convection, schematically

The Marangoni convection has three effects at CZ silicon growth: 1) It reduces the concentration of volatile elements - for example, of oxygen or antimony - in the melt near the crystal periphery. 2) It reduces the thickness of the diffusion layer in the melt at the solid-liquid interface near the crystal periphery. 3) It transports segregated impurities into the melt region below the crystal axis. All the three effects lead to a lower impurity concentration in the peripheral or to a higher impurity concentration in the axial region of the crystal.

2.10 Segregation and Evaporation

The ratio between the solubility in the solid and in the liquid is called "equilibrium distribution coefficient k_o", it is a constant over a wide concentration range. There is no element which has the same solubility in solid and liquid silicon at the freezing temperature. That means, all elements in silicon have distribution coefficients k_o different from one. The only element in silicon with a distribution coefficient greater than one is oxygen, with $k_o \approx 1.4$. All other elements have distribution coefficients smaller than one. This fact is very beneficial in the case of detrimental impurities like the transition metals, for instance, but it is adverse in the case of doping elements.

For CZ pulling, the law of "normal freezing" is valid, see Fig.13. The axial concentration of a non volatile impurity in a crucible pulled crystal varies according to the formula given in the diagram. In this formula, C_o is the starting impurity concentration in the melt, C_x is the concentration in the crystal at its freezing interface when the melt portion x has frozen and k is the effective distribution coefficient. The effective distribution coefficient has to be applied if the impurity segregation cannot be balanced out by impurity diffusion in the melt, what is given at higher resp. technical freezing velocities. In this case, a layer of higher (for $k_o < 1$) impurity concentration forms in the melt at the solid-liquid interface. By this effect, the crystal is in contact with a higher impurity concentration and it consumes a higher impurity quantity. In Fig.13, the axial distribution of five doping elements is shown. The most beneficial doping element with respect to axial distribution is boron, with an equilibrium distribution coefficient of $k_o = 0.8$. Second best to boron is phosphorus with $k_o = 0.35$. The doping elements Sb, Ga and In have very small distribution coefficients, so that most of the dopant remains in the melt, and only a small amount of the doping element is incorporated into the crystal.

The axial dopant resp. resistivity distributions shown in Fig. 13 are only valid if there is no evaporation of dopant during crystal pulling. This condition is fulfilled to a good approximation for B, Al and Ga. Considerable deviations are found with the other dopants and extremely strong deviations with Sb. Due to its high vapour pressure, Sb evaporates violently out of the silicon melt at low pulling pressures. The evaporation of Sb is so strong, that the enrichment by segregation can easily be compensated or overcompensated.

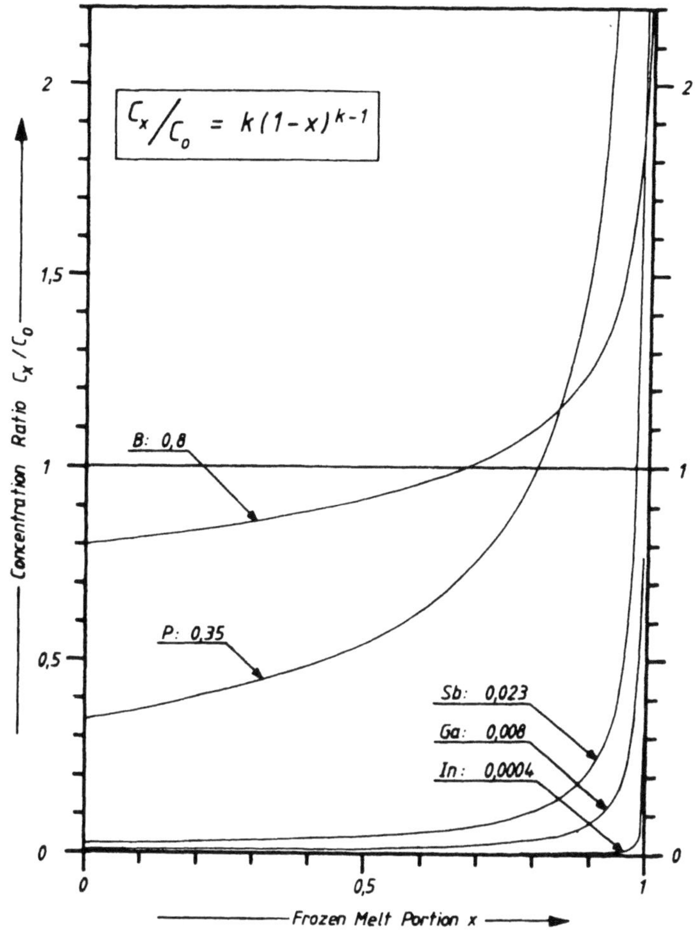

Fig. 13: Axial concentration variation with normal freezing for different distribution coefficients

2.11 Radial Impurity Distribution

Caused by segregation and unfavourable convective flows, especially the Marangoni flows, the radial impurity distribution is in no case uniform. The radial impurity variations are more or less strong, sometimes very strong. As shown in Fig.12 for the case of oxygen, the convective flows can exert a strong influence on the radial impurity distribution. This is also valid for dopants. Figure 14 shows the radial resistivity variation of two comparable crystals at the same position in the crystal. The pulling of the crystals differed only in the crystal rotation rate. As can be seen from the diagram, the crystal rotation rate influences strongly the radial dopant distribution.

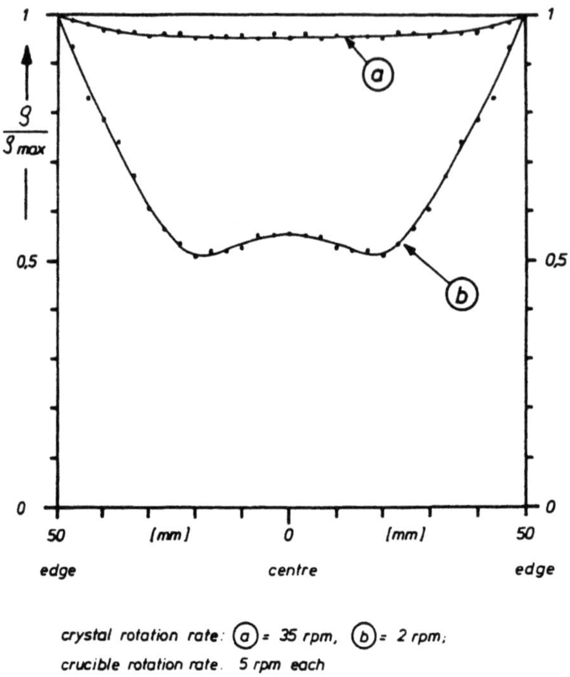

Fig. 14: Relative radial resistivity variation in two differently pulled CZ-Si crystals, <100>, 100 mm Φ, phosphorus doped

Next to the convective flows, the distribution coefficient and the crystal orientation influence the radial resistivity variation. This is shown in Fig. 15. The three crystals of Fig.16 are grown under the same conditions. They differ only in the dopant or in the crystal rotation. The upper diagram shows the difference between B and P or the distribution coefficients 0.8 and 0.35, resp.. The smaller distribution coefficient clearly generates the larger resistivity variation. The lower diagram shows the difference between the orientations <100> and <111>. With respect to radial resistivity variation, the <111> orientation is the most unfavourable one. Its resistivity variation differs by a factor of 2 from that of the <100> orientation.

2.12 Striations

The convective flows, the distribution coefficients and the crystal orientation do not only influence the macroscopical impurity distribution, they also have a strong influence on the microscopic impurity distribution - mostly called impurity striations. Due to temperature fluctuations in the melt, the crystal does not grow with a constant freezing velocity. On

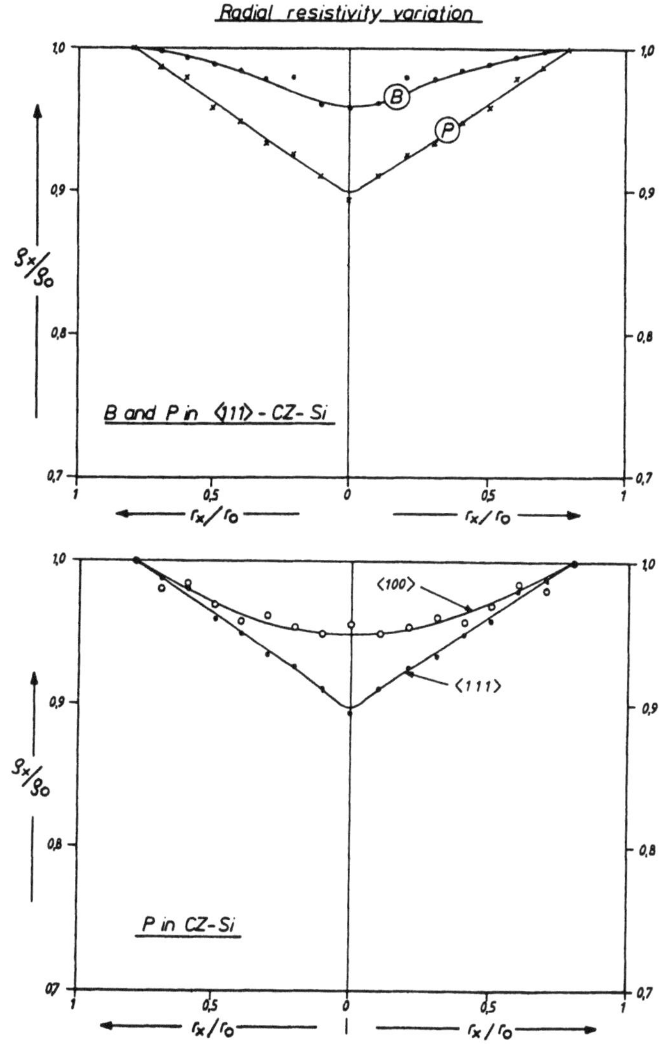

Fig. 15: Radial (relative) resistivity variation in CZ-Si crystals for different dopants and orientations

the other hand, the effective distribution coefficient k_{eff} depends on the growth velocity: It is the higher the higher the growth velocity is and becomes one at extremely high growth velocities. Therefore, the impurity concentration varies in the rhythm of the freezing velocity. As a consequence, the axial section through the crystal shows layers of varying impurity concentration. The thickness of the layers and the differences in impurity concentration depend on the growth conditions. The layer thickness can vary from 5 to 500 µm and the impurity concentration up to +/- 50 %, whereby the crystal rotation rate exerts the strongest influence. Figure 16 shows an example for a good and for a bad microscopic impurity distribution, measured at two phoshorus-doped CZ silicon crystals with the spreading resistance method. The pulling of the two crystals differs mainly in the crystal rotation rate, which was higher in the case b.

A comparable difference in striation strength can be observed between phosphorus (strong striations) and boron (weak striations). In the case of oxygen, the striations result partly from the varying oxygen content of the melt near the freezing interface. The melt flows are temporally not constant and have different oxygen concentrations. From this, an additional striation-generating effect is given.

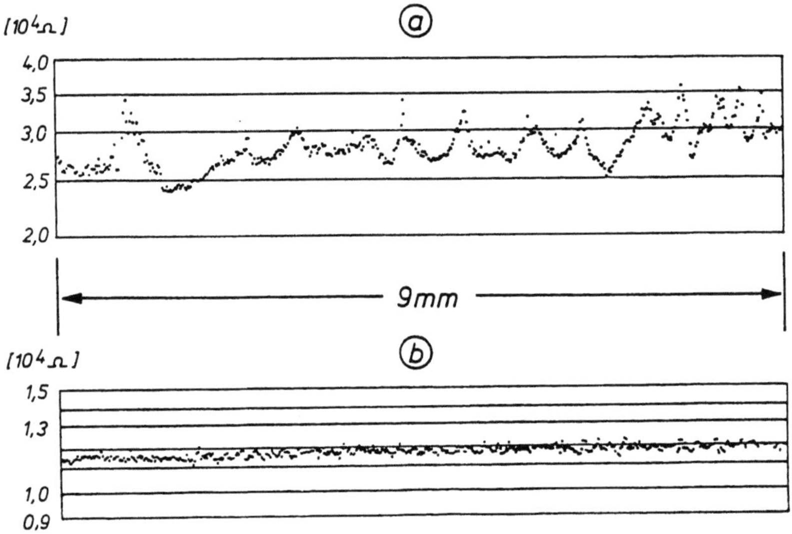

Fig. 16: Microscopic phosphorus distribution in CZ-Si, <100>, 100 mm Φ, ca. 5 Ωcm; spreading resistance measurement

2.13 Purity of CZ Silicon

There are several sources of impurities for CZ silicon: The chemically purified polycrystalline silicon, the handling of the poly-Si, the crucible material and the pulling furnace and atmosphere. The chemically purified poly-Si is of high purity, as is shown in Table 2. The table does not comprise the values for carbon and oxygen, which are about 100 ppba each. These two elements are the main impurities both in poly-Si and in CZ and FZ silicon. Carbon and oxygen are electrically neutral and not detrimental in this concentration range. The transition metals Fe, Ni and Cu with contents below 1 pp̧ba are more feared than carbon and oxygen. Except carbon and oxygen, the total impurity content of todays poly-Si is below 1 ppba. Table 2 also shows a typical analysis of CZ silicon. Despite the supply of impurities from the silica

Table 2: Repeated Neutron Activation Analysis (NAA) of CVD-polysilicon and CZ-monosilicon; concentrations in ppba

impurity	CVD-silicon	CZ-silicon
Ag	0,01	< 0,008
As	0,01	0,015
Au	10^{-5}	< 10^{-5}
Cd	< 0,006	0,008
Co	0,002	< 0,001
Cr	< 0,02	< 0,02
Cu	0,1	≤ 0,1
Fe	< 0,5	< 0,5
K	< 0,2	< 0,2
Mn	< 0,2	< 0,2
Na	< 0,3	< 0,3
Ni	0,4	< 0,2
Pt	< 0,01	< 0,01
Sb	0,0008	0,01
Ti	< 0,5	-
Zn	0,1	< 0,05
Zr	< 0,5	< 0,5

crucible, which is dissolved slowly by the melt, the impurity concentrations of the CZ silicon are mostly lower than that of the polysilicon. This results from the very small distribution coefficients of most of the elements in Si.

2.14 Design of the Pulling Apparatus

The basic design of a modern CZ crystal puller is shown in Fig. 17. A silica crucible, supported by a graphite susceptor, is surrounded by a cylindrical graphite heater and by a thermal insulation made of graphite, carbon felt and foam and of molybdenum. The machine is built as a vacuum system with water-cooled furnace walls of stainless steel, with vacuum flanges, mechanical pumps, vacuum meters, gas distribution systems and flow meters. Crystal and crucible can be rotated and lifted precisely. In most cases an optical system monitors the crystal diameter for automatic pulling of the crystal. A front opening door and an isolation valve allow removal of the crystal from the pulling machine with a hot furnace. This is especially necessary if recharging of the hot crucible with new poly-Si is required. A schematic drawing of such apparatus is shown in Fig. 18. In this case a poly-Si rod is dipped into the residual melt of the hot crucible and melted after the previous single crystal has been removed from the front opening chamber. In this apparatus the crystal is pulled by a chain or a rope instead of a shaft. Chain or rope pulling allows a cheaper and lower machine construction, but is more difficult in crystal pulling, because of swinging of the crystal.

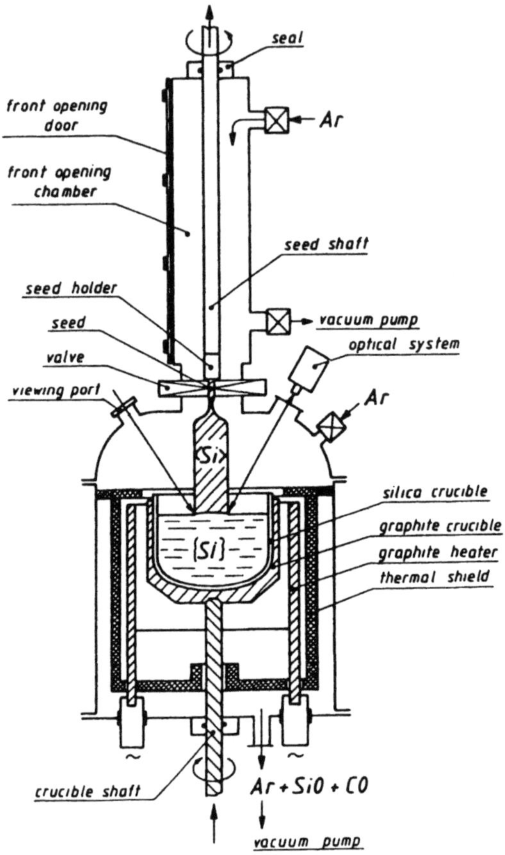

Fig. 17

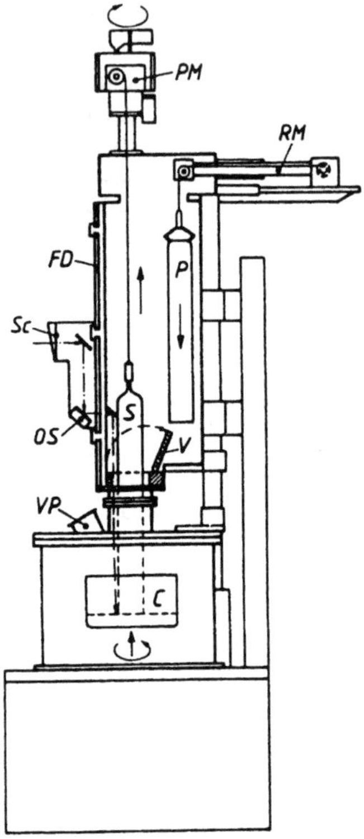

Fig. 18

Fig. 17: Schematic drawing of a CZ-Si crystal pulling apparatus with resistance heating and optical diameter control

Fig. 18: CZ crystal puller with recharging equipment for a silicon rod. The crystal is pulled by a rope instead by a shaft. S = single crystal, C = crucible, V = separating valve between furnace and front opening door, Sc = screen, OS = optical system for automatic diameter control, VP = view port, PM = pulling mechanism, P = poly-silicon rod, RM = recharging mechanism

The dimensions of the current commercial CZ Si crystals are: Diameters from 50 to 200 mm, lengths from 500 to 2200 mm and weights from 10 to 80 kg.

A schematic of a quasi-continuous recharging system is shown in Fig. 19, [77L]: In a separate chamber, poly silicon is melted and transmitted by a heated overflow pipe to the pulling crucible. The overflow is balanced by the gas pressure in the melting chamber.

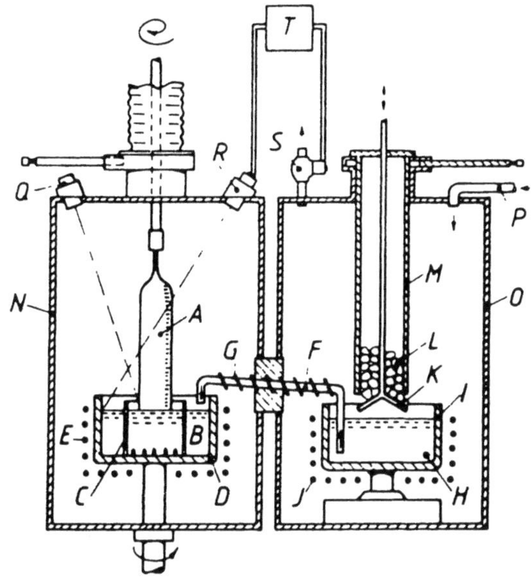

Fig. 19: A schematic of an automatic recharging of the crucible with melt during crystal pulling: A = crystal, B = melt, C = separating wall with holes, D = crucible, E = heating element, F = overflow tube, G = tube heating, H = reserve melt, I = crucible, J = heating element, K = store plug, L = polycrystalline silicon, M = storage bin, N = crystal pulling chamber, O = supply chamber, P = inert gas supply, Q = pyrometer for crystal diameter control, R = pyrometer for melt level control, S = drain valve, T = overflow control

References:

[1871M] Marangoni,C.: Poggendorf's Annal. Bd.CXLIII(1871) 337.

[17C] Czochralski, J.: Z.phys.Chemie 92 (1917) 219.

[50T] Teal, G.K., Little, J.B.: Phys.Rev. 78 (1950) 647.

[52T] Teal, G.K., Bühler, E.: Phys.Rev. 87 (1952) 190.

[59D] Dash, W.C.: J.Appl.Phys. 30 (1959) 459.

[67C] Carruthers, J.R.: J.Electrochem.Soc. 114 (1967) 959.

[68C] Carruthers, J.R. et al.: J.Appl.Phys. 39 (1968) 5205.

[74N] Nozaki, T. et al.: J.Radioanal.Chem. 19 (1974) 109.

[77L] Lorenzini, R.E. et al.: US-Patent 4,036,595 of 1977.

[82Z] Zulehner, W., Huber, D.: Crystals, Vol.8, ed. J. Grab=maier, Springer-Verlag, Berlin, Heidelberg, 1982, p.1.

Chemical and Physical Considerations in the Chemical Vapor Deposition Process

B.S. Meyerson

IBM Research Division, T.J. Watson Research Center,
Yorktown Heights, NY 10598, USA

1) Introduction

Chemical Vapor Deposition(CVD) is the premier method currently employed for the fabrication of silicon devices. CVD is a process wherein a gaseous source is decomposed, in order to form a film upon a substrate. There are numerous methods by which one may impart the energy required to decompose the source, but the most commonly employed method is pyrolysis at elevated temperatures. The variety of layers that may be deposited by this method is limited almost solely by the availability of appropriate gaseous precursors, and an exhaustive review of this field is thus well beyond what can be accomplished in this chapter. An extensive review of fundamental literature in this field has been published recently by Jasinski et. al.[1]. For the purpose of a brief survey of CVD processing, I will concentrate on one material system, silicon, and present an overview of one process employed in depositing that material, silicon epitaxy. The considerations applicable to CVD silicon epitaxy are frequently quite general in nature, and can be extended to the many other CVD processes practiced today. Common to all CVD processes are chemical and physical considerations due to hydrodynamics, gas phase(homogeneous) and surface(heterogeneous) chemistry. It is the relationship between these issues and the quality of the resultant layers that will be the subject of the following text.

2) Silicon Epitaxy

Silicon epitaxy by CVD was first performed employing a simple, atmospheric pressure deposition apparatus, as is shown in figure 1.

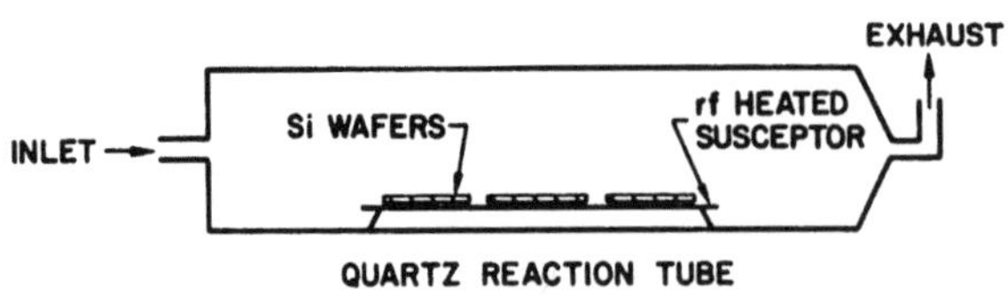

Fig. 1: An atmospheric pressure chemical vapor deposition apparatus

Springer Series in Materials Science, Vol. 13
Semiconductor Silicon Editors: G. C. Harbeke and M. J. Schulz

Atmospheric pressure chemical vapor deposition (APCVD) is principally employed to perform silicon epitaxy at high temperatures, typically in excess of 1100C[2]. A gaseous source, for example a silicon halide of the form $SiH_xCl_{4-x} 0<x<3$, is pyrolyzed upon the heated surface of a silicon substrate, and forms an epitaxial silicon layer. This process is known as homoepitaxy, where the film deposited upon a substrate exactly replicates both the substrates atomic ordering and chemical identity. In order for this to be successful, the initial growth interface must be highly perfect, accomplished in APCVD by the use of extreme processing temperatures. In the limit of high temperature processing, two physical phenomena work in favor of the maintenance of high quality silicon surfaces. The silicon surface is primarily contaminated by carbon and oxygen, resulting respectively in the formation of silicon carbides and oxides. At elevated temperatures, the silicon surface may be stripped of these contaminants, either by their evaporation, or their dissolution into the bulk of the silicon wafer due to their high solubilities. In the instance of silicon dioxide, it is known[3] to decompose to form the volatile species SiO, which then leaves the Si surface. There is no analogous mechanism for the evaporative elimination of carbon, but its reaction at high temperatures with the hydrogen ambient employed as a carrier gas in APCVD is one path for its removal. Alternatively, the silicon surface itself is evaporating at temperatures in excess of 1100C, presumably carrying off contaminating species.

The high temperatures employed in the APCVD epitaxial silicon process, when combined with viscous gas flow at atmospheric pressure, result in a phenomena known as autodoping[4]. This is the unintended incorporation of dopant atoms into the epitaxial layer being deposited, where the substrate itself serves as the source of dopant atoms. Typical of autodoping, the dopant profile has two distinct regions. At the extreme temperatures employed in this process, dopants diffuse by the conventional solid state process from the substrate into the deposited film, so that the dopant profile near the initial growth interface can appear to be the expected "ERFC" form. This is seen as region A in figure 2.

However, a second and distinct phenomena leads to the extended shallow tail of dopant present in region B. This is due to the evaporation of dopant from the growth interface into the gas stream, and its subsequent reincorporation into the growing layer. This phenomena is particularly damaging, in that the redistribution of dopant by this method occurs both in the vertical and lateral direction at the initial growth interface. This causes any preexisting dopant pattern on the substrate to be expanded and shifted anisotropically, as the gas stream causes the dopant layers to shift in the direction of flow. There are two methods by which one may reduce or eliminate this difficulty, those being either a reduction in processing temperatures, or a reduction in process pressure.

When process temperatures are reduced, solid state diffusion rates fall accordingly, greatly reducing that component of autodoping. Of greater benefit, the extent to which dopant is volatized into the gaseous source is also reduced, re-

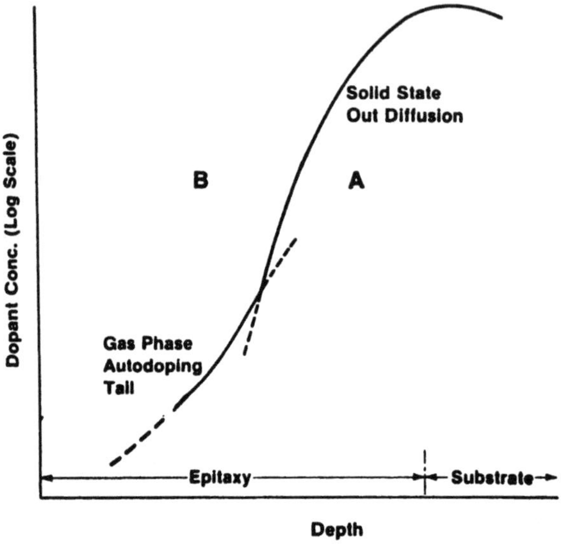

Fig 2: Schematic representation of dopant profile resultant from high temperature epitaxy of silicon on an arsenic doped substrate. After G. R. Srinvasan[4]

sulting in a more rapid falloff of the autodoping tail in region B. Although temperature reduction is desirable, it must be recalled that the high temperatures employed in this method serve to compensate for the presence of impurities which might otherwise contaminate and cause defects in the resultant epilayers. For this reason, the prime method developed for the reduction of autodoping was a reduction in processing pressures, leading to a process known imaginatively as Reduced Pressure Chemical Vapor Deposition(RPCVD)[5]

RPCVD differs little from APCVD, except for the addition of vacuum pumps which enable operation at reduced pressures. Operating at pressures in the range of 10-80 Torr, it is found that autodoping effects are reduced by as much as an order of magnitude over those seen in APCVD systems. Both the magnitude and width of autodoping are effected by this pressure reduction, the mechanism of which stems from the enhanced gas phase diffusivity of evaporated dopant species. This allows them to diffuse rapidly away from the growth surface, and thus be swept from the growth chamber rather than re-incorporated into the film. Apparatus employing this technique are at present the industrial standard for silicon epitaxy.

It was stated earlier that the quality of epitaxial Si material is intimately related to the growth temperatures employed. This results from the requirement that one prepare and maintain an atomically bare silicon surface prior to the commencement of epitaxy. This involves the control of several steps in the sample preparation process. When a Si wafer is received, it is packed in a "dust-free" plastic container, under ambient room atmosphere. One therefore starts with a Si wafer sealed inside a native oxide, which is in turn undoubtedly contaminated by a wide variety of hydrocarbon containing species. This presents a formidable task of sample preparation, and has been the subject of intense study[6].

The first step in sample preparation is a chemical cleaning procedure, a process which continues to evolve. In general, the chemical cleaning of silicon involves the repeated formation and stripping of chemical oxides from the silicon surface, where the chemical oxidant is hydrogen peroxide. A common example of this method is the "RCA" clean. In a version of this method, the Si wafer is first boiled in a mixture of hydrogen peroxide and ammonium fluoride, then in hydrogen peroxide and hydrochloric acid, and finally stripped of the oxide in a buffered HF solution. Hottier and Cadoret[7] conducted an excellent study of the efficacy of a number of variants on this method, taking freshly cleaned wafers and analyzing their residual level of surface contaminants employing Auger Electron Spectroscopy in a UHV apparatus.

Once a "clean" silicon surface is obtained, a wafer must be loaded into the processing chamber, at which time the process is begun. It is at this point where one must exercise extreme care to avoid the recontamination of the prepared silicon surface. Conventionally pumped epitaxy equipment, employing hydrocarbon based pump oils, or silicone rubber O-rings, or any other source of volatile contaminating species, will quickly cover the Si surface with a layer of carbon and oxygen containing adsorbates. Thus, even in the event that one prepares an atomically bare and stable Si surface by chemical means, it is lost at the inception of the epitaxy process, and must be recreated through the use of a high temperature bakeout process. A fine point here is that one need not maintain the high temperatures required for conventional epitaxy once the actual deposition process is underway. If a typical film growth rate is of the order 1 micron per minute, the silicon surface renews itself at the rate of approximately 100 times per second, as new layers of silicon atoms are incorporated. This dynamic surface is orders of magnitude less susceptible to contamination than the initial interface, as the initial Si surface is a single atomic layer which is exposed to the ambient for several minutes. This fact is employed in a variety of "Hi-Low" processes[8,9], where temperatures are greatly reduced once epitaxy commences. This is essentially the state of the art in commercial Si epitaxy, but the field is far from stagnant. A great number of papers appear in the literature each year exploring[10–13] techniques as "Low Temperature Epitaxy"(LTE) processes. I will describe, as an example, a method known as Ultra-High-Vacuum/Chemical Vapor Deposition(UHV/CVD)[14–16], and give details as to the methods it employs to achieve a true LTE process by a systematic approach to the fundamentals of epitaxy.

3) Low Temperature Epitaxy

A method by which one may eliminate problems associated with the redistribution of dopants during epitaxy is a radical lowering of the process temperatures. That this is possible is well established, in that molecular beam epitaxy(MBE)[10] serves as an existence theorem for the field of low temperature epitaxy. However, MBE has some similarities to the "HI-Low" processes just described, in that the Si samples are taken to 900C to desorb contaminants, and the deposition temperatures are then dropped to the 750C range for actual film growth. In order to achieve low temperature epitaxy by CVD, one must satisfy the key requirement, that the Si surface remain atomically bare up to the time epitaxy begins. A systematic approach to this is possible, in that equilibrium data for the maintenance of a bare Si surface in an oxidizing ambient is well known[17] from the literature. As an example, figure 3 presents equilibrium data for the instance of Si exposed to an oxidizing ambient, H_2O.

At either a high enough temperature or a low enough partial pressure of water vapor, the silicon surface remains bare. Figure 4 shows the UHV/CVD apparatus, the vacuum levels of which are designed to rapidly bring the silicon wafer into an environment consistent with the maintenance of a bare surface.

Unlike conventional epitaxy systems, it is configured as a hot walled furnace environment, with wafers stacked coaxially in the growth chamber. This places severe limits upon the reactivity of the gaseous source, in that a highly efficient source would yield highly non-uniform films in such an environment. Furthermore, in this geometry, where the growth source is isothermal with the furnace, significant homogeneous pyrolysis of the gaseous source might be anticipated. To reduce the possibility of this, deposition is performed at total pressures of 1 micron, from a 100% silane(SiH_4) source. In the limit of low pressure, silane's

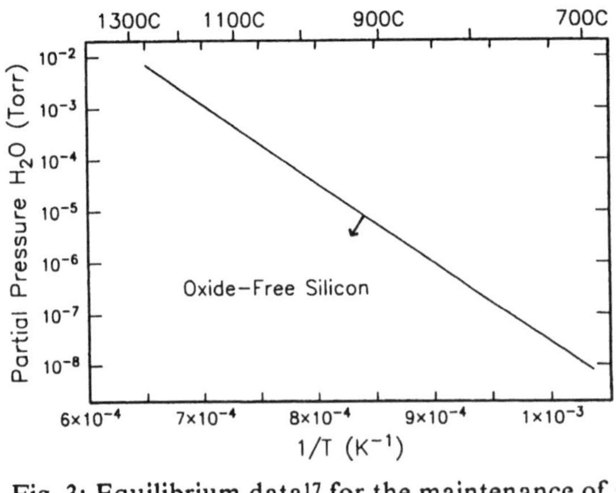

Fig. 3: Equilibrium data[17] for the maintenance of a bare Si surface in the presence of an oxidizing ambient, H_2O

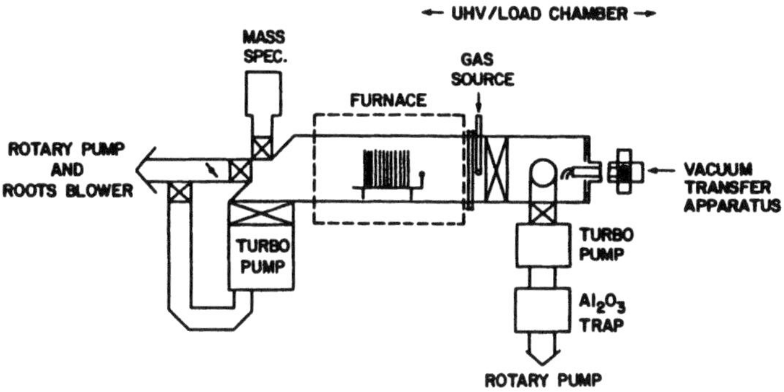

Fig. 4: Schematic of an Ultra-High Vacuum/Chemical Vapor Deposition(UHV/CVD) apparatus

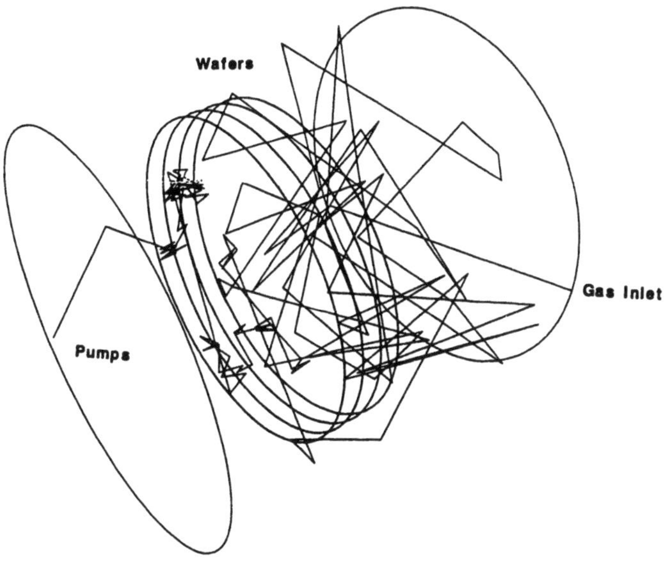

Fig. 5:Single molecule trajectory in UHV/CVD apparatus

rate of homogeneous pyrolysis is actually linear in total pressure, due to the inefficient manner in which molecules can transfer energy by collisions in this low pressure environment. Calculation[18] shows that silane pyrolysis is linear in total pressure below 1 Torr, so a three order of magnitude reduction in gas phase pyrolysis rates at a fixed temperature is achieved by simply operating at 1 micron versus 1 torr total pressure. pressure. This leads to a negligible rate of homogeneous pyrolysis, so that the depletion of the gaseous source due to gas phase chemistry is negligible in the 550-700C temperature range. It is therefore possible to employ this simple, isothermal growth environment, and still avoid uniformity problems associated with excessive gas phase source pyrolysis. Furthermore, at a pressure of 1 micron, one operates in the molecular flow regime, where molecules traverse the reactor primarily through gas-surface collisions. This eliminates completely the complex hydrodynamic considerations of viscous flow, in that recirculating flows, eddy currents, and all manner of turbulence is eliminated. One need only consider questions of the reactivity of the source species with various reactor surfaces in computing the time varying spatial distribution of the source. A preliminary model based upon a monte-carlo type molecular flow calculation is shown in figure 5, where Jensen and Meyerson have employed heterogeneous reaction rate data to model the path taken by a silane molecule in traversing the reaction zone at the furnace center.

After all modeling and optimization is performed, one must in the end produce an epilayer, and analyze its perfection, the topic of the next section.

4) Is it Epi?:

Establishing the quality of an epitaxial silicon layer is by far the most tedious of tasks in evaluating a new epitaxial Si deposition method, and in many cases the evaluation is most subjective. In the case of silicon on sapphire(SOS), where there exists a large lattice mismatch, a defect density of $1x10^6 cm^{-2}$ is an exceptionally good result. Furthermore, describing this result as "successful" epitaxy is very reasonable, given the state of the art. However, in the instance of silicon homoepitaxy, the standard is essentially perfection, in that it is quite common for defect densities of less than $100 cm^{-2}$ to be observed. Successful low temperature epitaxy is frequently reported, but layers can be laced with a near continuous pattern of dislocations, voids, inclusions, and precipitates, making such reports relatively meaningless. The defect density ranges reported for some "successful" methods can run as high as $1x10^{10} cm^{-2}$ [19], or as low as $40 cm^{-2}$ [20]. As long as these defect densities are reported, one may at least read the report with an appropriately critical eye, and evaluate its significance. More commonly, this data is lacking, in that evaluation techniques with minimal sensitivity are employed, so that a qualitative indication of epitaxy, rather than a quantitative one, is obtained. Methods such as Rutherford Back-Scattering(RBS), Low Energy Electron Diffraction(LEED), and X-Ray diffraction, are excellent techniques for monitoring the qualitative approach to epitaxy, but extremely misleading when employed to demonstrate the "quality" of silicon homoepitaxy. This is due to a profound lack of sensitivity in these methods, where they are transparent to defect densities of order one million per square centimeter. An example of this phenomena is given below.

Employing the UHV/CVD system above, a series of epitaxial layers were deposited at temperatures between 495C and 800C. RBS channeling was employed, where the sample is aligned such that a minimum backscattering yield is

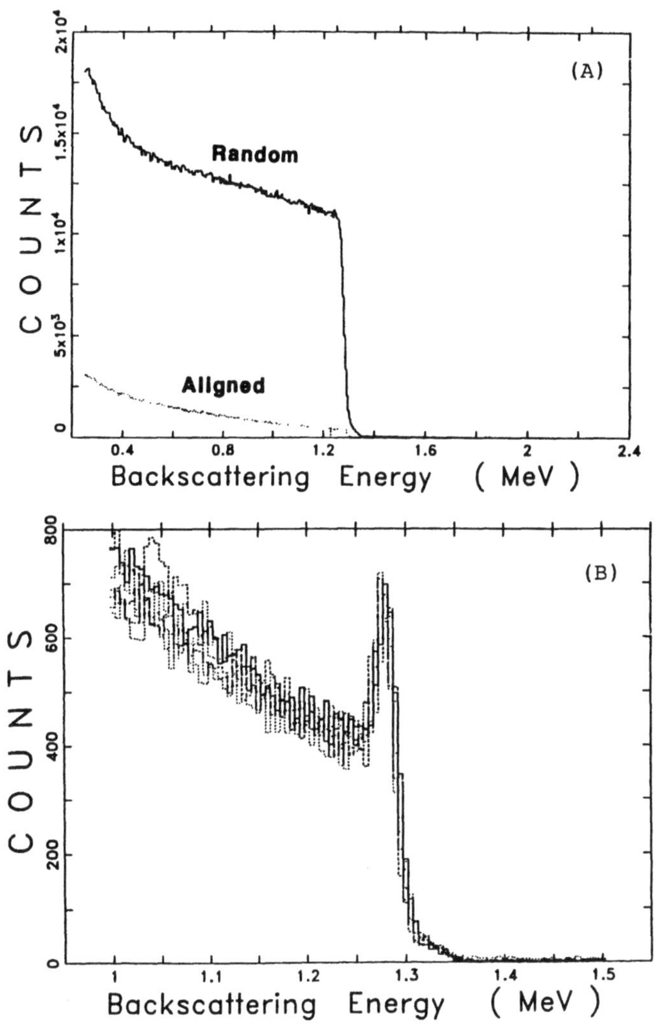

Fig 6A&B.:. RBS spectra from high defective and essentially perfect epilayers

obtained, and compared with the yield from a randomly oriented layer. The lower the yield ratio, aligned:random, the better the epilayer quality. Figure 6A displays the aligned and random spectra from a "perfect" epilayer, where the backscatter yield of the aligned spectrum is roughly 4.0% that of the random spectrum. In figure 6B, I show all the aligned spectra for the above samples plotted together. Although all channeled spectra are identical, the quality of these layers varies from defect free (< 100 cm^{-2} by cross-sectional Transmission Electron Microscopy (TEM) and Secco Etching) to $1x10^7$dislocations – cm^{-2}. This must not be taken as suggesting that RBS obtains the wrong answer, but is indicative of the limits of the techniques sensitivity in this mode of usage. Fortunately, there are numerous methods by which one may conveniently probe the quality of an epitaxial layer, and several are given below.

The oldest and best method for the analysis of epilayer crystallographic perfection is etching, where a preferential etch is employed to decorate defects in thick epilayers, which can then be counted using optical microscopy. This method is inexpensive, accurate, quantitative, and can in fact identify the specific nature of defects by the shapes of the etched features. The only caution when employing this method is that feature sizes will be on the order of the film thickness, so layers in excess of a micron thick are best employed. To examine the initial growth interface, a far greater effort is required, and cross-sectional TEM is the method of choice. Having atomic resolution, one can identify the source of a high defect level, in that a discontinuous oxide layer trapped at the initial growth interface is readily observed by TEM.

In addition to crystallographic defects, epilayers may contain a wide variety of electrically active defects which are not readily observed using the aforementioned methods. However, these defects are easily detected by the fabrication of electrical devices in the epilayer, in that even single electron events can now be detected in devices. When diodes, field effect or bipolar transistors, and/or metal-oxide-semiconductor capacitor structures are fabricated in an epitaxial layer, a wealth of quantitative quality data is available. Carrier lifetimes, interface and bulk Si trap densities, oxide fixed charge densities, reverse bias leakage currents in diodes, diode ideality factors, oxide breakdown voltages, AC and DC transistor characteristics, are all direct indicators of epilayer perfection, and are potentially more sensitive than any other method. Consider the fabrication of diodes of area several million microns square. In high reverse bias, it is possible to detect leakage due to a single stacking fault crossing the depletion layer, making this an effective means of quantifying the density of electrically active defects. This is a sensitive form of spectroscopy, in that the quantity being probed is readily measured with great precision. In general, electrical probes as enumerated above are the preferred methods of epitaxial layer characterization, as they yield the most direct indicator of the potential for any given epitaxial process.

5) Unique Materials:

Ultra High Vacuum/Chemical Vapor Deposition(UHV/CVD) is a highly non-equilibrium deposition method. Films containing in excess of $1 \times 10^{20} B - cm^{-3}$ are readily prepared, where the B remains in solid solution, and is not precipitated. Furthermore, layers whose carrier densities by Hall measurements exceed 1.5×10^{20} are also prepared at 550C. Plotting this data in figure 7, one observes that it exceeds equilibrium predicted active dopant concentrations by in excess of 3 orders of magnitude.

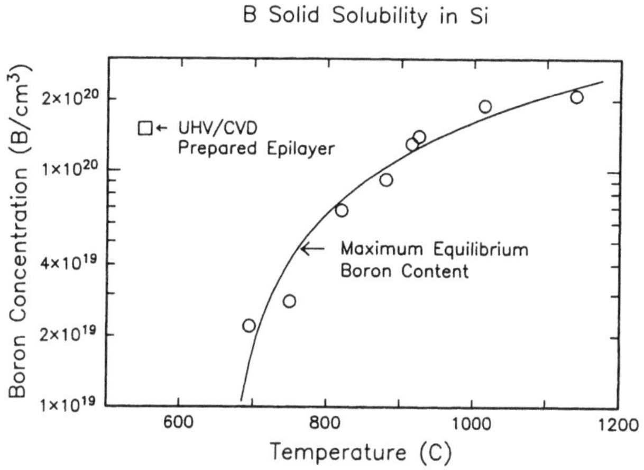

Fig. 7: As deposited carrier densities obtained in 550C B doped epilayers

Such materials are unique, in that they are so far beyond the norm of B content, that their behavior is difficult to predict, and much research is still required to fully characterize these layers. It has been found, however, that these layers remain stable up to temperatures comparable to those required to drive the diffusion of B in Si. If annealed extensively at temperatures in excess of 800C, precipitation takes place, as expected, and the properties of these layers then resemble those of conventional B doped materials. The impact of the existence of such novel materials upon the field of device technology has yet to be fully assessed, but initial results are promising. B doped single crystal emitters have been demonstrated[21], employing such supersaturated layers, in Si:Ge heterojunction bipolar PNP devices, which were found to have switching speeds(FT's) in the range 12-15 GHz. Put in perspective, these devices operate 250% faster than the fastest PNP Si devices yet reported, and demonstrate the promise of such novel materials. The minimum thermal cycle required to fabricate these devices was the primary reason for their success, in that conventional emitter formation requires high temperature excursions which result in defect generation in the Si:Ge layers, as well as broadening of the overall device profile as a consequence of diffusion.

In summary, there has been a radical change in the past two years in the field of epitaxy. Processing temperatures have fallen to a regime where solid state diffusion is no longer an issue, and the primary issues in material work has become the perfection that can be obtained at these reduced temperatures. I have outlined above several methods by which one may sensitively test materials, and have presented the results of such experimentation for the case of one Low Temperature Epitaxy method, Ultra-High-Vacuum/Chemical Vapor Deposition.

References:

1) J. M. Jasinski, B. S. Meyerson, and B. A. Scott, Ann. Rev. of Phys. Chem., 38, 109(1987).

2) W. A. P. Claasen and J. Bloem, J. Electrochem. Soc. 127, 194(1980).

3) G. Ghidini and F. W. Smith, J. Electrochem. Soc.131, 2924(1984).

4) G. R. Srinivasan, J. Electrochem. Soc. 127 1334(1980).

5) M. Ogirima, H. Saida, M. Suzuki, and M. Maki, J. Electrochemical Soc. 124, 903(1977).

6) J. L. Vossen, J. H. Thomas III, J. S. Maa, and J. J. O'Neill, J. Vac. Sci. Tech. A, 2, 212(1984).

7) F. Hottier and R. Cadoret, J. Crystal Growth 61, 245(1983).

8) G. R. Srinivasan, Solid State Technology 24, 101(1981).

9) J. F. Gibbons, C. M. Gronnet, K. E. Williams, Appl. Phys. Lett. 47, 721(1985).

10) R. People, J. C. Bean, D. V. Lang, A. M. Sergent, H. L. Stormer, K. W. Wecht, R. T. Lynch, and K. Baldwin, Appl. Phys. Lett. 45, 1231(1984).

11) J. H. Comfort and R. Reif, Appl. Phys. Lett. 51, 2016(1987).

12) R. A. Zuhr, B. R. Appleton, N. Herbots, B. C. Larson, and S. J. Pennycook, J. Vac. Sci. Tech A, 5, 2135(1987).

13) K. Baert, J. Symons, W. Vandervorst, J. Vanhellemont, M. Caymax, J. Poortmans, J. Nijs, and R. Mertens, Appl. Phys. Lett. 51, 1922(1987).

14) B. S. Meyerson, Appl. Phys. Lett. 48, 797(1986).

15) B. S. Meyerson, E. Ganin, D. A. Smith, and T. N. Nguyen, J. Electrochem. Soc. 133 1232(1986).

16) B. S. Meyerson, F. K. Legoues, T. N. Nguyen, and D. L. Harame, Appl. Phys. Lett. 50, 113(1987).

17) F. W. Smith and G. Ghiddini, J. Electrochem. Soc. 129, 1300(1982).

18) B. S. Meyerson and J. M. Jasinski, J. Appl. Phys. 61, 785(1987).

19) D. Meakin, M. Stobbs, J. Stoemenos, and N. A. Economou, Appl. Phys. Lett. 52, 1053(1988).

20) W. G. Townsend and M. E. Uddin, Sol. St. Elec. 16, 39(1972).

21) D. L. Harame, J.M. C. Stork, G. L. Patton, S. S. Iyer, B. S. Meyerson, G. L. Scilla, E. F. Crabbe, and E. Ganin, Proceedings of IEDM '88., pub. IEEE, (in press).

Silicon Molecular Beam Epitaxy (Si-MBE)

E. Kasper

AEG Research Center Ulm, Sedanstr. 10, D-7900 Ulm, Fed. Rep. of Germany

1. MOTIVATION

Today's microelectronics development is governed by the shrinkage of lateral dimensions. As a consequence the technology is forced also to reduce vertical dimensions, and to decrease process temperatures. Other material systems, like GaAs, are competing with Si, especially for high speed devices and for optoelectronic applications. The electronic functions are confined to a rather thin surface layer on a several hundred micron thick substrate. With increasing tendency this surface layer is fabricated by epitaxy - a single crystal deposition process at temperatures well below the melting point. The substrate serves as stable handling support which gives orientation information to the epitaxy layer grown on it. In a traditional way a uniform epitaxy layer is grown, which serves as a thin substrate to the following device definition processes. The well proven chemical vapour deposition (CVD) process fulfills nearly perfectly the requirements of this simple layer structure as long as the layer thickness is well above the micron.

The introduction of a new method such as silicon molecular beam epitaxy (Si-MBE) is justified if besides better solutions to remaining problems new opportunities for device physics and fabrication will be opened. Si-MBE offers submicron layers, extremely abrupt junctions, complex dopant profiles for complete realization of the vertical device structure, processing at temperatures low enough to totally suppress solid state diffusion of group III/V elements, and heteroepitaxy of Si-semiconductor, Si-metal and Si-insulator structures.

The reader is referred to three books /1-3/ which allow an in depth sight of the main topics treated here. Individual references are only given for very recent results. A good opportunity for more information is given to scientists with the 3rd Int. Symposium on Si-MBE (E-MRS Spring Meeting, May, 29th - June, 2nd, Strasbourg, 1989).

2. METHOD

A rather small community (roughly fifty groups around the world) from universities and industrial research laboratories define the technical status of Si-MBE.

2.1 Principle

MBE as a physical vapour deposition (PVD) process refines the evaporation method by condensation of the molecular beams on heated surfaces within an

Springer Series in Materials Science, Vol. 13
Semiconductor Silicon Editors: G. C. Harbeke and M. J. Schulz

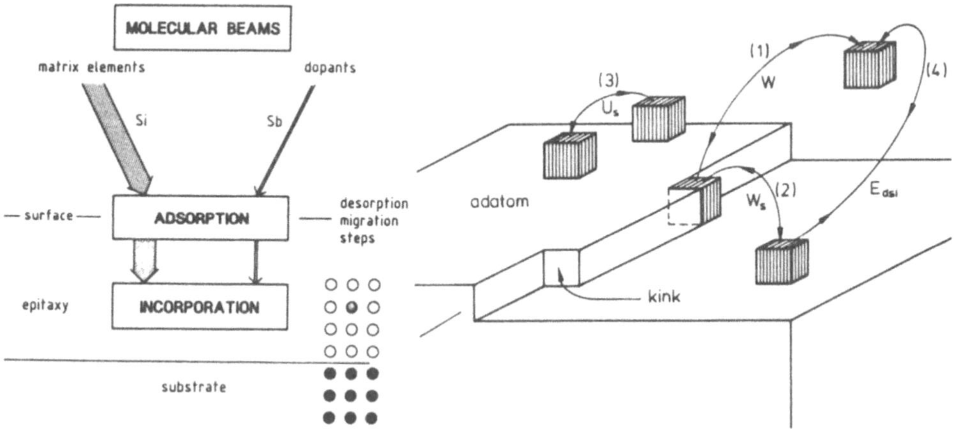

Fig. 1: Molecular Beam Epitaxy (MBE). Adsorption and incorporation of beam atoms (left side). Surface model (right side). Desorption energy E_{dSi} (4), binding energy W (1) and diffusion barrier U_s (3)

ultrahigh vacuum (UHV) environment (Fig. 1). Of essential importance are the cleanliness of the surface and an understanding of the adsorption of atoms on the growing surface. The cleanliness of the surface is determined by an in situ cleaning step before epitaxy and by the clean ultrahigh vacuum surrounding. Atoms from the molecular beam impinging on the surface are bonded to the surface with a desorption energy lower than the binding energy of the crystal. The adatoms diffuse rather easily on the surface creating a twodimensional adatom gas. Adatoms are incorporated at surface steps which move forward by this adatom catching. Vertical growth proceeds by lateral movement of atomic steps.

2.2 Apparatus

Within the ultrahigh vacuum pumped growth chamber are three subsystems (table 1). Matrix and dopant elements are evaporated from the sources

Table 1: Subsystems of an Si-MBE growth chamber

Subsystem	Sources	Wafer Heater	In situ Monitoring
Typ. examples	Effusion cell, E-gun evaporator, low energy ion implanter	graphite meander (heated by electrical current)	Quadrupole mass spectrometer (QMS) Reflection high energy electron diffraction (RHEED)

equipped with beam shutters. Fig. 2 shows an effusion cell used for dopant evaporation. As dopants from effusion cells Sb for n-doping and Ga, B for p-doping are used. Low energy ion implanters use preferably As for n-doping and B for p-doping.

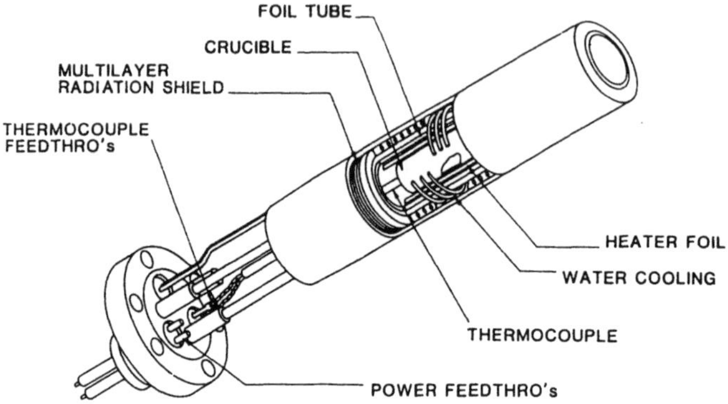

Fig. 2: MBE-equipment: Typical effusion cell design /1,2/

The growth chamber is or can be connected with further UHV-chambers, e. g. load lock, preparation chamber, analyses chamber. Many of the equipments are home made with differing design. Within the last years equipment manufacturers (table 2) improved their apparatus considerably.

Table 2: Si-MBE equipment manufacturers

Anelva	Atomika	Perkin Elmer	Riber	Varian	VG
(Japan)	(FRG)	(USA)	(France)	(USA)	(England)

We want to address the aspects of advanced materials in Si-MBE. Especially, three materials have featured the design of todays Si-MBE equipment. High purity graphite coated by a pyrolytic graphite film is used for construction of ultrahigh vacuum compatible large area heaters which allow efficient contactless radiation heating from the wafer backside (Fig. 3). Pyrolytic boron nitride (PBN) enabled the manufacturing of effusion cells for very pure molecular beams of a variety of materials. The most unique development in this area involved the usage of ultrapure single crystal silicon for construction of evaporators, of shutters, of apertures and heat shields. The purity of molecular beams, and the crystal quality of MBE layers improved considerably by using silicon commonly considered for semiconducting devices, not for constructions.

2.3 Growth kinetics

For modelling the silicon growth, we have to consider the special growth conditions typical for MBE. Especially important for understanding the growth behaviour are the following features:

(i) Growth at temperatures far below the melting point.
(ii) Clean surfaces without adsorption layers from residual and carrier gases, or chemical reaction products apart from the intentional dopant material.

(a)

high purity graphite coated with pyrolithic graphite

(b)

ultrapure single crystalline silicon

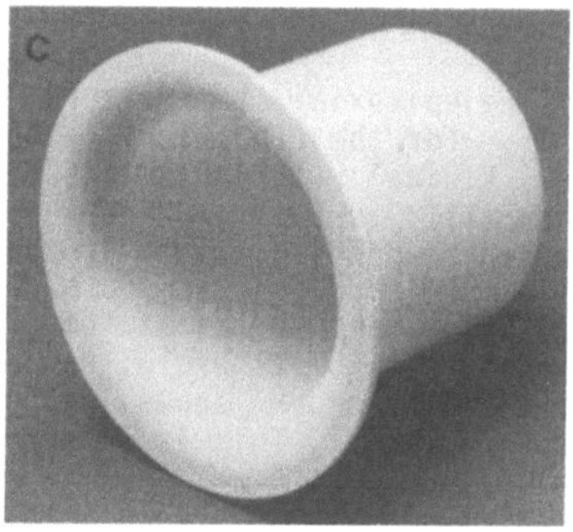

(c)

pyrolithic boron nitride (PBN)

Fig. 3: Advanced materials play an essential role in Si-MBE. As examples are shown: (a) UHV-compatible wafer heater. (b) crucible part for the electron gun evaporator. (c) effusion cell crucible

(iii) Irradiation of the surface by electrons, ions and electromagnetic radiation originating from the electron gun evaporator.

(iv) Complete condensation of the Si-beam. Negligible reflection or desorption of Si-atoms (condensation coefficient $\eta = 1$).

Standard chemical vapour deposition (CVD) of silicon takes place 200 - 350 K below its melting point (1690 K), whereas the MBE process for doped layers is usually performed at far lower temperatures (850 - 1050 K) near half the melting point, but well above the Debye temperature (660 K). At present, no consensus exists about absolute values of the lowest epitaxial temperatures. This is probably because of the variety of growth conditions used by various investigators (cleanliness of environment, irradiation conditions) and their different investigative methods (LEED, RHEED, Rutherford Backscattering). The most rigorous investigation was carried out by de Jong /4/ using LEED for the characterization of the ordering of MBE grown Si layers. The general results of his investigation are as follows:

(i) Epitaxial growth with ordered, reconstructed surfaces is obtained above a certain temperature T_{em} which means the minimum epitaxial temperature. Table 3 lists values of T_{em} found by different authors /4-8/.

Table 3: Epitaxial temperature T_{em} above which ordered bulk-like growth is obtained. Values obtained by different authors /4-8/ on (100), (111) and 4° vicinal (111) surfaces

Surface	T_{em} (K)	de Jong /4/	Shiraki /5/	Gossmann /6/	Gronwald /7/	Herzog /8/
(100)	T_{Epi}	470	440	570	-	510
(111)	T_{Epi}	870	-	790	970	760
vicinal (111)	T_{Epi}	770	-	-	-	-

(ii) The epitaxial temperature, T_{em}, depends on the surface orientations. T_{em} for the (100) surface is much lower than T_{em} for the (111) surface. Vicinal (111) surfaces, tilted a few degrees, exhibit a lower T_{em} than (111) surfaces, which were specified to deviate less than 0.5° from the corresponding crystallographic plane.

(iii) Below the epitaxial temperature, T_{em}, disordered growth occurs indicated by a LEED pattern with enhanced background and gradually decreasing fractional-order spots. With very thin layers some sort of ordering seems to be present. This is in accordance with the earlier observations of Jona /9/.

In our laboratory /8/ the following experiment was performed to investigate the absolute value of epitaxial temperature, T_{em}. The basic idea was to start growth at a temperature high enough to ensure epitaxial ordering. Thereupon the substrate temperature was gradually decreased to room temperature but maintaining a constant Si molecular beam flux of about 1×10^{15} atoms/cm^2s. Examination of the grown layer by electrical profiling and Rutherford backscattering yields the ordering at each depth corresponding to the accompanying substrate temperature (Fig. 4).

The low growth temperature regime of Si-MBE does not overlap with the high temperature growth regime (1200-1500 K) of conventional chemical vapour deposition (CVD) of epitaxial silicon layers. We have performed some MBE experiments with enhanced growth temperature up to 1225 K /10/ to elucidate such general properties as condensation behaviour of silicon epitaxy. The condensation coefficient, η, measures the portion of the incoming Si flux which condenses to create the epitaxial film. With CVD, the growth rate around 1200 K decreases with decreasing temperature /11/. This is partly ascribed to the desorption of Si atoms indicating a slow motion of Si adatoms to capture sites. R. Farrow /12/ deduced from mass spectrometric measurements of silane pyrolysis at 1200 K, a condensation coefficient $\eta = 0.15$ and an activation energy for surface diffusion of 1.6 eV (CVD). However, Si-MBE reveals a condensation coefficient near unity ($\eta = 1$) which indicates mobile adatoms with low activation energy for surface diffusion /13/. The different condensation behaviour of MBE and CVD grown films may be attributed to the adsorption of carrier gas and chemical reaction products on the growing surface.

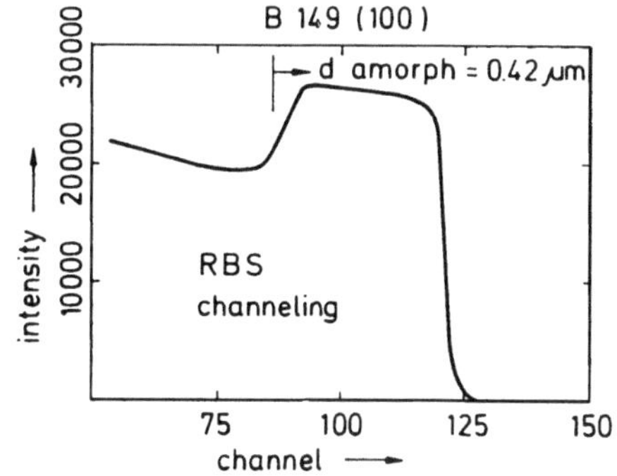

Fig. 4: Growth experiment with varying substrate temperatures. After a thermal anneal at 1175 K, growth was started at 1025 K and decreased to 300 K with increasing thickness of the (100) layer. Rutherford backscattering result for this sample showing the amorphous overlayer grown with temperatures $T < T_{em} = 510$ K

The surface of a solid is in equilibrium with its vapour if the flux of atoms adsorbed from the vapour equals the flux desorbed from the surface. The equilibrium flux density, F_o, of incident atoms is connected to the equilibrium vapour pressure, P_o, by

$$F_o\left[m^{-2}s^{-1}\right] = \sqrt{\frac{N_A}{2\pi MkT}}\; P_o = 8.33\cdot 10^{22}\,\frac{P_o\left[Pa\right]}{(MT)^{1/2}} \qquad (1)$$

where N_A is Avogadro s number, M the molecular weight, k Boltzmann's constant, and T is the temperature.

Table 4 gives numerical values of the equilibrium flux for silicon. This table shows the drastic decrease of desorption events on decreasing the temperature from the upper MBE temperature regime (1200 K) to the lower regime (750 K).

Table 4: Equilibrium flux densities F_o of Si atoms as a function of temperature T. All values are given in SI units

T (K)	723	823	923	1023	1123	1223
P_0 (P_a)	3.6×10^{-20}	2.5×10^{-16}	2.7×10^{-13}	7.1×10^{-11}	6.9×10^{-9}	3.2×10^{-7}
F_0 ($m^{-2}s^{-1}$)	6.6×10^{2}	4.4×10^{6}	4.3×10^{9}	1.1×10^{12}	1×10^{14}	4.6×10^{16}

Etching or growth of the solid proceeds from unsaturated or supersaturated vapours, respectively. A measure of the supersaturation, σ, under MBE conditions is the ratio

$$\sigma = (F_{Si}/F_0)-1 \qquad (2)$$

where F_{Si} is the actual Si flux density, and F_0 the equilibrium flux density calculated from (1).

Table 5 gives some numerical values of the supersaturation for typical MBE growth rates. Often, crystal growers prefer low supersaturation for growth of high quality material. As Table 5 shows, Si-MBE experiments are always performed under extremely high supersaturation, maybe the highest ever reported for high quality growth of single crystals.

Table 5: The supersaturation σ (2) as a function /14/ of the growth temperature T for a growth rate of 0.4 nm/s obtained with a Si flux density density $F_{Si} = 2 \times 10^{19} m^{-2} s^{-1}$

T (K)	723	823	923	1023	1123	1223
σ	$3x10^{16}$	$4.5x10^{12}$	$4.7x10^{9}$	$1.8x10^{7}$	$2x10^{5}$	$4.4x10^{3}$

Vertical growth by lateral motion of surface steps

A real surface contains surface steps at thermal equilibrium between vapour and solid. The origins of these surface steps at equilibrium are

(i) dislocations,
(ii) misorientation of the surface relative to the low index plane, and under supersaturation the mechanism of
(iii) two-dimensional nucleation which generates additional surface steps.

Introduction of a dislocation into a solid creates a surface step. The step runs along the intersection of the slip plane with the surface. Often, as an example, a screw dislocation penetrating the surface is sketched. During growth the steps originating from dislocations wind up in spirals creating growth pyramids (Fig. 5) as was elegantly demonstrated by liquid phase epitaxy (LPE) experiments /15/. We believe that steps from dislocations play a minor role in MBE because substrates are used with low dislocation densities.

Commercially available substrates are misoriented relative to the nominal low index plane by less than a degree. But this, nevertheless, would imply low index terraces separated by misorientation steps. In an idealized situation (Fig. 6), a periodic array of straight steps is given with step separation L_0

$$L_0 = h/\sin i \qquad (3)$$

where i is the angle of misorientation and h the step height.

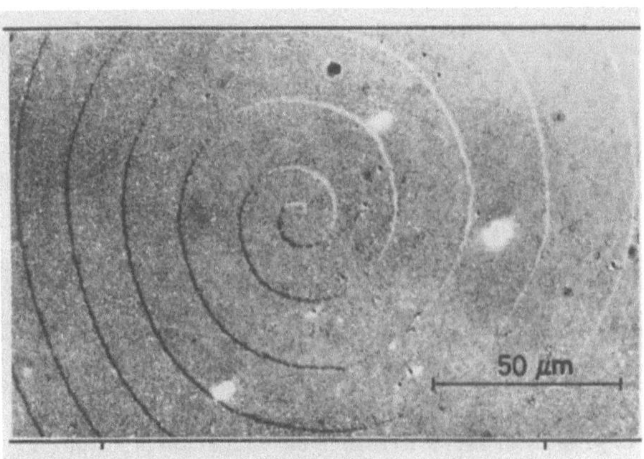

Fig. 5: Optical micrograph /15/ (Normanski differential interference contrast) of a growth spiral with atomic steps orginating from a dislocation intersecting the surface

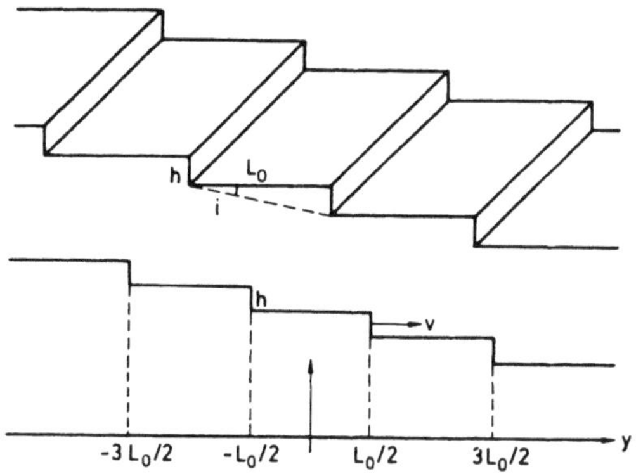

Fig. 6: Scheme of a misoriented surface (angle i) with low-index planes separated by misorientation steps of height h (step separation distance L_0)

Even from small misorientations, a high density of surface steps results, e.g., $1/L_0 = 10^7\ m^{-1}$ for i = 0.25°, h = 0.384 nm.

Adatoms migrating on the surface may join to form a two-dimensional nucleus. Under supersaturation, a nucleus above the critical size is stable and can grow further, but a nucleus below the critical size is unstable. The critical size is a function of supersaturation. The higher the supersaturation the lower is the critical size. Steps from nuclei annihilate each other after growth of a monolayer. Then a nucleation process restarts. In general, two-dimensional nucleation should be of minor importance on stepped surfaces (vicinal surfaces) whereas it must be the dominant source of steps on well prepared surfaces with very small misorientations.

The lateral motion of microscopic surface steps causes a macroscopic vertical growth rate, R, which is step flux times step height. Trains of steps progress along the surface.

The capture of adatoms by steps causes lateral step motion. A clear treatment of the capture of adatoms may be given within the framework of adatom diffusion theory as described below. Burton, Cabrera, and Frank considered the diffusion of adatoms towards steps driven by the supersaturation, σ. The step itself was considered as a perfect sink of adatoms, that means the adatom density n_s drops at the step to the equilibrium value n_{sequ}.

$$\lambda_s^2 \Delta n_s + n_s = n_{sequ}(\sigma+1) \tag{4}$$

with the boundary condition $n_s = n_{sequ}$ at step positions.

Burton, Cabrera, and Frank solved (4) for some step configurations. They neglected in their solution the movement of the step boundary, which is not allowed at high supersaturations. For a straight step at y = 0 they found

$$n_s = n_{sequ}\left[1 + \sigma\left[1 - \exp(-y/\lambda_s)\right]\right] \tag{5}$$

Recently, Voigtländer /14/ and Fuenzalida /16/ extended the solution to a parallel sequence of steps using a movable coordinate system. Large deviations from the classical BCF theory can take place for the adatom distribution, the step velocity and the condensation coefficient. The asymmetry of the adatom distribution is caused by the unidirectional movement of the step train.

LEED/RHEED Oscillations

A milestone in understanding MBE growth was reached with the discovery of LEED/RHEED oscillations from growing surfaces by the groups of Henzler /17/ (LEED) and Joyce /18/ (RHEED). The intensity of electron diffraction (ED) spots from the growing surface oscillates both in low energy (LEED) and reflection high energy (RHEED) experiments. The period of oscillations usually coincides with the growth of a monolayer. It is believed that the oscillations are a proof of (i) the two-dimensional growth mode and, (ii) the occurrence of two-dimensional nucleation processes (Fig. 7). Neither three-dimensional (3D) growth mode with rough surfaces nor two-dimensional (2D) growth mode via dislocation steps or misorientation steps would generate such intensity oscillations. Note that 3D growth would create a rough surface, 2D growth via misorientation steps would create the smoothest surface available at a given misorientation. It is not easy to observe electron diffraction intensities with high accuracy in a Si-MBE apparatus because of electrons from the electron gun evaporator. Henzler et al. used a special growth apparatus without an electron gun evaporator. Also they polished the substrate surfaces for accurate orientation whereas commercially available substrates are slightly misoriented ($i < 5 \times 10^{-3}$).

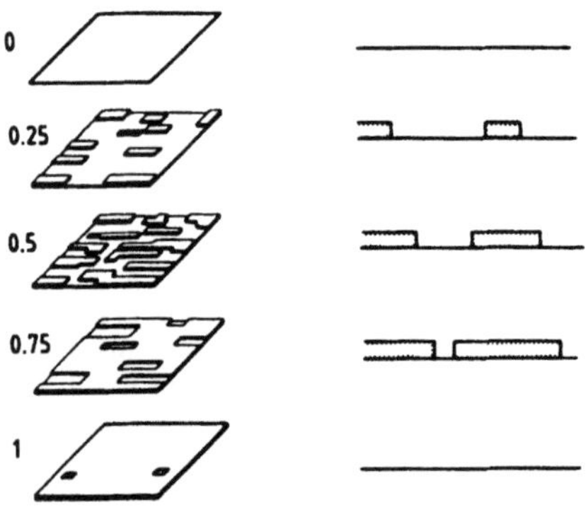

Fig. 7: Two-dimensional growth on a low index plane without misorientation steps. Growth proceeds by nucleation and lateral expansion of the nucleus via capture of adatoms at the steps. After completion of a monolayer, the process is repeated periodically leading to oscillations in the LEED and RHEED intensities /18/

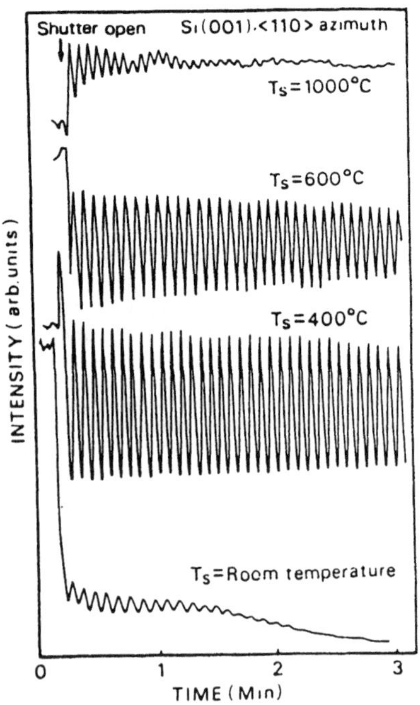

Fig. 8: Dependence of the amplitude of RHEED oscillations on growth temperature T_S. Within the epitaxial growth range ($T > T_{em}$) the amplitude increases with decreasing temperature /19/

Recently, Sakamoto /19/ observed clear RHEED oscillations in an Si-MBE apparatus. Two striking features of such oscillations (Fig. 8) are:

(i) the oscillations fade out with time, i.e. with layer thickness;
(ii) the strength of the oscillations is strongly dependent on growth temperatures.

As a rough rule for many material systems, Ishizaka /20/ claimed that growth occurs via two dimensional nucleation at temperatures $T_m/4 < T < T_m/2$ (T_m melting point) and via misorientation step flow at $T_m/2 < T < 3T_m/4$. Considering the foregoing sections, we can add that the exact transition temperature ($T_m/2$ in Ishizaka s rule) should be influenced by the particular surface properties, such as low index plane, misorientation, step configuration.

2.4 Impurity Incorporation into Si-MBE Layers

Though process temperatures of Si-MBE are far below temperatures usually needed for activation of solid state diffusion, fabrication of arbitrarily fine doping structures is not a simple matter. Initial experiments using coevaporated antimony for n-type doping have shown pronounced profile smearing. In addition, low sticking coefficients, i.e., ratios of dopant in the film to dopant evaporated onto the surface have been established. Except for boron, such or a similar behaviour was found for many of the common dopants such as As, Sb, Al, Ga, and In. Figure 9 illustrates schematically this behaviour. A dopant surface layer, deposited prior to MBE growth, remains almost totally in front of the growing surface, leaving behind only a weakly doped MBE film.

A first model to describe this unexpected impurity incorporation behaviour was suggested by Iyer et al. /21/. They assumed that a dopant atom entrapped on the Si surface, either desorbs or becomes incorporated after a mean residence time. Thus, an accumulated dopant adlayer is present at the surface which is assumed to obey a first order kinetic equation. τ_D and τ_i are the time constants for desorption and incorporation, respectively,

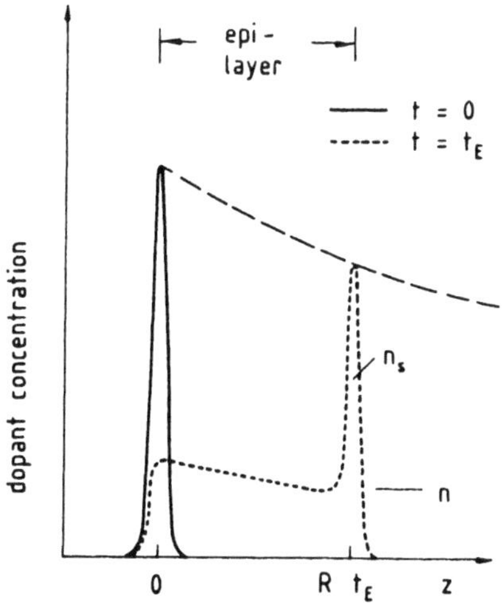

Fig. 9: Incorporation behaviour of common dopants in Si-MBE (schematically). A dopant surface layer (~ one monolayer) deposited onto the substrate, remains at the surface, leaving behind only a weakly doped MBE film (t_E epitaxy duration, R growth rate)

j is the incident dopant flux. Table 6 includes values of τ_D and τ_i for the system Sb on Si (111) for various growth temperatures. Both, τ_D and τ_i, are considerably larger than the time typically needed for one monolayer MBE growth ($t \simeq 1$ s). Particularly at higher temperatures, τ_i is significantly larger than τ_D and, accordingly, the sticking coefficient S, given by the ratio $S = \tau_D/\tau_i$, is very low at these temperatures. The flux density of dopant atoms into the growing MBE film corresponds to the term n_S/τ_i and so the bulk concentration n immediately below the surface is given by

$$n R = n_S/\tau_i \tag{6}$$

where R is the growth rate.

Table 6: Time constants for desorption (τ_D) and incorporation (τ_i), sticking coefficients S, and profile broadening Δ, for Sb on Si (111) at various growth temperatures T_S (growth rate $R \simeq 1$ Å/s)

T_S (C)	τ_D (s)	τ_I (s)	S	Δ (cm)
600	1.0×10^{5}	3.1×10^{6}	3.2×10^{-2}	1.0×10^{-3}
700	3.5×10^{3}	1.8×10^{6}	1.9×10^{-3}	3.5×10^{-5}
800	2.4×10^{2}	1.1×10^{6}	2.2×10^{-4}	2.4×10^{-6}
900	2.5×10^{1}	7.8×10^{5}	3.2×10^{-5}	2.5×10^{-7}

Equation (6) enables to simulate the doping profiles for different doping procedures (coevaporation, pre-deposition and flash-off) by use of the appropriate initial conditions. With coevaporation the profile smearing or broadening, Δ, is found to be

$$\Delta = R/(\tau_D^{-1} + \tau_I^{-1}) \tag{7}$$

In Table 6 values of Δ are shown for the system Si (111): Sb for some temperatures. The term Δ strongly decreases with increasing temperature. Unfortunately, the sticking coefficient simultaneously decreases drastically. By Sb adlayer pre-adjusting, however, while temporarily arresting silicon growth, arbitrarily sharp profiles can be grown even at lower temperatures, Since within the framework of the model, τ_i is still large, only low doping levels can be achieved at low dopant surface coverages, in both cases.

The segregation model predicts the existence of a transition from equilibrium to kinetically-limited segregation. In early experiments some evidence of this transition was found for Sb on Si (111) at temperatures below 650 °C where the sticking coefficient is significantly higher than the values predicted by the adlayer-model (Fig. 10). In recent experiments, where dopant incorporation was studied in a relatively low temperature regime, the transition was clearly observed for Sb on Si (100) (Fig. 11) /22/. These experiments revealed a well-defined transition temperature, T* = 560 °C at R = 3 Å/s growth rate. In addition, the transition temperature was found to shift to lower values when the growth rate is lowered /22/.

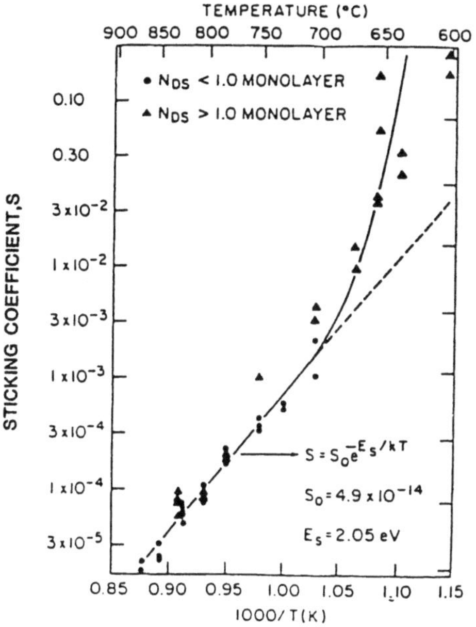

Fig. 10: Sticking coefficient of Sb on Si vs growth temperature. At T_S < 650° C the sticking coefficient significantly exceeds the adlayer-model prediction /21/

Secondary Implantation

The extremely high surface segregation, particularly of antimony on silicon, induces high surface coverages at simultaneously low bulk concentrations. For instance, an adlayer density of $n_s = 1 \times 10^{14}$ Sb atoms/cm^2 yields a bulk concentration of below 1×10^{16} cm^{-3} at T = 650 °C growth temperature. This offers new feasibilities to study secondary implantation effects, which are similar to well established processes of recoil implantation and collisional mixing (/23/ and references therein), at growing surfaces. Exciting features of this novel application of these techniques are the use of submonolayer dopant coverages and, accordingly, the use of very low energy projective coverages. Thus , multiple collisions between dopant atoms, as in ordinary recoil implantation experiments /23/, usually need not to be considered here. Dopant incorporation results from a single binary collision between the primary projectile and the dopant adatom.

Secondary implantation of Sb into Si-MBE layers has been demonstrated in a recent experiment /24/. With this experiment, low energy Si ions (200 eV), which were generated by partially ionizing the Si beam, were used as primary projectiles. The energy of the Si ions was adjusted by applying a proper voltage between substrate and grounded Si source. After growing a buffer layer, Sb was evaporated onto the growing surface. The MBE process was performed at T = 650 °C where extreme surface segregation (Fig. 11) restricts the corresponding bulk concentration to $n \approx 10^{16}$ cm^{-3}. When the growing film was irradiated by low energy Si ions, however, the bulk concentration increased by three orders of magnitude up to the 10^{19} cm^{-3} range. Subsequently, the adlayer was depleted by incorporation, causing an exponential doping decay. The amount of Sb incorporated is almost identical to the amount deposited initially. This and similar experiments /25,26/ have shown that

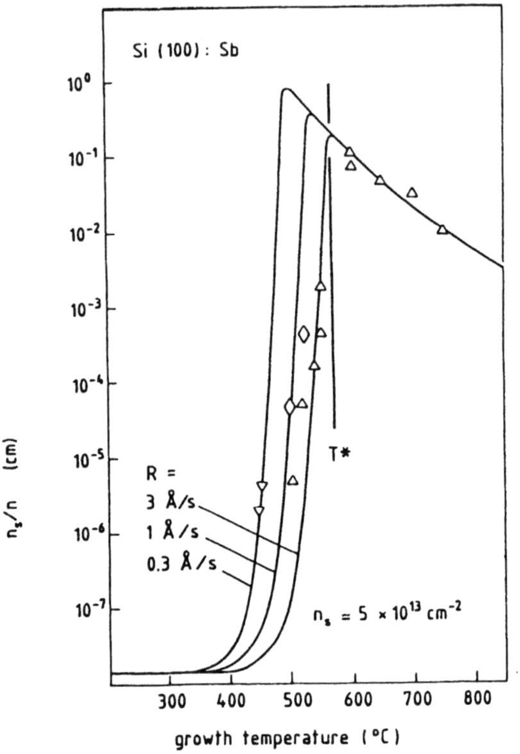

Fig. 11: Segregation coefficient, $r = n_S/n$ (solid line) versus growth temperature in the vicinity of the transition temperature, T*

(i) sputtering of Sb atoms is negligible at least up to 1500 eV Si ion energy.
(ii) Sb atoms are completely substitutionally incorporated.
(iii) Electron mobilities are comparable to bulk mobilities even at high doping levels.

2.5 Heteroepitaxy

Semiconductor-silicon, but also metal-silicon and insulator-silicon couples can be grown by Si-MBE (table 7). Silicides, calcium fluoride and silicon germanium are the most prominent partners of silicon based heterosystems.

Table 7: Si-based heteroepitaxy partners

material class	semiconductor	metal	insulator
partner	SiGe, Ge GaAs	$NiSi_2$, $CoSi_2$	CaF_2

A connection between Si-MBE and III/V-MBE and III/V-MBE equipment will allow growth of a wide variety of materials. SiGe is treated in the following section as model system for strained layer epitaxy.

3. STRAINED LAYER SUPERLATTICE

The cubic lattice cell of both materials, Si and Ge, and of the completely miscible alloy is of the diamond type. The room temperature lattice constant a_0 of Si amounts to 0.5431 nm. Ge is lattice mismatched to Si by 4.17 % (a = 0.56575 nm, T = 300 K). The lattice mismatch of the Si Ge alloy can be expressed in a first approximation by Vegard's law.

Nature gives two answers to uncracked accommodation of lattice mismatch. For layers thinner than a critical thickness t_c the accommodation is performed by strain ε distorting a cube with side length a into a tetragon with base length $a_\parallel$ and height $a_\perp$. In an isotropic material the height to base ratio is given by

$$a_\perp/a_\parallel = 1 - \frac{1+\nu}{1+\nu}\,\varepsilon \tag{8}$$

with Poisson's ratio ν. A treatment of the stress problems connected with anisotropy of diamond and zincblende type materials is given by W.A. Brantley /27/ and J. Hornstra et al. /28/. Strained Layer Superlattices (SLS) are created by stacking such thin strained layers (Fig. 12).

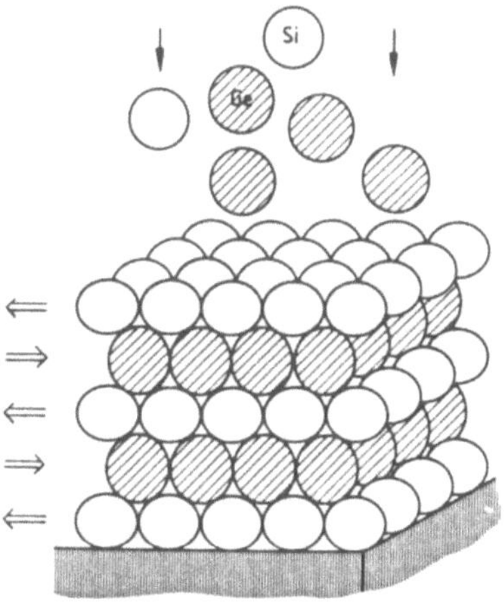

Fig. 12: Strained Layer Superlattice (SLS) grown from condensing Si and Ge molecular beams. The thin layers are strained because of lattice mismatch between Si and Ge

For layers thicker than the critical thickness t_c accommodation of mismatch takes partly place by generation of a misfit dislocation network lying in the interface. In the SiGe material system metastable strained layer growth /29-31/ shifts the critical thickness especially at growth temperatures below 900 K to higher values as compared with equilibrium.

3.1 Virtual Substrate

Let us consider a Si/SiGe superlattice on a Si substrate. The straight forward way would be to place the superlattice directly on the substrate /32/. Although this direct placing of the superlattice on the Si substrate has some advantages as high crystal perfection and ease of processing, it cannot be used for a general approach /30/ because of limited thickness range of the superlattice (e.g. 10 nm for a Si/Ge superlattice /33/) and inadjustable strain distribution (SiGe layers compressed, Si layers unstrained /30/).

The strain can be adjusted by a virtual substrate /34/ consisting of the Si substrate and a SiGe buffer layer between substrate and superlattice (Fig. 13). Often a rather thick and graded buffer layer is used which is inconvenient for MBE-growth methods and for device technology. We instead proposed a thin, homogeneous buffer and gave design rules for that buffer layer /35/.

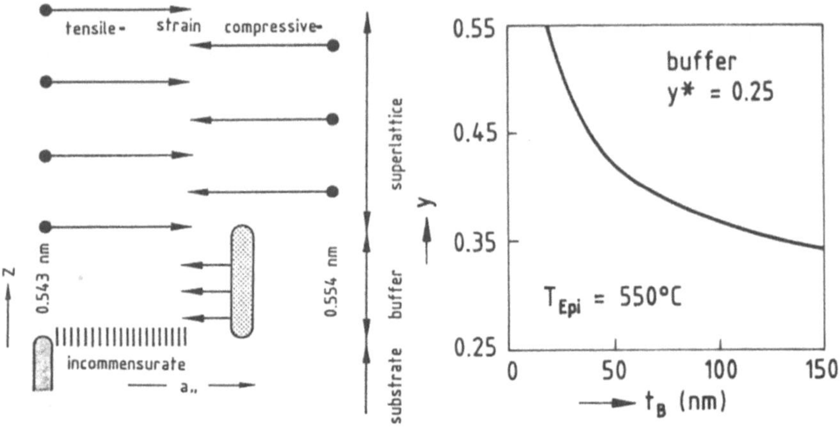

Fig. 13: Strain symmetrization in an SiGe/Si SLS by a virtual substrate consisting of an Si substrate with a covering incommensurate $Si_{1-y}Ge_y$ buffer layer. Left side: In plane lattice constants of the layers as influenced by chemical composition (dots) and strain (arrows). Right side: Design chart of the corresponding buffer layer with effective Ge-content of 25 % (y* = 0.25)

3.2 Stability of a Strained Layer Superlattice

Growing a SLS on different virtual substrates will result in different strain energies stored in the superlattice. Only the superlattice structure with the lowest energy content will be stable up to infinite thickness. All other structures will be unstable above a critical thickness t_{CS} of the whole superlattice, although the individual layers of the SLS are thinner than the critical thickness t_C for layer growth. Dislocations tend to move to the interface between virtual substrate and superlattice within these unstable SLSs.

The elastic energy E_h stored in a SLS with N periods and period length L is given by /34/

$$E_h = 2\mu \frac{1+\nu}{1-\nu} N (\varepsilon_1^2 t_1 + \varepsilon_2^2 t_2) \tag{9}$$

if ε_i, t_i are elastic strain and thickness of the individual layers of the superlattice period,

$$L = t_1 + t_2 \tag{10}$$

The strain values are connected by the lattice mismatch η between the layer materials.

$$\eta = 2 \frac{a_2 - a_1}{a_2 - a_1} \tag{11}$$

Neglecting second order terms we obtain

$$\eta = \varepsilon_1 - \varepsilon_2 \tag{12}$$

Minimizing the elastic energy, E_h, yields a condition for the stable SLS

$$\begin{aligned} \varepsilon_1 &= \eta \, (t_2/L) \\ \varepsilon_2 &= -\eta \, (t_1/L) \end{aligned} \tag{13}$$

Strain symmetrization is required for the important case of equally thick ($t_1 = t_2 = L/2$) layers

$$\varepsilon_1 = -\varepsilon_2 = \eta/2 \tag{14}$$

If we discuss strain symmetrization in context with SLS stability we always refer to (13) if unequally thick layers are used. Unsymmetrically strained SLSs are unstable above a critical superlattice thickness t_{CS}. This critical thickness is roughly the same as that for an alloy layer of the same integral composition.

3.3 Strain Adjustment

The strain in the SLS is adjusted by the choice of the virtual substrate with its in-plane lattice constant $a_{\parallel}$

$$\varepsilon_i = \frac{a_{\parallel} - a_i}{a_i} \tag{15}$$

with a_i lattice constant of the ith layer. Symmetrizing the strain in an Si/SiGe superlattice on Si substrate is obtained when the in-plane lattice constant of the buffer layer is chosen half between the lattice constants of the Si and SiGe layers.

3.4 Measurement of Strain Distribution

The strain distribution in these SiGe SLSs was determined by X-ray analysis /36/ Raman scattering /37/ and channeling RBS /38/. Fig. 14 gives an example of RBS measurement of a nearly symmetrically strained SiGe SLS.

Angular resolved channeling yields from Si and SiGe layers of the superlattice, respectively, prove the deviations of the (111) channel of the tetragonally distorted lattice cells of the superlattice layers from the undistorted substrate channel. The amount of the deviation (in the given example (Fig. 14) 0.°035 for the Si layers and -0.°043 for the SiGe layers) measures the strain distribution inside the superlattice.

3.5 Type II Superlattice

As detailed examples we choose two properties of the SiGe superlattice, which are essentially influenced by strain adjustment. Such properties are conduction band offset and zone folding effects.

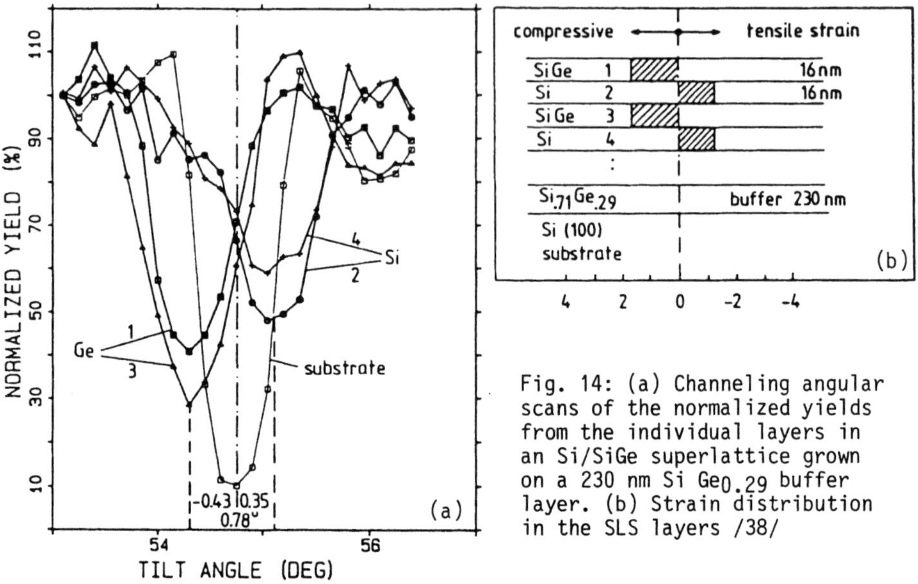

Fig. 14: (a) Channeling angular scans of the normalized yields from the individual layers in an Si/SiGe superlattice grown on a 230 nm Si $Ge_{0.29}$ buffer layer. (b) Strain distribution in the SLS layers /38/

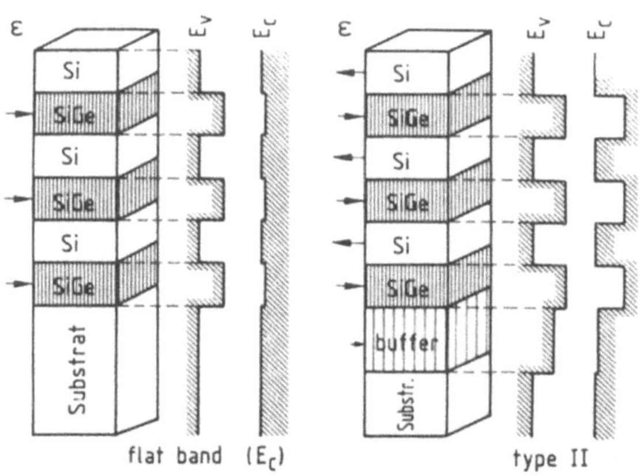

Fig. 15: Strain adjustment influences essentially the electronic properties of SiGe/Si superlattices. Left side: Unsymmetrically strained SLS results in nearly flat conduction band condition. Right side: Strain symmetrization by a buffer layer shifts the band ordering toward large conduction band offsets of a type II superlattice (electron energies in the small band semiconductor SiGe are higher than those in the large bandgap semiconductor Si)

P-channel MODFETs were successfully fabricated by a Bell group /39/ from an unsymmetrically strained Si/SiGe heterostructure. N-channel MODFETs could not be fabricated from that structure because of near flat conduction band conditions. On the other hand our group has fabricated N-channel MODFETs from symmetrically strained Si/SiGe heterostructures /40/, which was explained by a type II band alignment at the symmetrically strained

Si/SiGe interface /37/. Type II means a staggered band alignment with the conduction band edge of the small band gap semiconductor SiGe above that of Si (Fig. 15). This behaviour can be explained by the strain induced splitting of the six fold degenerate conduction band minimum /37/.

3.6 Zone Folding

Folding of the Brillouin zone by the superlattice period is predicted to be able to change the indirect character of the Si-type band gap /41/. The zone folding scheme is shown in Fig. 16. More detailed calculations of the zone folding effects in the Si/Ge superlattice system /42/ underline the essential effect of strain adjustment for this material system. New direct transitions are generated by the one-dimensional periodicity of the superlattice, which are in competition with the indirect transitions of the strained material. The material will change its character from the indirect band gap of Si-type semiconductors to a direct band gap superlattice

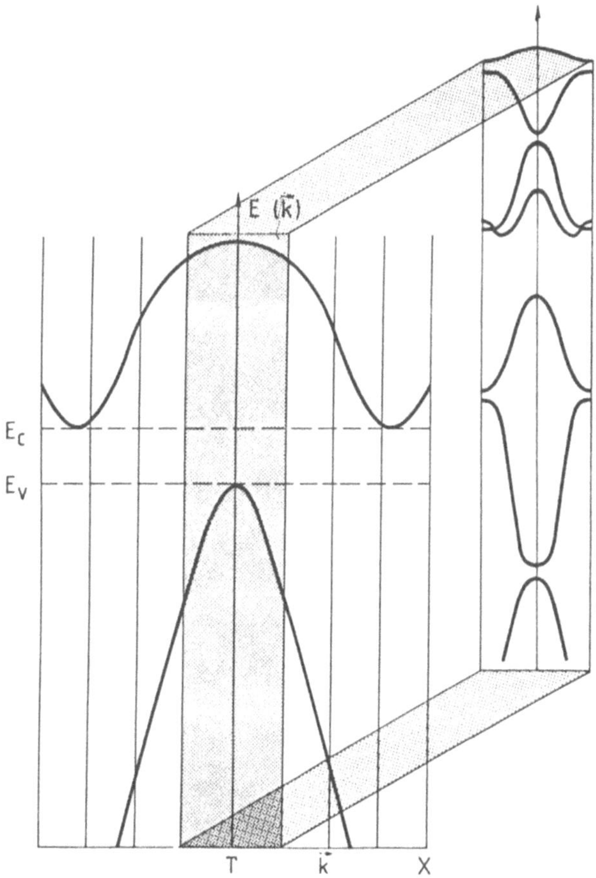

Fig. 16: Scheme of Brillouin zone folding by a superlattice ($L = 4a_0$). Minibands are generated in the new first Brillouin zone

material if the lowest transition is direct. This direct band gap is only predicted /42/ for strained superlattices from the symmetrical point to a Ge-rich virtual substrate. An Si/Ge superlattice on a Si substrate is of indirect character following these calculations.

4. DEVICES

Device classes are selected which are not only DC characterized.

4.1 Integrated Bipolar Transistors

In one approach the simple uniformly doped layer of roughly one micron thickness was varied systematically in thickness. With the given layout and process sequence a maximum transit frequency of 7 GHz was obtained with 1.2 µm layer thickness. The transit frequency decreased to 6 GHz and 4 GHz for layer thicknesses of 0.9 µm and 1.7 µm, respectively. 2.5 GHz frequency dividers were fabricated with high yield from Si-MBE layers.

A more interesting approach demonstrated lateral definition of device structures by growth into oxide windows. This growth mode is called differential epitaxy (Fig. 17). The first results were encouraging resulting in transit frequencies comparable to the above mentioned data. There was to my knowledge no real attempt to fabricate the whole structure of a high frequency transistor by Si-MBE.

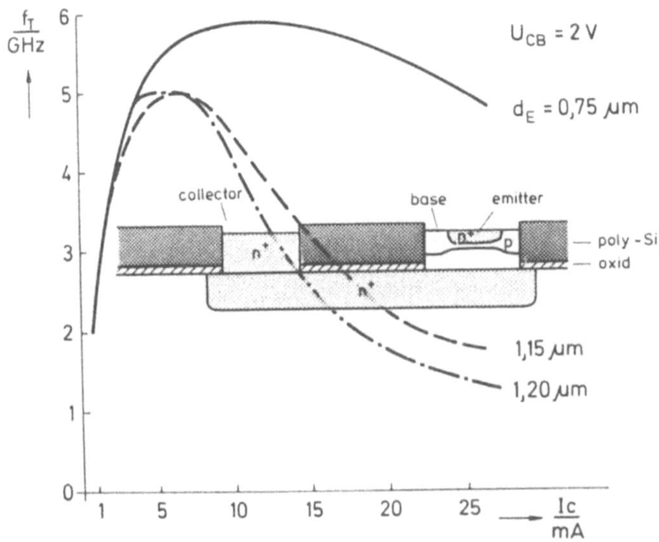

Fig. 17: Integrated bipolar transistor from differential epitaxy layers. Frequency f_T versus current I_c. Inset: transistor structure

4.2 Silicon mm-wave integrated circuits (SIMM WIC)

Up to now the brightest success with Si-MBE structures was obtained for discrete mm-wave (> 30 GHz) devices. The first commercial Si-MBE device - a 90 GHz-Impact Avalanche Transit Time (IMPATT) diode - is the existing most powerful semiconductor device in this frequency range, and has the potential to operate in oscillators from 60 GHz to 300 GHz.

A real break through was the monolithic integration of the IMPATT into a 75 GHz-oscillator /43,44/. In the meanwhile mm-wave receivers and transmitters with integrated antennae were realized (Fig. 18) pushing Si-technology to the 100 GHz regime and beyond this border.

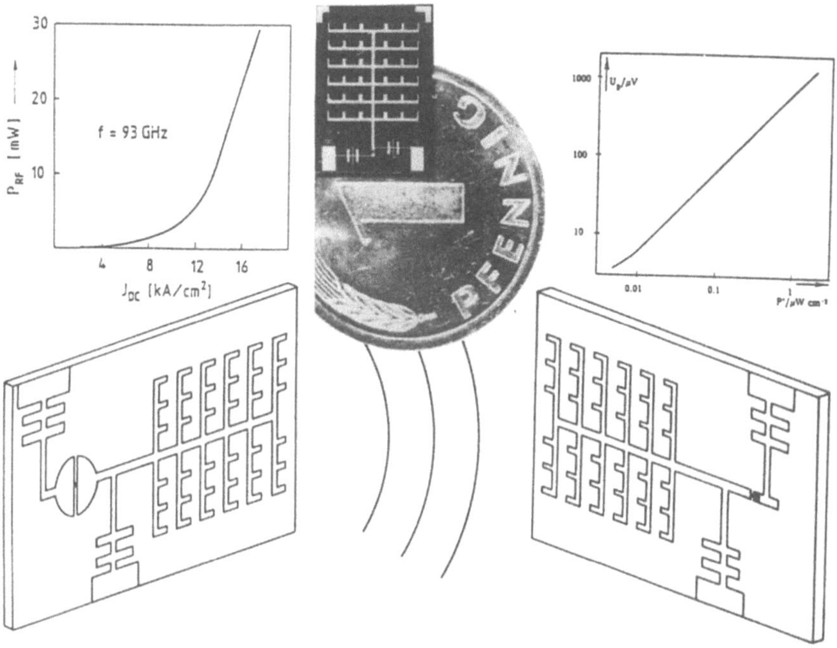

Fig. 18: Si-based integration of 100 GHz devices with microstrip oscillators, receivers and antennae (SIMMWIC). Scheme (below), output power (above left) and a receiver chip (above, middle) are shown

4.3 SiGe Heterostructure Devices

Heterostructures offer the designer a new freedom because of the unique properties not available in homogeneous materials. A nice example is J.F. Luy's /45/ work on a MITATT diode with improved noise properties. But the classical examples are the n-channel MODFET and the waveguide APD optical detector (Fig. 19), fabricated by H. Dämbkes and T.P. Pearsall from E. Kasper's and J. Bean's material, respectively. A review of the recent work on SiGe heterostructure devices is given in /46/ (table 8).

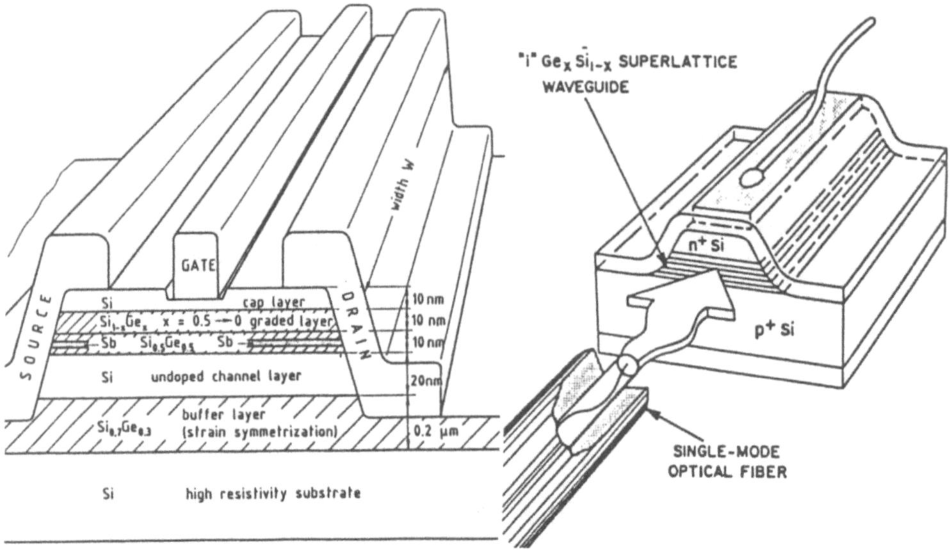

Fig.19: N-channel MODFET (left side) and optoelectronic (1.3 µm) APD-receiver (right side) based on SiGe heterostructures on Si substrates

4.4 Silicon on Insulator (SOI)

The creation of silicon on insulator (SOI) structures does not belong to the original strengths of Si-MBE. But now, at least three ways are offered to create SOI-structures:

- Epitaxy on porous silicon. The porous oxide can be oxidized with high lateral oxidation rates from windows etched into the epitaxy layer.

- Lateral solid phase epitaxy (L-SPE). High quality solid phase epitaxy is obtained by a slight variation of the MBE-process. At room temperature the impinging molecular beams condense to an amorphous layer which can be either n- or p-doped up to levels of 10^{20}/cm . This amorphous layer recrystallizes from the interface at temperatures around 600 °C. Lateral solid phase epitaxy (L-SPE) is obtained if the deposition takes place on a prepatterned SiO_2 layer with windows (seed area) to the substrate.

- CaF_2/Si heteroepitaxy. From several groups encouraging results /47/ were obtained with the CaF_2/Si heterosystem. CaF_2 is a good insulator with a small lattice mismatch to silicon.

4.5 Heterobipolar transistor (HBT)

It is predicted, and in the III/V material system already proven, that bipolar transistors would benefit from heterotransitions between the base and emitter regions. For silicon based transistors two concepts are now under actual consideration.

Table 8 : Essential steps in silicon based heterostructure device work from SiGe-MBE material

Year	Organization	Device	Remarks
1984	AT&T	Ge-PIN Detector	incommensurate Ge
1985	AT&T	p-MODFET	commensurate SiGe
	AEG	n-MODFET	symmetrical strain
1986	AT&T	Waveguide PIN-Detector	Si/SiGe superlattice
	AT&T	Waveguide APD Detector	Si avalanche region+superlattice det.
1987	IBM *	Heterojunction Bipolar Transistor (HBT)	SiGe base
	AEG	SiGe/Si-MISFET	low temperature growth (<450 °C)
	NEC	Collector Top-HBT	two step differential epitaxy
1988	NRC, Canada	Resonant Tunneling Diode	hole tunneling
	AEG	Hetero-MITATT	SiGe avalanche region
		QWITT	quantum well injection

*AT & T and AEG reported about HBT work at the same meeting (2nd Int. Symp. Si-MBE)

(i) The low band gap base HBT (Fig. 20) with a thin, strained SiGe base layer between Si collector and Si emitter. Since end of 1987 at least five companies (AEG, AT&T, IBM, NEC, British Telecom) are developing this transistor type /48,49/ with predicted over 40 GHz operation of a silicon based device.

(ii) The metal base transistor utilizes ultrathin metal layers cladded by Si layers as vertical transistor structures. As metal layers $NiSi_2$ and $CoSi_2$ are used. Both silicides exhibit low lattice mismatch to silicon.

5. MONOLITHIC INTEGRATION OF SILICON BASED SLS WITH CONVENTIONAL IC

Completely new and tailorable properties are expected from classes of semiconductor devices, which are based on superlattice structures. A superlattice structure is produced by arranging alternate layers of different elements with a period length, which superposes the natural period of the lattice. New electronic and optical properties arise from this synthetic semiconductor if the period of the superlattice is small (three to hundred ato-

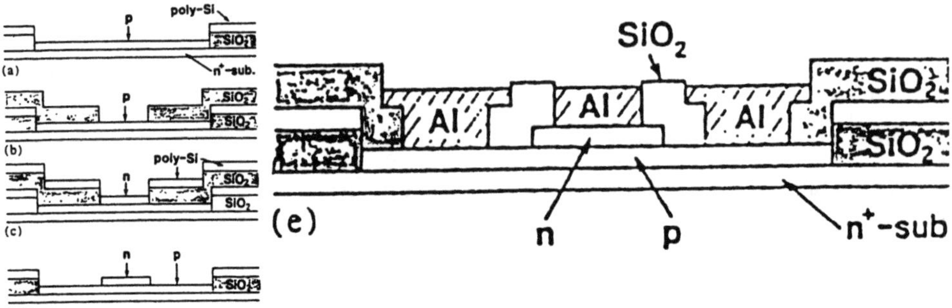

Fig. 20: SiGe base HBT. Structure and fabrication steps of the NEC transistor /48/

mic layers). The properties of the synthetic semiconductor are controllable by the geometrical dimensions, the chemical composition and the strain inside of the superlattice structure. Alternating layers of lattice mismatched materials can be grown with high perfection if the individual layer thickness is small. With very thin layers the lattice mismatch is accommodated by tetragonal distortion of the lattice cell leading to strained layer superlattices (SLSs). Since this discovery the SLSs gained increasing interest for several reasons. By preparation of mismatched material couples a wider range of superlattice systems is available. The strain can be used as additional factor influencing electronic and optical properties not available in unstrained material. The technologically important silicon based superlattices are strained because the silicon lattice constant is smaller than that of the most other semiconductors.

Todays microelectronics has reached a very high level of performance and complexity mainly by applying techniques for shrinkage of the lateral dimensions of integrated circuit (IC) structures down to the submicron level. The vast majority of high complexity ICs is manufactured with high yield and astonishingly low costs on silicon (Si) substrates. The performance of these conventional ICs is furtheron increasing with no reliable prediction of saturation effects. Future circuits based on new, more general concepts should utilize the tailorable properties of heterostructure - or superlattice devices for high performance cores, high speed interlinks and novel sensor integration on the chip. But the tremendous complexity of conventional ICs and the technical, economical and environmental advantages of silicon substrates should not be missed. A material concept combining silicon based superlattices with conventional ICs (Fig. 21) can fulfill the requirements if progress in mismatch accommodation, low temperature stacking of different device levels and production technology continues.

Silicon substrates are favoured for a broad industrial application because of the superior properties regarding wafer size, crystal perfection, thermal conductivity, handling and price. The large resources (silicon is the 2nd most frequent element of earths crust) and the environmental harmlessness are additional factors favouring a broad usage.

A realization of this concept could start with the conventional IC on the silicon substrate. The high complexity conventional part covers the main part of the chip with the exception of core regions, interlink region and sensor regions. The conventional IC-part is nearly completely processed

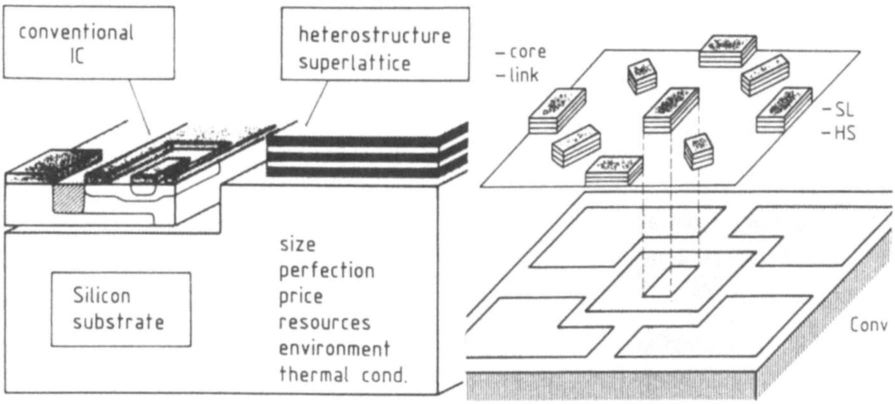

Fig. 21: Si-based monolithic integration of conventional IC with superlattices/heterostructure devices (left). A small part of the chip area (core regions, intrachip links) is occupied by the heterostructures (right).

up to the last metallization level. Then the superlattice structure is defined by a low temperature process within the predetermined regions and connected by the last metallization level.

Acknowledgement: Long time cooperation with H. Kibbel and H.-J. Herzog is acknowledged.

6. LITERATURE

1. The practice and technology of MBE, ed. by E.H.C. Parker, Plenum Press, New York, 1987
2. Silicon molecular beam epitaxy, ed. by E. Kasper and J.C. Bean, CRC Press, Boca Raton, Florida, 1988
3. Proc. 2nd Int. Symp. Si-MBE, ed. by J.C. Bean and L. Schowalter, Electro-Chem. Soc., Pennington, USA 1988
4. T. de Jong, W.A.S. Douma, L. Smit, V.V. Korablev, F.W. Saris: J. Vac. Sci. Technol. B 1, 808 (1983)
5. Y. Shiraki, Y. Katayama, K. Kobayashi, K.F. Komatsubara: J. Crystal Growth 45, 287 (1978)
6. H.J. Gossmann, L.C. Feldman: Appl. Phys. A 38, 171 (1985)
7. K.D. Gronwald, M. Henzler: Surf. Sci. 117, 180 (1982)
8. H.-J. Herzog, E. Kasper, P. Eichinger, H. Kibbel: unpublished
9. F. Jona: Appl. Phys. Lett. 9, 235 (1966)
10. E. Kasper: Appl. Phys. A 28, 129 (1982)
11. J. Bloem: J. Crystal Growth 50, 581 (1980)
12. R.F.C. Farrow: J. Electrochem. Soc. 121, 899 (1974)
13. H.C. Abbink, R.M. Broudy, G.P. McCarthy: J. Appl. Phys. 39, 4673 (1968)
14. K. Voigtländer, H. Risken, E. Kasper: Appl. Phys. A 39, 31 (1986)
15. D. Kass, M. Warth, W. Appel, H.P. Strunk, E. Bauser: Proc. 1st Int. Symp. Si-MBE, Toronto, ed. by J.C. Bean, Vol. 85-7 (The Electrochem. Soc., Pennington, NJ, USA 1985) p. 250
16. V. Fuenzalida, I. Eisele: J. Crystal Growth 74, 597 (1986)
17. M. Henzler: Appl. Surf. Sci. 11, 450 (1982) or Appl. Phys. A 34, 205 (1984)

18. B.A. Joyce, H.J. Neave, P.J. Dobson, P.K. Larsen: Phys. Rev. B 29, 814 (1984)
19. T. Sakamoto, N.J. Kawai, T. Nakagawa, K. Ohta, T. Kojima, G. Hashiguchi: Surf. Sci. 174, 651 (1986)
20. A. Ishizaka: Jap. J. Appl. Phys. 27, 883 (1988)
21. S.S. Iyer, R.A. Metzger, F.G. Allen: J. Appl. Phys. 55, 831 (1981)
22. H. Jorke: Surf. Sci. 193, 569 (1988)
23. J.J. Grob, N. Mesli, A. Grob, P. Siffert: Appl. Phys. A 35, 161 (1984)
24. H. Jorke, H.-J. Herzog, H. Kibbel: Appl. Phys. Lett. 47, 511 (1985)
25. H. Jorke, H. Kibbel: J. Electrochem. Soc. 133, 774 (1986)
26. H. Jorke, H.-J. Herzog and E. Kasper: Superlattices and Microstructures 4, 319 (1988)
27. W.A. Brantley: J. Appl. Phys. 44, 534 (1973)
28. J. Hornstra and W.J. Bartels: J. Crystal Growth 44, 513 (1978)
29. R. People and J.C. Bean: Appl. Phys. Lett. 47, 322 (1985)
30. E. Kasper: Surf. Sci. 174, 630 (1986)
31. J.Y. Tsao, B.W. Dodson, S.T. Picraux, and D.M. Cornelison: Phys. Rev. Lett. 59, 2455 (1987)
32. E. Kasper, H.-J. Herzog and H. Kibbel: Appl. Phys. 8, 199 (1975)
33. T.P. Pearsall, J. Bevk, L.C. Feldman, J.M. Bonar, and J.P. Mannaerts: Phys. Rev. Lett. 58, 729 (1987)
34. E. Kasper, H.-J. Herzog, H. Dämbkes, and G. Abstreiter: Mat. Res. Soc. Proc. Vol. 56, ed. by J.M. Gibson, G.C. Osbourne and R.M. Tromp, p. 347, Mat. Res. Soc., Pittsburgh (1986)
35. E. Kasper, H.-J. Herzog, H. Jorke and G. Abstreiter: Superlattices and Microstructures 3, 141 (1987)
36. B. Ofner, H. Krenn, G. Bauer: pers. comm.
37. G. Abstreiter, H. Brugger, T. Wolf, H. Jorke, and H.-J. Herzog: Phys. Rev. Lett. 54, 2441 (1985)
38. S. Mantl, E. Kasper and H. Jorke: Mat. Res. Soc. Symp. Proc. Vol. 91, 305 (1987) and
H.-J. Herzog, H. Jorke, E. Kasper, and S. Mantl: Proc. 2nd Int. Symp. Si-MBE, ed. by J.C. Bean and L. Schowalter: Electrochem. Soc., Pennington, N.Y. (1988), p. 58
39. T.P. Pearsall and J.C. Bean: IEEE EDL-7, 308 (1986)
40. E. Kasper and J. Dämbkes: Inst. Phys. Conf. Ser. No. 82, p. 93, ed. by D.F. Moore, IOP Publishing, Bristol (1987)
41. U. Gnutzmann and K. Clausecker: Appl. Phys. 3, 9 (1974)
42. A. Zunger: 2nd Int. Symp. Si-MBE, Honolulu, Oct. 1987 and S. Froyen, D.M. Wood and A. Zunger, Phys. Rev. B 36, 4547 (1987)
43. J.F. Luy, K.M. Strohm, J. Büchler: Microwave and Optical Technology Letters (MOTL) 1, 117 (1988)
44. J. Büchler, E. Kasper, J.F. Luy, P. Russer, K. Strohm: Electronics Lett. 24, 977 (1988)
45. J.F. Luy: will be published
46. E. Kasper: Heterostructures on Silicon: A Step Further with Silicon, ed. by Y. Nissim, K. Ploog, and E. Rosencher: will be published
47. J. Phillips, J.L. Batstone and J.C. Hensel: MRS Spring Meeting 1988, Symposium A
48. T. Tatsumi, H. Hirayama, N. Aizaki: Appl. Phys. Lett. 52, 895 (1988)
49. S. Iyer, H. Temkin, H. Dämbkes: in 3.

Part II

Processing

Simulation of Silicon Processing

M.E. Law, C.S. Rafferty, and R.W. Dutton

Integrated Circuit Laboratory, Stanford University, Stanford, CA 94305, USA

1 Introduction

In the fabrication of silicon integrated circuits, there are two main purposes to most fabrication steps. The first is the controlled introduction of dopant atoms into the silicon lattice. The other is the patterning of various material interfaces. It is difficult to calculate dopant profiles without the material shape information, since materials are used to confine and constrain the doping process. By controlling the materials and doping, the conductivity of the silicon is altered and transistors, diodes, and other electronic devices can be fabricated.

Integrated circuit (IC) fabrication processes are designed to control the quantity and location of the dopant atoms in the silicon substrate as well as the material properties. In designing such a process, it is important to understand the effect of each step on the materials and doping profiles. The goal of process simulation is to predict the device structure as it changes throughout the IC fabrication process.

Process simulation is able to provide several useful services. The first is to allow computer aided design of the process sequence. The engineer can examine the result of a fabrication sequence without having to make a costly test fabrication run. Second, the engineer is allowed to gain understanding of the physical processes which occur in the design process.

Figure 1 shows the cross section of a N Channel Lightly Doped Drain (LDD) structure.

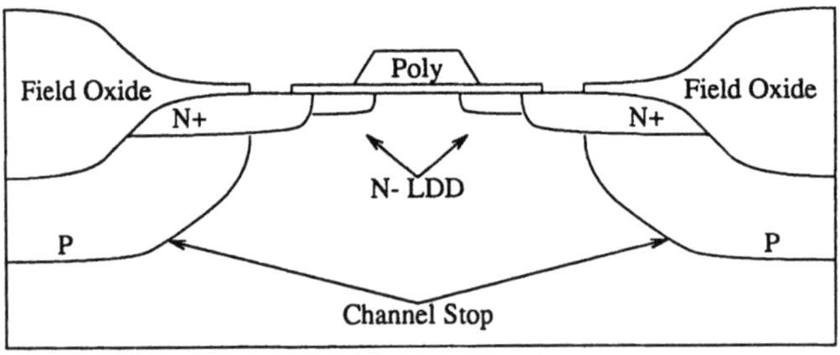

Figure 1 - Section of a N Channel LDD device

This illustrates the device of interest for modern technology, which is formed with heavily doped source and drain regions. These are adjacent to lightly doped regions which reduce the peak field near the gate and increase the reliability of the device. The device is surrounded by field oxide regions which provide isolation from neighboring devices.

This device shows several of the important process steps involved in IC fabrication. First, oxidation of silicon is important in forming isolation and gate insulators. Dopants are introduced by ion implantation, and these dopants redistribute by diffusion during high temperature anneals. Various materials are deposited and etched on the silicon, including

Springer Series in Materials Science, Vol. 13
Semiconductor Silicon Editors: G. C. Harbeke and M. J. Schulz

polysilicon, oxide, nitride, and photoresist. A process simulator should be able to handle these process steps.

2 Etching, Deposition, and Lithography

Etching, deposition, and lithography simulation can be lumped together because all these process steps are simulated using the same algorithm. This scope of this chapter does not allow a full discussion of the physics of these processes. However, the computation technique is powerful and applicable to all material simulation in IC processing.

Oldham, *et al.* [1,2] applied the string algorithm to model deposition, etching and lithography. The essential portion of the algorithm is to treat the interfaces between materials and the top surface as a line in two dimensional models and a planar surface in three dimensional models. During the time evolution of the surface, the line representing the surface is evolved to indicate the changes in the material boundaries.

The physical model can be quite complex in computing the evolution of the line. Each point on the line is given a single velocity as determined by the physics of the process involved. For reactive ion etching, it is necessary to calculate the incident quantity and velocity distribution of the incoming ions at each point. Given this data, each point on the line is given a velocity and updated.

In updating the line, there are several special cases which can cause problems. First, care must be taken when etching through a material. The etch rate may change substantially when moving from one material to another and this must be taken into account by correcting the etch rate in the new material. This action can also split a line in two. Figure 2a demonstrates this phenomena.

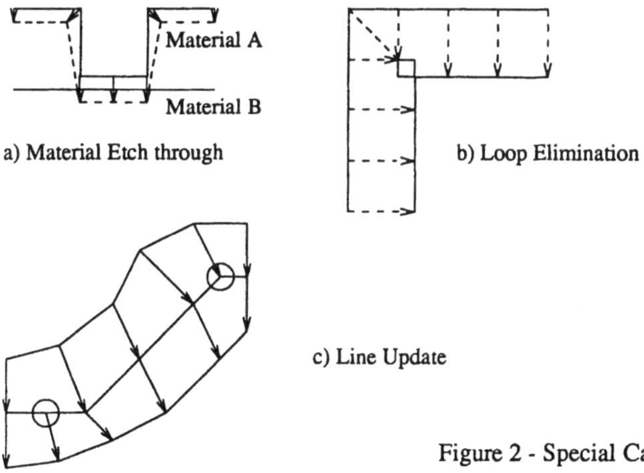

a) Material Etch through

b) Loop Elimination

c) Line Update

Figure 2 - Special Cases in String Algorithms

A second special case is the problem of looping, illustrated in Fig. 2b. Around a corner, there is the potential of the string looping. This must be taken into account and corrected for. The final case is shown in Fig. 2c. During a long etch or deposition, the surface may need to be rediscretized to account for changes occurring. This is typically handled by updating the line through removal or addition of points on the line.

This model has had a very successful past in modeling etching and deposition processes. For example, see Ohgo *et al.* [3] or Pelka *et al.* [4] for some recent work. Pelka *et al.* have excellent agreement with SEM cross sections of ion milling of gold.

3 Ion Implantation

The physics of ion implantation are covered elsewhere (see Poate) in this book. Therefore, the discussion here will be limited to techniques of calculation of ion implanted profiles of dopants. Most of the calculations of impurity profiles resulting from ion implantation have been based on statistics.

Gibbons *et al.* [5] compiled range statistics for use with Gaussian profiles. These have formed the basis for much of the range statistics modeling used in ion implantation. The ion profile is represented by a Gaussian distribution characterized by a projected range and standard deviation. Furakawa *et al.* [6] demonstrated how to extend this technique to multiple material layers. Runge [7] extended this treatment for arbitrary shaped mask edges. Ryssel *et al.* [8] extended the statistical approach by using the Pearson IV distribution and demonstrated that this gave better fits to boron implantation data.

The approach of characterizing profiles with range statistics suffer from several problems. First, range statistics data is all derived from infinitely thick material layers. In modern device structures, this is rarely true. Several thin layers are present and the ions may travel through one or more materials before coming to a stop. Second, calculations of damage to the material layer are not possible. Third, the layers are assumed amorphous and there is no accounting for ions that may scatter into the relatively unobstructed channels in the crystal. Finally, the work is primarily based on one dimensional calculations. This makes calculation of lateral profiles more difficult.

Monte Carlo calculations can provide information about damage and distribution functions in arbitrary structures. The algorithm for Monte Carlo involves tracing each incident ion and determining the stopping position. Each ion is traced through all scattering events and the final position is recorded in the doping profile. These calculations take a great deal of time, especially in two dimensions, because of the necessity of following a statistically significant number of ions striking the surface. In two dimensions, ions have to be incident along the entire length of the surface which can greatly increase the computation time.

Hobler *et al.* [9] used Monte Carlo techniques to compute range statistics for use with distribution functions. They extracted a depth dependence of the lateral standard deviation that is important in two dimensional calculations. Most previous range statistics work extracted only a single lateral standard deviation parameter. Hobler and Selberherr [10] also performed damage calculations and extracted moments. This allows simple calculations of the damage profiles. These types of calculations are quite useful since the accuracy of Monte Carlo calculations can be accessed without the computational overhead.

4 Oxidation Kinetics and Modeling

The basis of oxidation modeling was first described by Deal and Grove [11] in 1965. They proposed that oxidation occurred by diffusion of an oxidizing species through the oxide and a reaction with silicon at the interface. Figure 3 shows the fluxes through the oxide. Deal and Grove assumed that the flux through the oxide is equal to the oxidant consumed at the

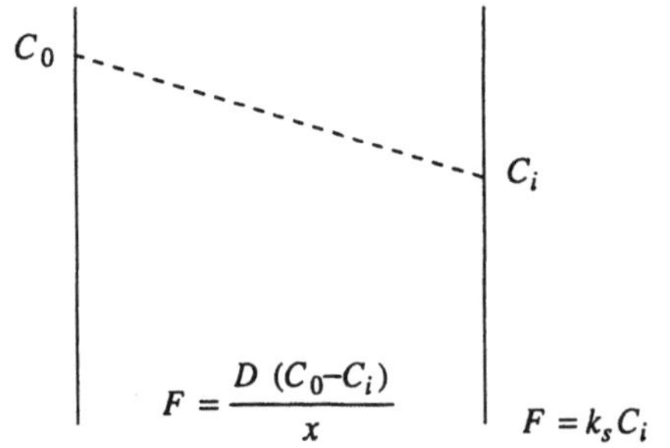

Figure 3 - Deal-Grove Oxidation Model

interface. By equating the fluxes in steady state, Deal and Grove derived the well known linear-parabolic growth law of oxide.

$$x^2 + Ax = Bt$$

where x is the oxide thickness, t is the oxidation time elapsed, and B and A are constants related to the surface concentration of oxidant C_0, the diffusivity D, and the reaction rate k_s.

Since the introduction of the LOCOS process [12] in 1970, modeling of non-planar oxide growth has been of interest. Experimental data indicated that the growth of the oxide layer can not be solved by simple extension of the one dimensional Deal-Grove model to two dimensions. On non-planar surfaces, the Deal-Grove model predicts the oxidation rate at the interface, but has nothing to say regarding the flow of the oxide above the interface. Wilson [13] assumed that the oxide grew only normal to the wafer, but found that this was insufficiently accurate. Chin [14] introduced a viscous fluid model of oxide in an attempt to correctly predict oxide shapes in two dimensions, and performed the first successful simulations of the bird's beak.

Many complications have arisen in analyzing oxide growth and subsequent flow since Chin [14] first introduced his viscous fluid model. The silicon step structures of Marcus and Sheng [15] showed that growth is retarded at both inside and outside corners. The cylindrical oxide structures of Kao [16] helped quantify this phenomenon, and showed that the retardation was greater at inside corners than outside corners. These works show that there is a feedback path from the stress to the oxidation rate coefficients, introducing a nonlinearity which must be solved self-consistently during simulation. This effect is significant even for planar oxides. Taft [17], Irene [18], and Landsberger [19] have observed that the diffusivity of oxidant in SiO_2 films depends not only on the temperature of the oxidation but also on the thermal history of the films. The variations of diffusivity with film history are believed to be due to density variations. Low temperature oxides grow in a state of compressive stress [20, 21] and are denser than oxides grown at high temperature [22].

The experimental work of Kao is important since it gives quantitative predictions of the effect of stress on oxide growth rates. The process sequence is shown in Fig. 4. Silicon rings are etched in the wafer and are oxidized to the same thickness at several temperatures. Polysilicon is deposited for contrast and the wafers are lapped back. The oxide thickness can be measured as a function of the radius of curvature of the silicon ring.

The data from these experiments is shown in Fig. 5. The vertical axis is the oxide thickness normalized by the value grown on a planar wafer. The final thickness is plotted against the inverse of the radius of the cylinders. Thickness reduction begins to appear at a radius of curvature of 1.0μm on convex corners, but at around 10.0μm on concave corners. The compressive stress at an inside corner is more effective at reducing the growth rate than the tensile stress on an outside corner.

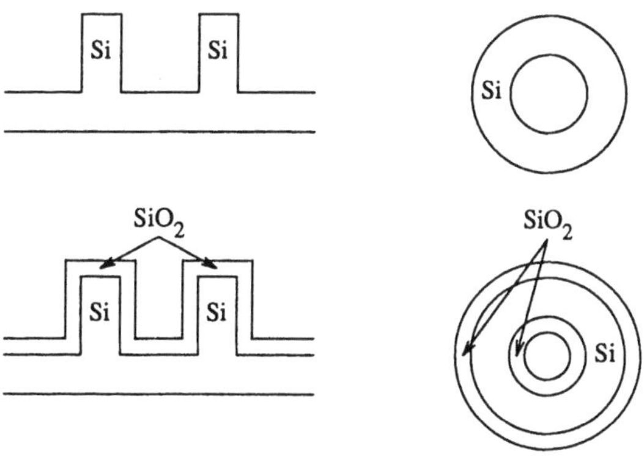

Figure 4 - Process Sequence for Kao's Experiment

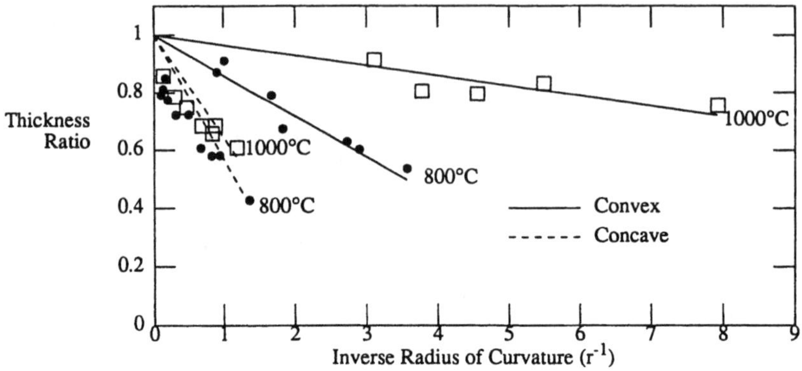

Figure 5 - Experimental Data for Wet Oxide from Kao

A nonlinearity can be inferred in the oxide flow itself by analyzing the stresses. If the measured viscosity of SiO_2 [23] is substituted in a linear model (constant viscosity), the predicted stresses around silicon corners are large enough to cause gross mechanical failure in the silicon. Such failures were not observed, so it must be that the viscosity is (substantially) reduced by the flow. Again there is a supporting evidence of this notion in planar oxidation. The stress relaxation studies of Landsberger [24] show that density relaxation in thermal oxides is much faster than the decaying exponential law predicted by a linear model. The increased rate of relaxation indicates a reduced resistance to flow, that is, a reduced viscosity.

The challenges facing two-dimensional modeling of oxide flow are to experimentally determine the stress dependence of the growth coefficients (B and B/A), the low-stress viscosity (μ_0) and its modification under high stress, and to incorporate these nonlinearities in an oxide flow simulator. In addition, work is needed to extract the viscous and/or elastic coefficients of silicon nitride in order to make meaningful predictions of stress in LOCOS structures. A third important unknown is the correlation of stress with dislocations in the substrate.

Nonlinear models have been implemented for oxide flow and growth in SUPREM-IV. The nonlinear flow model is based on the theory of Eyring [25]. His celebrated work on viscosity and diffusion predicted that the viscosity should drop exponentially when the work done by an applied shear force during a molecular hop becomes comparable with the activation barrier for flow. He derived the equation

$$\mu(\tau) = \mu_0 \frac{\tau/\sigma_c}{\sinh(\tau/\sigma_c)}$$

where τ is the maximum shear stress, and σ_c is a critical stress above which nonlinearities appear. This prediction has been verified experimentally for many different fluids, ranging from polymers [26] to iron-nickel and lead glasses [27] soda-lime glass [28] rubidium oxide glass [29] and last but not least, silicon dioxide [30]. In spite of its widespread use in other materials, only recently has the model been adopted for thermal SiO_2 growth [31]. It is reasonable to assume that nonlinear effects come into play since the threshold at which nonlinear behavior appears ($\approx 10^9$ dynes/cm^2) [30] is considerably less than the stresses predicted by the linear model ($>10^{10}$ dynes/cm^2) [16].

The net effect of the model is that the predicted stresses during oxidation are dramatically reduced. For a given rate of strain ε, instead of a linear stress of $2\mu_0\varepsilon$, the nonlinear stress is $2\sigma_c \sinh^{-1}(\mu_0\varepsilon/\sigma_c)$. For small strain rates, the expressions reduce to the same value, but for large strain rates the nonlinear stress increases only logarithmically with strain rate. At 800 °C, the nonlinear stress is three orders of magnitude less than the linear prediction.

Kao's model for the reduction of the oxidation coefficients due to stress has been tested. It predicts an exponential decrease of the rate coefficients due to stress:

$$B = B_0 e^{-PV_d/kT}$$

$$B/A = B/A_0 e^{\sigma_n V_r/kT}$$

where V_d and V_r are unknown constants, P is the hydrostatic pressure and σ_n is the stress component normal to the oxide/silicon interface. Because modifications to these coefficients alter the solution of the oxidant diffusion equation, the diffusion equation must now be solved self-consistently with the stress equations. To carry this out reliably, a Newton algorithm has been implemented to solve the diffusion and flow equations simultaneously. Although the Newton algorithm converges quadratically at the root, convergence problems are often seen far from the root. That is, once the solution variables are correct to within 1%, the updates decrease very rapidly, and an error of 10^{-6} is easy to achieve. To get to the point where the updates have decreased to 1%, it was found necessary to damp the Newton update to prevent overshoot.

By choosing values of V_d, V_r, $\mu_0(T)$, and $\sigma_c(T)$ at each temperature in order to fit Kao's data, one can in principle calibrate the simulator. However, with twelve parameters in a nonlinear model, the confidence level is not very high and it is desirable to seek additional data supporting the models. To that end, a new analysis of existing data and several new experiments have been carried out. Annealing experiments where no growth takes place are useful to estimate the parameters of the flow model, and oxidation of samples with a measured built-in stress can be used to estimate the effect of stress on the growth rate coefficients.

It has already been mentioned that oxides grown at low temperatures have a built-in stress. Their refractive indices are also higher than fused quartz or high temperature oxides, indicating higher density material. It is believed that the higher density is the origin of the built-in stress. If such oxides are annealed for long periods of time, the refractive index falls and the oxide thickness increases. Given enough annealing, the refractive index eventually reaches a value consistent with measurements on bulk oxide samples. This process does not involve oxide growth; if high-temperature (relaxed) oxides are annealed together with low-temperature (unrelaxed) oxides, the relaxed oxides show no change in thickness in an Ar ambient [19]. The density relaxation process is an internal reorganization of the oxide that involves viscoelastic flow, and therefore proceeds at a rate which depends on the instantaneous viscosity. In a linear model where the oxide had a constant viscosity μ_0, the normalized density excess $\rho' = (\rho - \rho_\infty)/(\rho_\infty)$ would follow an exponential decay law of the form [22]

$$\rho' = \rho_0' e^{-t/\tau}$$

where t is the anneal time, ρ_0' is the density excess of the as-grown oxide, and τ is a relaxation time. Landsberger [24] observed that an exponential decay does not fit the relaxation of dry oxides well. Instead, there is a very rapid initial relaxation, occurring in less than 15 minutes, followed by a slower transient which depends on the anneal temperature. This is consistent with observations in bulk samples. Mackenzie [32] found empirically that a $\log t/\tau$ dependence fitted his data better than $\exp(-t/\tau)$.

Given the Eyring prediction regarding viscosity reduction, it is interesting therefore to inquire what effect a nonlinear viscosity model might have on the relaxation time. The Eyring viscosity model was introduced into a viscoelastic model of the dilation, and the relaxation curve calculated [33]. The full solution is of the form

$$\rho' = \frac{2g\sigma_c}{B}\tanh^{-1}\left\{\tanh\left[\frac{B}{2g\sigma_c}\rho'_0\right]e^{-Bt/f\mu_{b0}}\right\}$$

B is the bulk modulus of the oxide, μ_{b0} the bulk viscosity. The constants g and f involve the Poisson ratio and are close to unity. If the exponentials are expanded, the solution can be approximated, except at very short and very long times, as

$$\rho' = -\frac{g\sigma_c}{B}\ln\left[\frac{Bt}{f\mu_{b0}}\right]$$

Instead of a change of relaxation time, the shape of the relaxation curve changes. The relaxation curve is logarithmic, exactly as observed by Mackenzie. The log law also fits the thin film data well, as shown in Fig. 6.

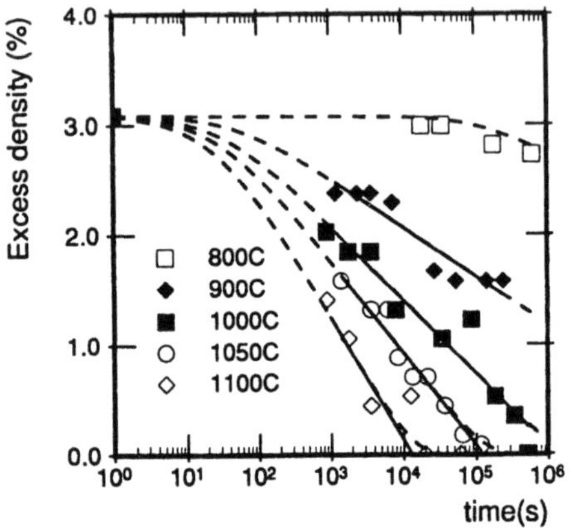

Figure 6 - Density Relaxation of Dry Oxides

From the above formula, it can be seen that the slope of the density relaxation line versus log time gives an estimate of the critical stress σ_c, and the intercept an estimate of the low-stress viscosity μ_0. The extracted data is given in Fig. 7.

Temperature (°C)	800	900	1000	1050	1100
σ_c (10^9 dynes / cm ^{2})	0.43	0.43	0.66	0.82	0.88
μ_{b0} (Poise)	10^{23}	10^{20}	$2.9 \cdot 10^{17}$	$2.9 \cdot 10^{16}$	$5.7 \cdot 10^{15}$

Figure 7 - Extracted values of critical stress and low stress viscosity

The low-stress viscosities have an activation energy of about 7.4eV in reasonable agreement with Hetherington's measurements on I.R. Vitreosil, a form of SiO_2 with low water content which is comparable to dry thermal oxide. The critical stress is found to increase at higher temperatures, in qualitative agreement with bulk data [30]. Quantitative agreement is not expected since the bulk samples were not prepared thermally, and appear to have a high water content.

To investigate the effect of stress on the growth rate coefficients, an experiment was carried out using the built-in stress as a probe. The built-in stress has the advantages that it can be measured, by wafer-bending experiments, and controlled, by selecting the oxidation temperature. The procedure was to grow the same primary thickness of oxide at 5 different temperatures from 800°C to 1100°C. Selected wafers were removed and stress measurements performed. Other wafers were selected and reoxidized long enough to grow an extra 5% at 800°C. This is is illustrated schematically in figure 8.

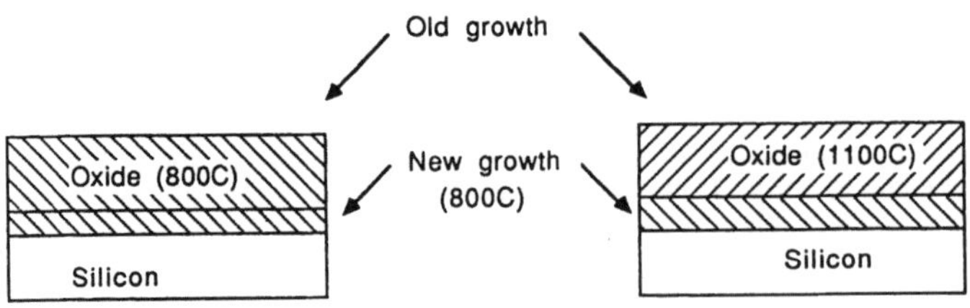

Figure 8 - Reoxidation experiment.

The experiment was repeated using primary oxide thicknesses between 300 Å and 9600 Å. The time to grow the extra thickness was much less than the days needed for relaxation at 800°C. Therefore the stress in the films was essentially constant for the duration of the experiment, and depended only on the temperature at which the film was created. The reoxidized films all began with the same thickness and were reoxidized together at the same temperature. Nevertheless, the amount of oxide grown in the second step (the "extra" oxide) varied by up to 50%. The direction of the differences was that more oxide grew under films that were created at high temperature than under films created at low temperature. This is due to the built-in stress in the low-temperature films. By comparing the different growth rates with the different measured stresses, the stress dependence of the rate coefficients can be extracted.

Stress measurements were made using a wafer curvature measurement instrument developed by Flinn *et al.* [34] A scanning laser beam is sampled at 20 locations across the wafer. The variation in beam deflection allows the curvature to be calculated. The substrate curvature is proportional to the force applied to it, so the stress in the oxide film can be deduced. By comparing curvature before and after oxidation, any initial curvature effects can be corrected for. A stress difference of about 3×10^9 dynes/cm^2 was found between films oxidized at 800°C and films oxidized at 1100°C.

Figure 9 shows reoxidation results. Each data point represents the *ratio* of extra oxide grown under a film created at a high temperature, to the extra oxide grown under a film created at 800°C, when both films are reoxidized together. In every case, more oxide grows under a high temperature film than under a lower temperature film. The differences appear to be more significant for thick films than for thin films, but it is striking that even for very thin films, the amount of oxide grown under a relaxed film (1100°C) is 30-40% greater than under an unrelaxed film (800°C). This suggests that there may be a history effect on B/A as well as B.

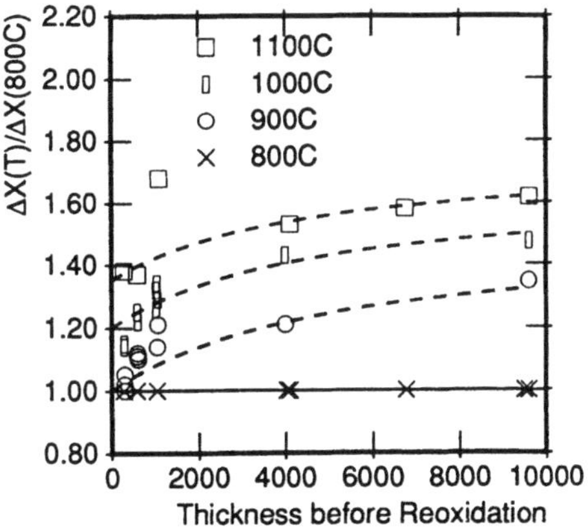

Figure 9 - Ratio of extra oxide grown as a function of film history and thickness

Since the amount of oxide grown in the second step changes the total film thickness by less than 5%, the growth rate is approximately constant and is given by

$$\frac{\Delta x}{\Delta t} \approx \frac{dx}{dt} = \frac{1}{1/(B/A)' + 2x/B'}$$

where the primes signify that the Deal-Grove coefficients depend on the upper layer. In the case where the upper layer is also formed at 800°C, the growth rate is simply $1/(1/(B/A) + 2x/B)$. Since the growth rates are lower in the latter case, $(B/A)'/(B/A) \equiv r_l > 1$ and $B'/B \equiv r_p > 1$. Multiplying through, the ratio of growth rate under a relaxed film to that under an unrelaxed film is related to the ratio of the linear and parabolic coefficients by

$$r_\Delta = \frac{A + 2x}{A/r_l + 2x/r_p}$$

Fitting the measured ratios for the 1100°C data against thickness, r_l is estimated to be ≈ 1.35 and $r_p \approx 1.75$. The activation volume V_d in the expression $B = B\exp(-PV_d/kT)$ then is of the order of $40A^3$, and that for B/A is $15A^3$.

In summary, silicon oxidation is a strongly nonlinear problem, with stress dependencies in both growth rate coefficients and in the viscosity. The stress-dependent oxidation model predicts the reduction in oxide thickness that occurs at silicon corners, for instance in trench oxidation. Experimental investigations were carried out to independently estimate the parameters of the stress-dependent models. Very little data is available for the stress-dependence of SiO_2 viscosity, even in bulk samples. By analyzing relaxation curves, the temperature dependence of the low-stress viscosity and critical stress for thin films was obtained. By comparing differential growth rates at constant, measured levels of stress, the dependence of the oxide diffusivity on stress was estimated.

5 Diffusion Kinetics and Modeling

The modeling of dopant diffusion is a complex problem due, in part, to the effect of the surface boundary conditions and the non-linearities of the diffusion rates of the dopants. In attempting to explain these effects, it is important to proceed from a sound physical basis so that the predictive capability of the process simulation is as accurate as possible. Most dopant atoms dissolve almost exclusively on lattice sites in a silicon crystal. As a consequence, most dopant diffusion takes place through interaction with point defects. Point defects can take three forms: 1) vacancies - the absence of atoms on lattice sites, 2) interstitials - an excess of atoms in between lattice sites, 3) interstitialcies - two atoms sharing lattice sites.

An example of the vacancy mechanism is shown in Fig. 10a. If a vacancy is adjacent to an impurity, the impurity and vacancy can exchange lattice sites. The vacancy then must diffuse away and approach from a different direction for the process to continue. Clearly, an impurity will diffuse more rapidly if there are more vacancies present in the lattice. An interstitial type mechanism is shown in Fig. 10b. This is an interstitialcy mechanism, where two atoms share a single atom site. Once again, the diffusivity of the dopant will be higher if there are more interstitials to interact with.

Diffusion in silicon is complicated by the fact the defects can be charged and will be effected by the Fermi level changes introduced by high concentration doping. Watkins [35] showed that vacancies exist in four charge states; positive, neutral, negative, and double negative. Since the vacancies are charged, they will be effected by changes in the Fermi level and electric field.

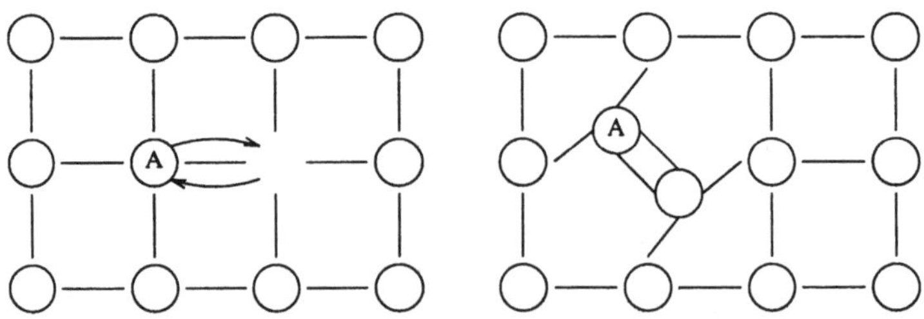

a) Vacancy Mechanism a) Interstitialcy Mechanism

Figure 10 - Defect Mechanisms for Impurity Diffusion

Charged flaws in a semiconductor were treated by Shockley and Last [36]. They considered the relative concentrations of charged defects as a function of the Fermi level in equilibrium conditions. The ratio of the concentration of charged defects to the concentration of neutral defects is constants times a simple function of the electron concentration. This result is important to diffusion modeling, since it states that at equilibrium the number of charged defects will be a function of the Fermi level. Since dopants diffuse by interaction with defects, as the number of defects change so will the diffusivity of the dopant.

Shockley and Last's work on the relative concentrations of defects is the basis for Fair's models of diffusion [37,38]. These models are based on the assumption that impurities in silicon diffuse by means of interaction with charged defect species. The total diffusion of a dopant is computed by summing the contributions from each of the pairs made by the dopant and defects.

$$\frac{\partial C_A}{\partial t} = \nabla \left[\sum_{X,c} d_{AX^c} \left\{ \nabla C_{AX^c} + Z_{AX^c} \frac{q}{kT} C_{AX^c} \nabla \psi \right\} \right]$$

where C_{AX^c} refers to the concentration of impurity A - charge state c vacancy pairs, d stands for the diffusivity of the quantity in the subscript, Z is the charge state, and ψ is the potential. The sum is over all the charge states and both defect types. This can be simplified using local thermal equilibrium, *i.e.*, $C_{AX^c} = k(T) C_A C_{X^c}$. If this is true at every point, the continuity equation can be written [39]:

$$\frac{\partial C_A}{\partial t} = \nabla \left[\sum_{X} D_{AX} \left\{ \nabla C_A \frac{C_X}{C_X^*} + C_A \frac{C_X}{C_X^*} \nabla \log \frac{n}{n_i} \right\} \right]$$

where the summation is over both defect types, n and n_i are the electron and intrinsic electron concentration, and D_{AX} can be written:

$$D_{AX} = \sum_c D_{AX^c} \frac{C_{X^c}}{C_{X^c}^i}$$

where the sum is over all charge states of defect X, and the superscript i refers to inert surface conditions with intrinsic doping. The ratio of defect concentrations can be simplified by using the work of Shockley and Last [36].

Of practical interest is the effect of the surface treatment on the distribution of point defects in the wafer. Oxidation, nitridation, and oxynitridation all are known to perturb the defect concentrations and therefore the impurity diffusivities. The best known of these effects is oxidation enhanced diffusion (OED). Oxidation of silicon injects silicon interstitials which causes increased diffusion of boron, arsenic and phosphorus as well as the growth of stacking faults. In two dimensions, the interstitials diffuse away from the oxidizing interface into the bulk and recombine at neighboring surfaces. This gives rise to a two dimensional interstitial profile. Lin [40] designed experiments to try and understand the two dimensional nature of the interstitial profile. The experimental structure is shown in Fig. 11. This experiment varied the width of stripes of masked oxide separated by exposed oxide. Diffusion was carried out in an oxidizing ambient. The injected interstitials flow laterally, enhancing the diffusion of boron in the masked areas. By plotting the junction depth at the center of the masked stripes as a function of the width of the stripe, an estimate for the lateral decay of OED can be obtained.

Fahey [41] used ammonia ambients to investigate diffusion. Bare silicon when exposed to ammonia reacts and becomes thermal nitride. This process is thought to inject vacancies. Pad oxide when exposed to ammonia reacts and forms oxynitride which injects interstitials. These regions were used side by side with inert areas to investigate diffusion further. Fahey concluded that boron and phosphorus diffuse primarily by interaction with interstitials. Both thermal nitridation and oxynitridation produce two dimensional defect distributions which perturb the dopant diffusivity from equilibrium.

Hu [42,43,44] proposed a standard form of the boundary condition to use for defects. This boundary condition included both a term for the recombination and generation at the surface. This boundary condition is general enough to represent the effect oxidation, nitrida-

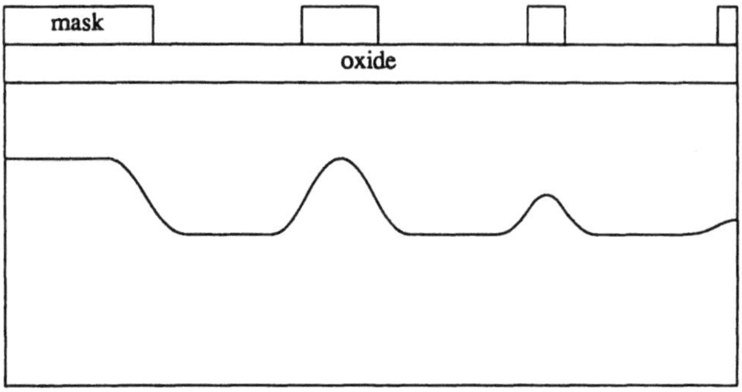

Figure 11 - Lin's Experiment to Determine the Lateral extent of OED

tion, and thermal nitridation have on diffusion. The flux balance of defects at the surface can be expressed:

$$-D_X \frac{\partial C_X}{\partial y} + K_X (C_X - C_X^*) = g_X$$

The first term is the diffusion of the X type defect away from the surface, the second, $K_X (C_X - C_X^*)$ is a surface recombination term, and g_X represents the generation at the surface. C_X is the concentration of X defects, and C_X^* is the equilibrium concentration of X defects, which can be either interstitials or vacancies.

In the bulk the defects are known to obey a standard continuity equation of the form:

$$\frac{\partial C_X}{\partial t} = \nabla \left[\sum_c \left\{ D_{X^c} \left[\nabla C_{X^c} + C_{X^c} \nabla \log \frac{n}{n_i} \right] \right\} \right]$$

where D_{X^c} is the diffusivity of the defect X in the c charge state. It is important to include the effect of the field on the defects, since they are a charged species.

To model diffusion in two dimensions, it is necessary to extract the defect parameters. For an experiment similar to Lin's, there are several parameters that must be determined. Figure 12 illustrates these parameters. At the oxidizing interface, it is necessary to extract both a surface recombination and generation rate. The nonoxidizing interface has a characteristic recombination velocity. The bulk diffusion and recombination must also be modeled.

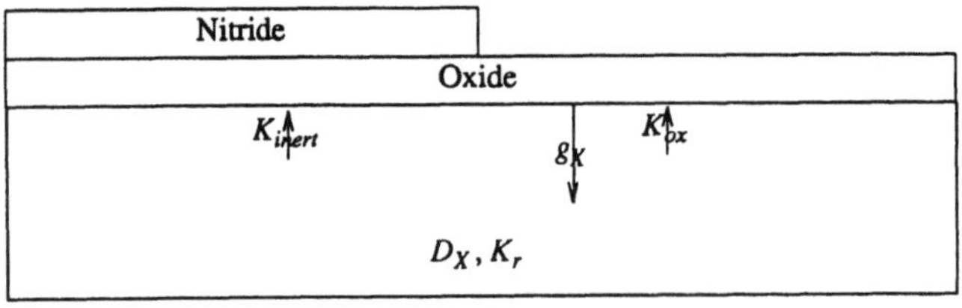

Figure 12 - Boundary Conditions and Parameters for Defects

Griffin and Plummer [45] performed a series of experiments to investigate the defect parameters. Stacking fault experiments have shown that oxidation injects interstitials. Therefore, oxidation experiments can be used to monitor the interstitial behavior. These experiments were similar to the stripe experiment of Lin. Griffin was able to extract diffusivities and recombination rates uniquely based on the two dimensionality of the structure and the time dependence of the enhanced diffusion. A large spread in the data has been reported for the interstitial diffusivity in literature [46,47,48,49]. Several trends can be noticed. First, the diffusivities from gettering experiments are higher than those reported from dopant diffusion studies. Second, the diffusivity in dopant diffusion studies is different for epilayers, float-zone, and Czochralski materials. The diffusivity is fastest for epi and slowest in Czochralski silicon. Griffin explained this with an interstitial trap mechanism which depends on the silicon material. Most experiments measure an effective diffusivity and ignore the effects of bulk recombination. The interstitial trap mechanism allows the wide variation in data to be explained. Griffin extracted the following expressions for surface recombination and diffusivity:

$$D_I = 2.70 \cdot 10^{-4} \exp\left(\frac{-1.57}{kT}\right)\frac{cm^2}{s}$$

$$K_{ox} = 7.93 \cdot 10^{-11} \exp\left(\frac{0.81}{kT}\right)\frac{cm}{s}$$

Novel structures can be designed using surface properties to enhance or retard the diffusion through point defects. N-well processing is a good example of the type of options available. A standard process uses a phosphorus implant to dope the well n-type followed by a long drive-in (i.e., 30 hours at 1125° C). This produces a deep junction depth (3.5μm), and also produces considerable lateral diffusion(2.8μm).

Fahey [41] has shown that oxynitridation injects interstitials and nitridation injects vacancies. Wong and Ekstedt [50] made use of this fact to design a new well process which showed the same junction depth as the standard process but could be realized with only a 15 hour cycle when diffused in ammonia instead of 30 hours. In addition, the lateral diffusion of the well is reduced from 2.8μm to 2.3μm. The area in the non-well regions is stripped and a pad oxide is left on the well regions. When exposed to ammonia, the bare surface forms a thermal nitride and injects vacancies. The pad oxide is converted to oxynitride and injects interstitials.

Since phosphorus primarily diffuses via an interstitial mechanism [41], the phosphorus diffusion is enhanced in the well region. This allows the same vertical junction depth to be obtained with half the drive-in time. The nitridation that occurs in non-well areas provides a source of vacancies for bulk recombination as well as a surface sink for interstitials. Therefore, the enhancement factor drops as the phosphorus diffuses under the mask edge. This results in a reduced lateral junction, which is measured as 2.3μm. Simulation showed that the important factor determining the lateral diffusion is the surface recombination [51]. Further improvements are possible after the analysis of the surface and bulk effects. The bulk recombination is of little importance in determining the junction depth in either the lateral or vertical direction. The surface recombination can be exploited to improve the junction shapes.

The observed variation in surface recombination rate for point defects with surface materials (oxide, nitride, and oxynitride) suggest possible new applications to lateral junction engineering. Figure 13 shows the qualitative descriptions of the surface material's effects on defect concentrations. One improved structure can be fabricated by replacing the nitride with a pad oxide. Since the pad oxide has a higher surface recombination, it should further reduce the lateral diffusion. This structure will still feature reduced drive-in times because oxynitridation is occurring.

Surface	Recombination	Injects
Nitride	Weak	Vacancies
Pad Oxide	Moderate	Nothing
Oxynitride	Strong	Interstitials

Figure 13 - Qualitative Surface Description

Other structures can also be designed which make use of antimony as the well dopant. Antimony is typically not used for a diffused dopant because of its low diffusivity. However, Fahey [41] showed that antimony diffuses via a vacancy mechanism. The enhancement from a thermal nitridation for antimony is a factor of five at 1100° C and is approximately independent of time. Under a thermal nitridation condition, antimony has a diffusivity which is suitable for well diffusion. Two new processes can be proposed which make use of a nitride in the well regions and antimony as the well dopant. The first process uses a pad oxide in the non well areas and the second uses an oxynitride.

The simulated processes for well fabrication are summarized in Fig. 14. All of these processes were simulated using SUPREM-IV. All of the processes are straightforward modifications to the standard process, mainly involving new surface treatments for the well and non-well areas. All of the new process drive-ins are performed in an ammonia ambient. The drive-in times were chosen so that the wells would all have the same vertical junction depth. It is clear that the material on the surfaces adjoining the well region have a strong influence on the final lateral junction.

Name	Ion	Ambient	Well Surface	Adjacent Surface
Standard	P	H_2	Oxide	Oxide
Wong & Ekstedt	P	NH_3	Oxynitride	Nitride
Phosphorus 1	P	NH_3	Oxynitride	Oxide
Antimony 1	Sb	NH_3	Nitride	Oxide
Antimony 2	Sb	NH_3	Nitride	Oxynitride

Figure 14 - Descriptions of Well Processes

By simple variation of the surface materials, the shape of the junction can be radically altered. Each of the processes proposed has the same vertical junction depth but the lateral junction is strongly influenced by the surface materials. This type of processing can be done for other process steps to control the lateral profile separately from the vertical profile.

Simulation Results			
Name	Drive In	Vertical	Lateral
Standard	30 h	3.54μm	2.65μm
Wong & Ekstedt	15 h	3.48μm	2.45μm
Phosphorus 1	15 h	3.49μm	2.31μm
Antimony 1	32 h	3.52μm	2.00μm
Antimony 2	32 h	3.49μm	1.68μm

Figure 15 - Simulated Results for Junctions

6 Numerical Solutions

The underlying assumption of most numerical simulators is that the local defect concentration determines the local diffusivity of the dopant. Many simulators [52, 53, 54, 55] have used the Fermi level to determine the defect concentrations. This approximation is fine for simple cases. However, both for phosphorus diffusion and oxidation the defects are perturbed away from the equilibrium conditions. Therefore, many simulators have begun to solve for the defect concentrations explicitly [56, 57, 58, 59]. This is a complicated problem due to the large difference in typical defect and dopant diffusivities. It is also complicated by the fact the during some of the most interesting cases, oxidation is occurring and the boundary is moving. This requires accurate solutions and modeling of the oxide growth rate as well as the effect of the boundary on the diffusion.

The numerical solution of these equations can be broken into three main parts. The first part is the time integration of the partial differential equations. The second uses Newton's method to solve the non-linear system of equations produced by finite element techniques. The third part involves the solution of the resulting linear system.

The choice for the time integration is strongly influenced by the moving boundary problem. Since the grid can potentially change at every time step, higher order time integration schemes would require several interpolations onto the latest grid. The interpolation error can then affect the accuracy of the time integration or influence the estimates of the time step size. The moving boundary requires that a fairly simple low order integration scheme be used. However, since the problem is stiff, stability is an issue.

The time step scheme that has proven most stable for integration of the dopant defect equations is the combined trapezoidal rule - second order backward difference (TRBDF) scheme proposed by Bank, *et al.* [60] This scheme takes two time steps in the interval $t = t_n$ to $t = t_n + h_n$. The first time step is a trapezoidal rule step from $t = t_n$ to $t_n + \gamma h_n$, where γ is a factor between 0.0 and 1.0. The second time step is a second order backward difference step to the final value of $t_n + h_n$. This scheme is L-stable and A-stable and provides a convenient means to estimate the time integration error to allow the choice of a time step. Bank *et al.* show that by choosing γ to be equal to $2.0 - \sqrt{2.0}$, the time terms in the Jacobian are the same. Divided differences or a Milne's technique [57] can be used to estimate error and for time step size control.

The continuity equations for dopants and defects as well as the oxidation flow and growth equations are strongly nonlinear. Newton's method can be used to linearize the differential equations. This is a simple extension of the well known one dimensional Newton's method to work with operators and multiple variables. This produces a system of linear equations representing the Jacobian. The resulting linear equations can be converted to an algebraic system by a straightforward application of the finite element method [61,62]. The diffusion solutions typically make use of first order elements and the oxidation solution makes use of second order elements.

Finite element techniques are based on approximation of functions theory. Each element, *e.g.* a triangle, has the function of interest defined over its space. The function is defined to be the sum of several nodal shape functions in the element. For a triangle, each of three nodes would be represented by a shape function. These functions are chosen so that they are one at the associated vertex and zero at the other vertices. In addition, the function is set to zero outside the element. These functions are usually simple polynomials. In SUPREM-IV, either linear or quadratic elements are used.

The resulting linear algebraic system is solved using an iterative scheme. This iteration is a conjugate residual technique based on that proposed by Elman [63] with a preconditioner consisting of the inverses of the diagonal blocks. Each diagonal block is the Jacobian of the continuity equation of a impurity with respect to that impurity concentration. The inverses are computed using a complete LU decomposition, or can be computed using some other iterative technique. By relying on a block iterative technique, the size of each factorization is equal to the mesh size, not the number of differential equations times the mesh size. Since factorization tends to be exponentially related to the matrix size, savings of time of approximately an order of magnitude can be achieved over factoring the entire system.

Most of the iterative methods have shown to be more costly then direct factorization of the matrix. The direct factorization makes use of auto Newton-Richardson techniques. The matrix is refactored only when the Jacobian changes significantly -- in practice, the Jacobian is factored once for every time integration. By choosing γ appropriately in the TRBDF method as discussed in section 2.2, the Jacobian has the same form over both time integration steps [60]. Using TRBDF, the matrix is factored only once each time step. The total CPU time is approximately a linear function as a function of the number of mesh points. There are two reasons for this. First, the number of LU decompositions for each time integration is small. Less than 20%, on average, of the CPU time is spent performing the numerical factorizations of the Jacobian. The bulk of the time is spent in performing forward/backward solves. The forward/backward solve is more closely linear in the number of nodes. Second, the vectorizing compiler decreases the power dependence of most algorithms on mesh points. It is for these two reasons that solution time is essentially linear in the number of nodes.

7 Summary

The challenges to process simulation have been to model dopant redistribution and material deposition, etching and growth. Progress has been made in understanding the physical processes of oxide growth and dopant diffusion. Experiments have been carried out which help quantify the various model coefficients. Further work needs to be done on high concentration effects, although some promising work has been done [64, 65]. Additional work also needs to be done on the complex problem of the oxidation of polysilicon gate material.

References

1. W.G. Oldham, S. Nandgaonkar, A. Neureuther, and M. O'Toole, "A General Simulator for VLSI Lithography and Etching Processes: Part I Applications to Projection Lithography," *IEEE Trans. on Electron Devices*, vol. 26, p. 717, April 1979.
2. W.G. Oldham, S. Nandgaonkar, A. Neureuther, and M. O'Toole, "A General Simulator for VLSI Lithography and Etching Processes: Part II Application to Deposition and Etching," *IEEE Trans. on Electron Devices*, vol. 27, p. 1445, August 1980.
3. S. Yamamoto, T. Kure, M. Ohgo, T. Matsuzawa, S. Tachi, and H. Sunami, "A Two Dimensional Etching Profile Simulator: ESPRIT," *IEEE Transaction on Computer Aided Design*, vol. 6 No. 3, p. 417, May 1987.
4. J. Pelka, K.P. Muller, and H. Mader, "Simulation of Dry Etch Processes by COMPOSITE," *IEEE Transaction on Computer Aided Design*, vol. 7 No. 2, p. 154, February 1988.
5. J.F. Gibbons, W.S. Johnson, and S.W. Mylroie, *Projected Range Statistics*, Dowdon, Hutchinson and Ross, 1975.
6. S. Furakawa, H. Matsumura, and H. Ishiwara, *Japan J. Appl. Phys.*, vol. 11, p. 134, 1972.
7. H. Runge, *Phys. Status. Solidi*, 39(a), p. 595, 1977.
8. H. Ryssel, K. Haberger, K. Hoffman, G. Prinke, R. Dumcke, and A. Sachs, "Simulation of Doping Processes," *IEEE Transactions on Electron Devices*, vol. 27 No. 8, p. 1484, Aug., 1980.
9. G. Hobler, E. Langer, and S. Selberherr, "Two-Dimensional Modeling of Ion Implantation with Spatial Moments," *Solid State Electronics*, vol. 30 No. 4, p. 445, 1987.
10. G. Hobler and S. Selberherr, "Two-Dimensional Modeling of Ion Implantation Induced Point Defects," *IEEE Transactions on Computer Aided Design*, vol. 7 No. 2, p. 174, February, 1988.
11. B.E. Deal and A.S. Grove, "General Relationship for the Thermal Oxidation of Silicon," *J. Appl. Phys.*, vol. 36(12), pp. 3370-3378, 1965.
12. J.A. Appels, E. Kooi, M.M. Paffen, J.J.H. Schatorje, and W.H.C.G. Verkuylen, "Local Oxidation of Silicon and its Application in Semiconductor-Device Technology," *Philips Research Reports*, vol. 25, p. 118, 1970.
13. Lynn O. Wilson, "Numerical Simulation of Gate Oxide Thinning in MOS Devices," *Electrochemical Society Journal*, vol. 129, no.4, p. 831, Apr 1982.
14. D. Chin, "Two-Dimensional Oxidation," PhD dissertation, Department of Electrical Engineering, Stanford University, June 1983.
15. R.B. Marcus and T.T. Sheng, "The Oxidation of Shaped Silicon Surfaces," *J. Electrochm. Soc.*, vol. 129(6), 1982.
16. D-B. Kao, "Two-dimensional Oxidation Effects in Silicon - Experiments and Theory," PhD Dissertation, Department of Electrical Engineering, Stanford University, June 1986.
17. E.A. Taft, "Diffusion of Oxygen in Silicon Thermal Oxides," *J. Electrochem. Soc.*, vol. 132(10), p. 2486, 1985.
18. J.K. Srivastava and E.A. Irene, "A Measurement of the Effect of Intrinsic Film Stress on the Overall Rate of Thermal Oxidation of Silicon," *J.Electrochem. Soc.*, vol. 132, p. 2815, 1985.
19. L.M. Landsberger and W.A. Tiller, "Indicators of Stress Relaxation in Two-Step Thermally Grown Silicon Dioxide Films on Silicon," in *Proceedings of the Symposium on*

Silicon Nitride and Silicon Dioxide Thin Insulating Films, vol. 87-10, p. 333, San Diego, Oct 20-24, 1986.

20. E.P. Eernisse, "Stress in Thermal SiO_2 during growth," *Appl. Phys. Lett.*, vol. 35(1), 1979.
21. E. Kobeda and E.A. Irene, "A measurement of intrinsic SiO_2 film stress resulting from low temperature thermal oxidation of Si," *J. Vac. Sci. Technol*, vol. B 4, p. 720, June 1986.
22. E.A. Irene, E. Tierney, and J. Angilello, "A Viscous Flow Model to Explain the Appearance of High Density SiO_2 at Low Oxidation Temperatures," *J. Electrochem. Soc.*, vol. 129(11), p. 2594, 1982.
23. G. Hetherington, K.H. Jack, and J.C. Kennedy, "The viscosity of vitreous silica," *Physics and Chemistry of Glasses*, vol. 5 No. 5, p. 130, Oct 1964.
24. L.M. Landsberger and W.A. Tiller, "Refractive Index, Relaxation Times and the Viscoelastic Model in Dry-Grown SiO_2 films on Si.," *Appl. Phys. Lett.*, vol. 51(18), p. 1416, Nov. 1987.
25. H. Eyring, "Viscosity, Plasticity, and Diffusion as Examples of Absolute Reaction Rates," *J. Chemical Physics*, vol. 4, p. 283, Apr 1936.
26. Ferry, J.D., *Viscoelastic Properties of Polymers*, Wiley, New York, 1970.
27. D.G. Ast and D.J. Krenitsky, "Stress relaxation in coiled ribbons of $FE_{40}Ni_{40}P_{14}B_6$ and $FE_{29}Ni_{49}P_{14}B_6S_2$," *J. Materials Science*, vol. 14, p. 287, 1979.
28. J.H. Simmons, R.K. Mohr, and C.J. Montrose, "Non-Newtonian Viscous Flow in Glass," *J. Applied Physics*, vol. 53, p. 4075, Jun 1982.
29. J.H. Li and D.R. Uhlmann, "The flow of Glass at High Stress Levels," *Journal of Non-Crystalline Solids*, vol. 3, p. 127, 1970.
30. P.P. Donnadieu, O. Jaoul, and M. Kleman, "Plastic Behaviour of Silica Glass Near the Glass Transition," *Philosophical Magazine A*, vol. 52, p. 5, 1985.
31. P. Sutardja, W. Oldham, and D-B. Kao, "Modeling of Stress-Effects in Silicon Oxidation including the non-linear Viscosity of Oxide.," in *IEDM Tech. Digest*, 1987.
32. J.D. Mackenzie, "High-Pressure Effects on Oxide Glasses: II, Subsequent Heat Treatment," *Journal of the American Ceramic Society*, vol. 46, no. 10, p. 470, Oct 1963.
33. C.S. Rafferty, L.M. Landsberger, R.W. Dutton, and W. Tiller, "Nonlinear viscoelastic dilation of SiO_2 films," *Appl. Phys. Lett.*, (submitted)..
34. P.A. Flinn, D.S. Gardner, and W.D. Nix, "Measurement and Interpretation of Stress in Aluminum-Based Metallization as a Function of Thermal History," *IEEE Trans. Electron Devices*, vol. ED-34(3), p. 689, Mar 1987.
35. G.D. Watkins, "EPR Studies of the Lattice Vacancy and Low Temperature Damage Processes in Silicon," in *Lattice Defects in Semiconductors 1974*, ed. F.A. Huntley, Inst. Phys. Conf. Ser. 23, London, 1975.
36. W. Shockley and J.T. Last, "Statistics of the Charge Distribution for a Localized Flaw in a Semiconductor," *Phys. Rev.*, vol. 107 No. 2, p. 392, July 15, 1957.
37. R.B. Fair, "Concentration Profiles of Diffused Dopants in Silicon," in *Impurity Doping Processes in Silicon*, ed. F. F. Y. Yang, p. 315, North-Holland, New York, 1981.
38. R.B. Fair, "Recent Advances in Implantation and Diffusion Modeling for the Design and Process Control of Bipolar ICs," in *Proceedings of the 3rd International Symposium on Silicon Materials Science and Technology*, ed. H.R. Huff and T. Abe, p. 968, May 1977.
39. M.E. Law, "Two Dimensional Numerical Simulation of Dopant Diffusion in Silicon," PhD dissertation, Department of Electrical Engineering, Stanford University, Jan. 1988.
40. A.M. Lin, R.W. Dutton, and D.A. Antoniadis, "The Lateral Effect of Oxidation on Boron Diffusion in <100> Silicon," *Appl. Phys. Lett.*, vol. 35 No. 10, p. 799, Nov. 15, 1979.

41. P. Fahey, G. Barbuscia, M. Moslehi, and R.W. Dutton, "Kinetics of Thermal Nitridation Processes in the Study of Dopant Diffusion Mechanisms in Silicon," *Appl. Phys. Lett.*, vol. 46 (8), p. 784, April 1985.

42. S.M. Hu, "Kinetics of Interstitial Supersaturation During Oxidation of Silicon," *Applied Physics Letters*, vol. 43 No. 5, p. 449, Sept., 1983.

43. S.M. Hu, "Kinetics of Interstitial Supersaturation and Enhanced Diffusion in Short-Time/Low-Temperature Oxidation of Silicon," *J. Appl. Phys.*, vol. 57, p. 4527, 1985.

44. S.M. Hu, "On Interstitial and Vacancy Concentrations in Presence of Injection," *J. Appl. Phys.*, vol. 57, p. 1069, 1985.

45. P.B. Griffin and J.D. Plummer, "Process Physics Determining 2-D Impurity Profiles in VLSI Devices," *International Electron Devices Meeting*, p. 522, Los Angeles, Dec., 1986.

46. P.B. Griffin, P.M. Fahey, J.D. Plummer, and R.W. Dutton, "Measurement of Silicon Interstitial Diffusivity," *Appl. Phys. Lett*, vol. 47 No. 3, p. 319, Aug., 1985.

47. T.Y. Tan and U. Gosele, "Point Defects, Diffusion Processes, and Swirl Defect Formation in Silicon," *Appl. Phys.*, vol. A37 No. 1, p. 1, 1985.

48. K. Taniguchi, D.A. Antoniadis, and Y. Matsushita, "Kinetics of Self Interstitials Generated at the Si/SiO_2 Interface," *Appl. Phys. Lett.*, vol. 42 no. 11, p. 961, June 1983.

49. G.B. Bronner and J.D. Plummer, "Gettering of Gold in Silicon: A Tool for Understanding the Properties of Silicon Interstitials," *Journal of Applied Physics*, vol. 61 No. 12, p. 5286, June 15, 1987.

50. S.S. Wong and T.W. Ekstedt, "CMOS Well Drive-In in NH_3 for Reduced Lateral Diffusion and Heat Cycle," *IEEE Electron Device Letters*, vol. EDL-6(12), pp. 659-661, Dec. 1985.

51. M.E. Law, C.S. Rafferty, and R.W. Dutton, "New N-Well Fabrication Techniques Based on 2D Process Simulation," *International Electron Devices Meeting*, p. 518, Los Angeles, Dec., 1986.

52. C.P. Ho, J.D. Plummer, S.E. Hansen, and R.W. Dutton, "VLSI Process Modeling - SUPREM III," *IEEE Transactions on Electron Devices*, vol. 30, No. 11, pp. 1438-1453, Nov., 1983.

53. R. Tielert, "Two Dimension Numerical Simulation of Impurity Redistribution in VLSI Processes," *IEEE Transactions on Electron Devices*, vol. 27 No. 8, p. 1479, Aug., 1980.

54. B.R. Penumalli, "A Comprehensive Two-Dimensional VLSI Process Simulation Program, BICEPS," *IEEE Transactions on Electron Devices*, vol. ED-30(9), p. 986, Sept. 1983.

55. C.D. Maldanado, F.Z. Custode, S.A. Louie, and R.K. Pancholy, "Two Dimensional Simulation of a 2 micron CMOS Process Using Romans II," *IEEE Transactions on Electron Devices*, vol. 30(11), p. 1462, Nov. 1983.

56. P. Pichler, W. Jungling, S. Selberherr, E. Guerrero, and H. Potzl, "Simulation of Critical IC Fabrication Steps," *IEEE Transactions on Computer Aided Design*, vol. 4 No. 4, p. 384, Oct., 1985.

57. H.R. Yeager and R.W. Dutton, "An Approach to Solving Multi-Particle Diffusion Exhibiting Nonlinear Stiff Coupling," *IEEE Transactions on Electron Devices*, vol. ED-32, pp. 1964-1976, Oct. 1985.

58. M.E. Law and R.W. Dutton, "Verification of Analytic Point Defect Models using SUPREM-IV," *IEEE Transactions on Computer Aided Design*, p. 181, Feb., 1987.

59. M.R. Kump and R.W. Dutton, "The Efficient Simulation of Coupled Point Defect and Impurity Diffusion," *IEEE Transactions on CAD*, vol. 7 No.2, p. 191, Feb. 1988.

60. R.E. Bank, W.M. Coughran, Jr., W. Fichtner, E.H. Grosse, D.J. Rose, and R.K. Smith, "Transient Simulation of Silicon Devices and Circuits," *IEEE Trans. on Electron Devices*, vol. ED-32, p. 1992, Oct 1985.

61. K.A. Salsburg and H.H. Hansen, "FEDSS-Finite Element Diffusion Simulation System," *IEEE Transactions on Electron Devices*, vol. 30 No. 9, p. 1004, Sept., 1983.

62. A. Poncet, "Finite Element Simulation of Local Oxidation of Silicon," *Summer Course on VLSI process and Device Modeling*, Heverlee, Belgium, June 7-10, 1983.

63. H.C. Elman, "Preconditioned Conjugate Gradient Methods for Nonsymmetric Systems of Linear Equations," *Research Report* , vol. 203, April, 1981.

64. F.F. Morehead and R.F. Lever, "Enhanced "Tail" Diffusion of Phosphorus and Boron in Silicon: Self Interstitial Phenomena," *Appl. Phys. Lett.*, vol. 48 No. 2, p. 151, Jan., 1986.

65. B.J. Mulvaney and W.B. Richardson, "Model for Defect-Impurity Pair Diffusion in Silicon," *Appl. Phys. Lett.*, vol. 51 No. 18, p. ACO 1439, Nov. 2, 1987.

Simulation of Laser-Assisted Doping of Silicon – The Temperature Distribution

W. Zittel

Fraunhofer-Institut für Festkörpertechnologie,
Paul-Gerhardt-Allee 42, D-8000 München 60, Fed. Rep. of Germany

Doping of silicon driven by laser thermal heating is a challenging technological variant with future applications. In this contribution, we report on semianalytical solutions of the heat conduction equation. These allow a fast, but still rather accurate calculation of melting width and depth. The results are used to demonstrate the effects of different laser intensity distributions.

1. Introduction

Doping of silicon by laser heating offers several advantages. First, for SOI-structures recrystallized by a laser beam, it provides a natural extension to add a dopant precursor into the reaction chamber which reacts with the substrate during recrystallisation /1/. This process is without photographic, masking, implantation or etching steps and therefore, at least, compensates its shortcoming due to the low process speed of several cm/s. Secondly, for VLSI a reduced step size merits increased attention. Here laser doping may become a virtual tool for doping in the sub-μ range, if the process parameters can be controlled adequately /2/.

Modelling of the problem contains two major ingredients. First, the temperature distribution has to be calculated. This results in a knowledge of the melting regime. Secondly, the dopant penetration into that region has to be modelled. This usually is done by solving the diffusion equation according to proper boundary conditions /2/.

In this contribution, we concentrate on the calculation of the temperature distribution for bulk material. It is done analytically as far as possible. The generality of the method allows to transfer the results, if reasonably adjusted, to other heating problems, or to use them as a cross check for more sophisticated computer codes. Furthermore, we compare two different laser modes on their applicability for homogeneous doping.

2. Theory

The basic equation to be solved is given by the general heat conduction equation /3/:

$$\left(c_p(T)\,\frac{\partial}{\partial t} - \nabla \cdot K_{th}(T)\,\nabla\right) T(\vec{r},t) = \alpha e^{-\alpha z}\, I(\vec{r},t). \qquad (1)$$

Here, c_p gives the specific heat per unit volume, K_{th} the heat conduction coefficient, and ∇ the Nabla operator. The source term on the right hand side, $I(\vec{r},t)$, stands for the laser power density weighted with the absorption

Springer Series in Materials Science, Vol. 13
Semiconductor Silicon Editors: G. C. Harbeke and M. J. Schulz

path. The double Nabla operation can be transformed into the Laplace operator with the Kirchhoff transformation /4/:

$$\theta(\vec{r},t) = \frac{1}{K_{th}(T_0)} \int_{T_0}^{T} K_{th}(T')\, dT' \;. \qquad (2)$$

For silicon, where K_{th} is well modelled with $K_{th}(T)=299/(T-99K)$ W/cm /5/, the transformation can be carried out analytically with the result

$$T(\vec{r},t) = T_0 + (T_0-99K)\, \exp(\theta_0\, N(\vec{r},t)/(T_0-99K)). \qquad (3)$$

The transformation function θ has been split into a dimension carrying factor $\theta_0=P(1-R)/2\sqrt{\pi}K_{th}(T_0)w)$, and a scaling function $N(\vec{r},t)$. The closed form solution of the transformed equation is derived by integrating over the proper Greens function:

$$\theta(\vec{r},t) = \int\int d\vec{r}'dt'\; G(|\vec{r}-\vec{r}'|,|t-t'|)\alpha e^{-\alpha z'} I(\vec{r}',t')/c_p \qquad (4)$$

with

$$G(r,t) = \frac{1}{|4\pi Dt|^{3/2}} e^{-r^2/4Dt} \quad \text{and} \quad D = K_{th}(T)/c_p(T). \qquad (5)$$

This solution, of course, cannot account for latent heat and explicit melting. However, there are arguments that these effects do not significantly alter melting depth and width /5/.

For a Gaussian type source term most of the integrals can be carried out analytically and one ends up with a 1-dimensional integral whose evaluation is left for the computer. This solution has been studied in detail by several authors /5/ for a laser operating in the TEM_{00}-mode:

$$I_{00}(\vec{r},t) = \frac{P(1-R)}{\pi w^2} \exp(-((x-vt)^2 + y^2)/w^2). \qquad (6)$$

The width of the beam is defined with w, surface reflection of the laser with R, and the whole laser power is P; its scan velocity along the x-direction is v.

Experimentally, also the TEM_{01}*-mode with doughnut like intensity structure can be selected by properly choosing the laser diaphragm /6/. This mode, e.g., is the preferable one for simultaneous recrystallisation of SOI-structures, since it gives a temperature profile with a central minimum. Mathematically, it can be traced back to the TEM_{00}-mode according to

$$I_{01}{*}(\vec{r},t) = ((\tfrac{x-vt}{w})^2 + (\tfrac{y}{w})^2) I_{00}(\vec{r},t) = -\frac{d}{d\mu} I_{00}(\sqrt{\mu}\vec{r},\sqrt{\mu}t)|_{\mu=1} \;. \qquad (7)$$

In Fig.1 we present results for the calculated melting width and depth. The curves are derived solving (1) with v=1 cm/s and w=3.25 µm. The penetration depth is chosen as $1/\alpha$ = .6 µm which corresponds to the Argon laser at λ = 514 nm. Operating in the TEM_{00}-mode, the dependence of the melting width on the beam power is nearly linear. Melting starts somewhat below 1.4 Watt. On the other hand, the TEM_{01}*-mode starts abruptly with a width corresponding to the width of its intensity maximum and further increases similar to the TEM_{00}-mode. The melting depth (see Fig. 1b) rises for the higher order mode much faster. This feature makes it harder to control its depth in an experiment.

To study the qualitative differences in more detail, in Fig. 2 solutions for the scaling function N are drawn. At v = 0 cm/s and for surface absorp-

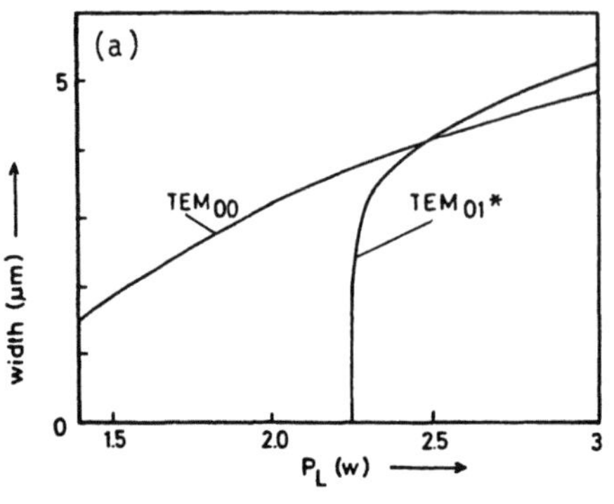

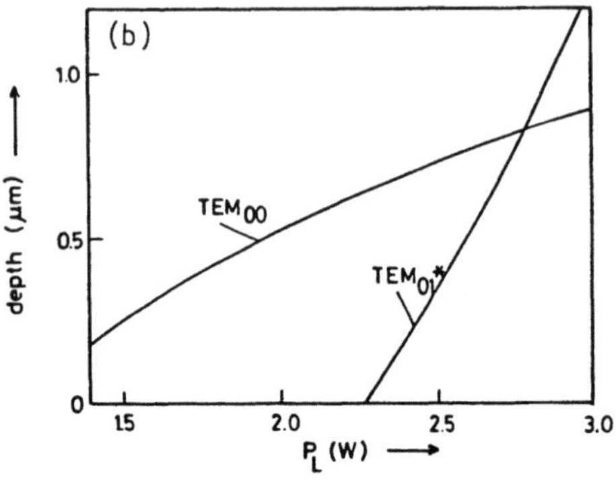

Fig. 1a : Calculated melting width for TEM_{00}- and TEM_{01}*-mode; parameters are w = 3.25 µm and T_0 = 300 K

Fig. 1b : Calculated melting depth for TEM_{00}- and TEM_{01}*-mode; parameters are w = 3.25 µm and T_0 = 300 K

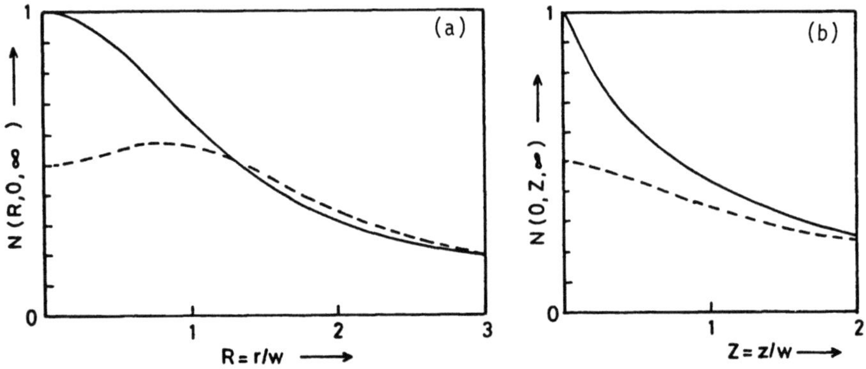

Fig. 2 : Distribution of the dimensionless temperature function N(r/w, z/w) in radial direction at the surface (a; left part) and at the center along the z-direction (b; right part); broken line is for TEM_{01}*-mode, straight line for TEM_{00}; the solutions are derived for surface absorption ($\alpha \to \infty$)

tion ($\alpha \to \infty$) these can be figured out analytically in terms of modified Bessel and error functions. For the TEM_{00}-mode the results are given in /7/. The corresponding formulas for the higher order mode can be derived by consistently applying the parametric differentiation introduced in (7). For constant heat conduction, K_{th}, N(R,Z) corresponds directly to the temperature distribution; in the general case it is intensified with the transformation function (3).

Figure 2a gives the radial temperature distribution at the surface (Z=0) for the two investigated modes. At the same laser power both modes roughly lead to the same temperature in the outer region (r > w). However, close to the center the Gaussian mode heats twice as much as the higher order mode. Figure 2b shows the central temperature change along the z-direction. The profiles clearly exhibit the stronger sensitivity of the TEM_{01}*-mode on laser power fluctuations due to its less steep profile.

3. Conclusion

We have calculated the temperature profile arising from the TEM_{00}- and $TEM_{01}*$ laser mode. The calculations are done for a semiinfinite bulk silicon. For the higher order mode the calculated melting depth is more sensitive to laser power fluctuations. On the other hand, its more homogeneous temperature profile is amendable, since it serves for more homogeneous doping.

4. References

1. H. Sigmund, J. Ernst: In Proceedings of the E-MRS Meeting, Vol. XV (Strasbourg, June 1987) p.193
2. T.W. Sigmon, P.G. Caray, K.H. Weiner: In Proceedings of the E-MRS Meeting, Vol. XV (Strasbourg, June 1987) p. 143
3. H.S. Carslaw, T.C. Jaeger: Conduction of Heat in Solids, 2nd.ed. (Oxford U.P., New York 1959) reprinted 1988
4. M. Lax: Appl. Phys. Lett. 33, 786 (1978)
5. H.E. Cline, T.R. Anthony: J. Appl. Phys. 48, 3895 (1977)
   Y.I. Nissim, A. Lietoila, R.B. Gold, J.F. Gibbons: J. Appl. Phys. 51, 274 (1980)
   J.E. Moody, R.H. Hendel: J. Appl. Phys. 53, 4364 (1982)
6. C. Stumpff, H. Sigmund: In Proceedings of the E-MRS Meeting, June 1988, to be published
7. M. Lax: J. Appl. Phys. 48, 3919 (1977)

Ion Implantation

J.M. Poate

AT&T Bell Laboratories, Murray Hill, NJ 07974, USA

1. Introduction

Ion implantation doping has become one of the dominant processing technologies in Si integrated circuit fabrication. The ability to implant electrical dopants into the surface layers of semiconductors with precise spatial and numerical control is the reason for the success of this technology. In these lecture notes we will firstly introduce the concepts and give some historical perspective. We will then discuss the ballistic aspects of an ion traversing a solid. The lattice damage produced by the impinging ion must be removed by an anneal before its electrical activity (donor or acceptor) is achieved. Finally we will discuss some recent trends in materials synthesis by implantation and fundamental studies of phase transitions in amorphous Si.

Planar silicon technology involves the fabrication of many circuit elements in a two-dimensional fashion on a flat silicon wafer. This configuration permits exact control over the fabrication steps required to make and connect the many individual circuit elements. The processing steps and materials involved can be exemplified through the field effect transistor shown in Figure 1. The goal of any electronic device is to control the flow of charge carriers - i.e., electrons or holes. In the field effect transistor the flow of carriers from the source to the drain is controlled by means of the gate electrode. The device is fabricated in the following fashion. Perfect silicon crystals are cut into wafers that are then polished to give mirror-like and defect-free surfaces. To control and improve the electrical properties of the surface, an additional layer of silicon is often grown on top of the wafer by an epitaxial growth technique such as chemical vapor deposition (CVD).

Many different processing steps are required to build up the structure and these steps must be accurately confined to specific regions of the wafer. This control is effected through pattern transfer techniques that are used to transfer geometric shapes on a mask to the surface of a silicon wafer. For example, the silicon wafers are heated in oxygen to form insulating SiO_2 layers for forming the field oxide. Lithographic techniques are used to transfer polymeric masks (photoresists) onto the regions that need to be protected in subsequent etching steps. The SiO_2 can be etched chemically or by plasma-assisted etching. In this way the unmasked SiO_2 is removed and source-drain regions are exposed. Pure silicon at room temperature has high resistivity, and dopants must be introduced to give the silicon its conducting properties. The dopants, such as arsenic, are injected into the silicon by means of ion implantation. The whole wafer is exposed to the energetic arsenic beam, but only the exposed or unmasked regions are implanted. This process is known as a self-aligned technology. A complex integrated circuit may have from 6 to 12 separate implantation steps.

The first studies of implantation in semiconductors appear to be due to Ohl [1] at AT&T Bell Laboratories. In 1952 he described the improvement in the electrical

Springer Series in Materials Science, Vol. 13
Semiconductor Silicon Editors: G. C. Harbeke and M. J. Schulz

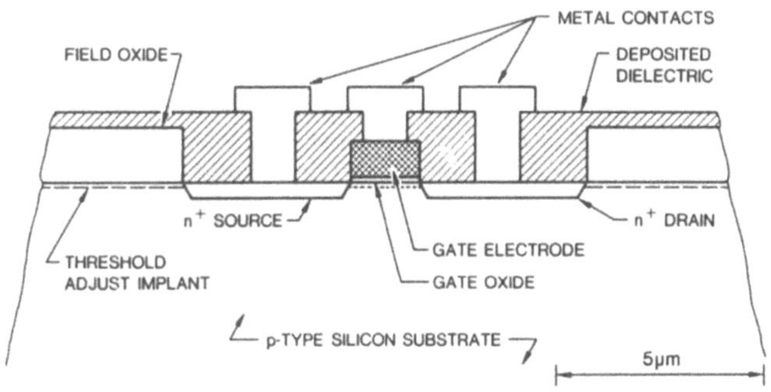

Figure 1. Cross-sectional view of a metal oxide semiconductor field effect transistor (MOSFET) structure showing the patterning and different components required to fabricate such a device structure

characteristics of Si point contact diodes produced by bombardment with hydrogen, helium, nitrogen and argon. The improvements were, in fact, caused by surface damage and not specific chemical dopings. The concept of chemical doping by implantation is spelled out in the 1954 patent by Shockley [2]. This patent entitled "Forming Semiconductive Devices by Ionic Bombardment" has proved remarkably prophetic and it is interesting to quote some of the sections of this patent.

"In accordance with this process, a semiconductor body of one conductivity type is bombarded with a monoenergetic beam of ions of a significant impurity element characteristic of a conductivity type opposite to that of the body in a manner to convert the conductivity type of a thin layer in this interior of the body...

In the process of the invention, the energy level of the bombarding beam is adjusted so that the projected beam will penetrate into the interior of the semiconductor body and be localized there for converting that region to opposite conductivity type. By utilization of a monoenergetic beam the ions are all made to penetrate to a fairly uniform depth whereby the thickness of the region of converted conductivity type is kept small. The thickness of this region can be controlled by variations in the energies of the bombarding ions. Thereafter, the semiconductive body is treated to repair the radiation damage done to the surface region penetrated. This is done advantageously by annealing at a temperature sufficiently low that there results little migration from the region of deposition of the significant impurity ions introduced by bombardment."

This patent statement still covers the subject as it exists today but the path from invention to commercial acceptance took many years. Groups in Europe and the United States in the 1960's studied the basic science of the process. Understanding the ranges of the implanted atoms and the means of removing the lattice damage were key developments. However, it was not until the early 1970's that the process gained commercial acceptance when it was recognized that only implantation doping could be used to shift threshold voltages in field effect transistors in a reproducible and reliable fashion. Today, every integrated circuit employs at least one implantation step. It is difficult to envision the remarkable growth in integrated circuit technology without the use of ion implantation.

2. Energy Loss, Range and Channeling

There are now many reviews and books on ion implantation [3,4,5]. Here we will emphasize the physical concepts and give an idea of some of the numbers involved. An ion traversing a solid loses energy by way of the Coulomb interaction with the atoms and electrons of the solids. It is convenient to mathematically separate these two components of energy loss per unit length i.e.

$$\left[\frac{dE}{dx}\right]_{total} = \left[\frac{dE}{dx}\right]_{nuclear} + \left[\frac{dE}{dx}\right]_{electronic} \tag{1}$$

The nuclear term refers to the elastic scattering event between the incoming ion and the atoms of the solid. Much energy loss and trajectory deviation can be imparted in these elastic collisions. The electronic term refers to the inelastic energy loss to all the electrons of the solid. Such energy loss will not cause the particle to deviate in its trajectory. Figure 2 shows calculations [6] of these components for As impinging on Si. The nuclear term peaks at energies less than 100 keV whereas the electronic term will peak at many MeV. Bohr [7] was able phenomenologically to divide the electronic energy loss processes into close, or single-particle and distant, or resonant collisions. The resonant collisions can be identified with the excitation of plasmons in the solid. The problem was placed on a more accurate quantum mechanical basis by Bethe and Bloch [8]. The Bethe-Bloch formula is good to within a few percent of predicting energy loss at high energies. Stopping power or $\left[\frac{dE}{dx}\right]_{electronic}$ is inversely proportional to the square of the incident ion velocity at the high energies, however, with decreasing energies the stopping power passes through a maximum as the traversing ions capture electrons and become neutralized. According to Bohr the maximum should occur at velocities of approximately $v_o Z_1^{2/3}$ where v_o is the Bohr velocity of 2.2×10^8 cm/sec, corresponding to 25 keV/nucleon. At lower energies the whole problem of the inelastic electronic interactions not only becomes more complex but elastic or nuclear collisions begin to play

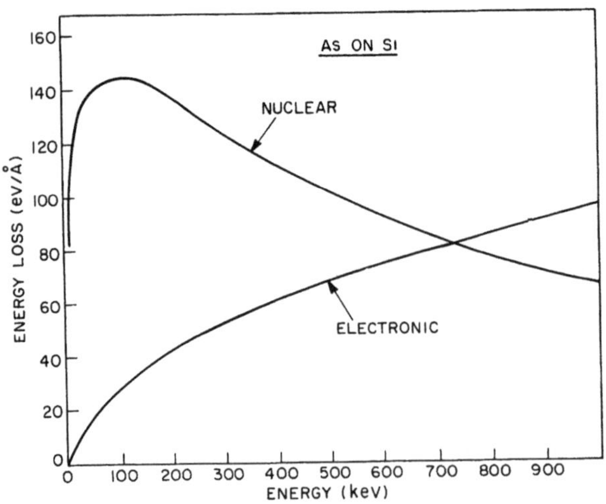

Figure 2. Energy loss as function of incident energy for As on Si. From Smith [6]

a strong role in the energy loss process. The theoretical treatment in this low velocity region is largely due to Lindhard and his collaborators. The elastic scattering cross sections are calculated from screened Coulomb potentials. The first calculation of the electronic energy loss in this region is due to Fermi and Teller [9] in 1947. They showed that the stopping power is proportional to velocity. The unified approach to the low energy stopping by Lindhard, Scharff and Schiott [10] - referred to as LSS Theory - is successful in predicting stopping and ranges. Implant energies can range from 10 keV - 1 MeV. As seen from Figure 2, the energy given to the lattice atoms is very high (~100 eV/Å) compared to binding energies. The energy given to the electronic systems is also high but will not result in lattice damage in Si. This electronic energy will eventually be dissipated as heat.

The parameter of interest in implantation is the range which can be obtained by integrating (1) i.e.

$$R = \int_0^E \frac{dE}{(dE/dx)_{total}} \tag{2}$$

where E is the incident energy. The calculations of the ranges using LSS theory [6,10] are good to within 10%. Figure 3 shows the ranges of As in Si and As in Au. Conventional implantation machines have energies ~100 keV so that ions can be implanted to depths ~1000Å in Si.

Energy loss is a statistical process which means that there will be a spread in the range distribution. To first approximation the range distribution will be Gaussian with standard deviations or straggling as shown in Figure 3. The standard deviations in the ranges are

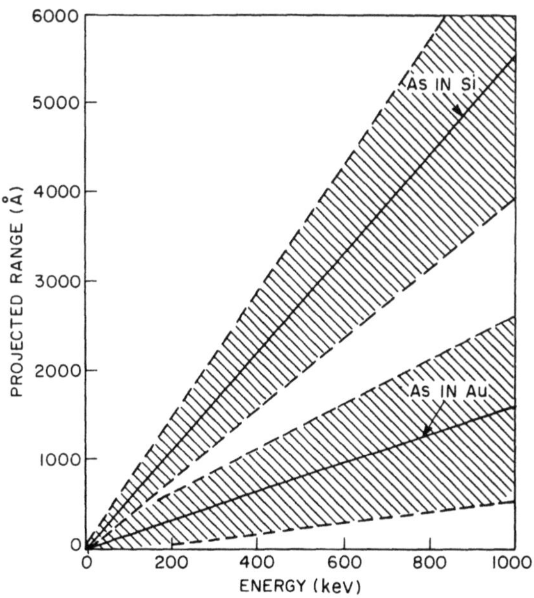

Figure 3. Projected ranges and straggling of As and Au in Si. From Smith [6]. The solid lines are the projected ranges and the hatched areas are the straggling

large because, for most heavy ion-target combinations, nuclear stopping predominates and the path of the ion through the solid will resemble a random walk process. This behavior is contrary to the ideas expressed by Shockley where he assumed that the ions would be deposited in extremely thin layers. Figure 3 gives an idea of the magnitude of the ranges and straggling. These dimensions are quite compatible with junction depths required in existing Si technology. A more exact fit to the range distribution involves a four-moment (Pearson-IV) approach from statistical theory. Figure 4 shows such a fit to B ranges [11] in fine-grain, polycrystalline Si. Gaussians only fit the ranges at low implantation energies or at the maxima of the distributions. The skewness on the shallow part of the profile can be understood in terms of B backscattering off the heavier Si atoms. Arsenic, which is heavier than Si, shows a skewness toward the deep side of the distribution. Several compilations [6,12-14] give the ranges and moments for most projectile-target combinations. Lateral straggling has to be considered when evaluating the depth distribution of ions implanted through a mask.

There is an energy loss phenomenon, channeling [15], which is not so amenable to range computations. An ion incident along a major crystallographic direction will be channeled between the strings of atoms and will experience a smaller energy loss than ions moving in a random direction in the crystal. Its range will therefore be greater than an ion of the same energy moving through a random array of target atoms. Channeling is usually manifested by a deeply penetrating tail in the range distribution. There is a certain critical angle for the incident ion, typically 5° for 100 keV ions, beyond which channeling will not initially occur. Single crystal Si wafers are usually misaligned 7° from the <100> direction to minimize channeling. Even in that situation, however, channeling effects can occur because the ions can get scattered into channeling directions after penetrating a certain thickness of the solid. The way to obviate channeling is to implant into an amorphous or fine-grain polycrystalline target. The optimum way to amorphize Si is by bombardment with Si ions.

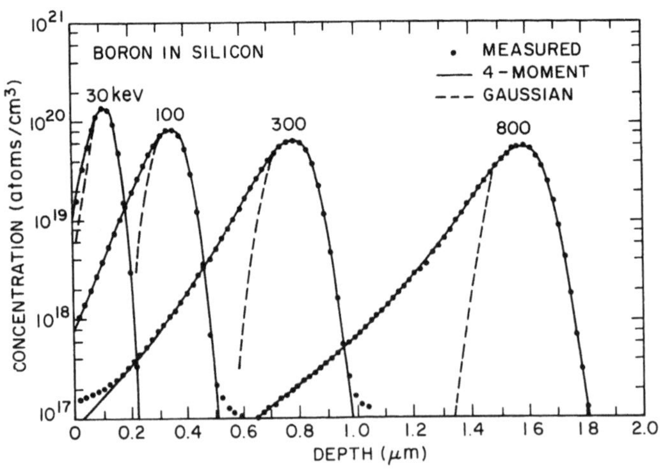

Figure 4. Boron range distribution in Si for 30, 100, 300 and 800 keV respectively. The B was implanted into fine-grain polycrystalline Si to obviate channeling effects. From Hofker [11]. Calculated ranges are four-moment or Gaussian

3. Sputtering

A noticeable effect of energy deposition from ion bombardment is sputtering, the process by which the incident beam removes surface atoms. Surface morphologies can be changed markedly by sputtering. Measurements of the sputtering yields of monoelemental targets are in good agreement [16] with theoretical predictions. The agreement between theory and experiment is shown in Figure 5 for Ar sputtering of Si. The sputtering yield, S, of Si (S = number of atoms removed per incident ion) is shown as a function of Ar ion energy. The sputtering yields for most ions incident on Si, for the energies described here, are in the range 1-5. The calculations [17] are based upon the collision cascade theory of sputtering. The incident ion makes multiple collisions with target atoms, which in turn collide with other target atoms. Momentum can thus be returned to the surface atoms and, as a result, those atoms which obtain sufficient energy to overcome the surface binding energy will be sputtered. Consequently the sputtering yield is proportional to the nuclear energy loss at the surface and is inversely proportional to the surface binding energy (E_b) i.e. $S \approx \frac{1}{E_b}\left(\frac{dE}{dx}\right)_{nuclear}$. The sputtering yield as a function of energy therefore shows the same energy dependence as the nuclear energy loss curve of Figure 2.

The data of Figure 5 show that sputtering must be considered when implantation experiments are carried out. For example, Ar implanted in Si at a dose of $10^{16}/cm^2$ and energy of 20 keV will sputter away 1.8×10^{16} Si/cm^2 or approximately 20 monolayers of Si. For high dose or very low energy implants, the amount of sputtered host material can equal or exceed the range of the incident ion. Range distributions can therefore be grossly changed by sputtering. Moreover sputtering ultimately imposes the limit on the maximum concentration of material that can be implanted. At steady state the maximum surface concentration of an implanted atom will be approximately equal to the inverse of the sputtering yield. For example if the sputtering yield of A or B is 5, then the maximum retained concentration of A in B will be 20 at.%.

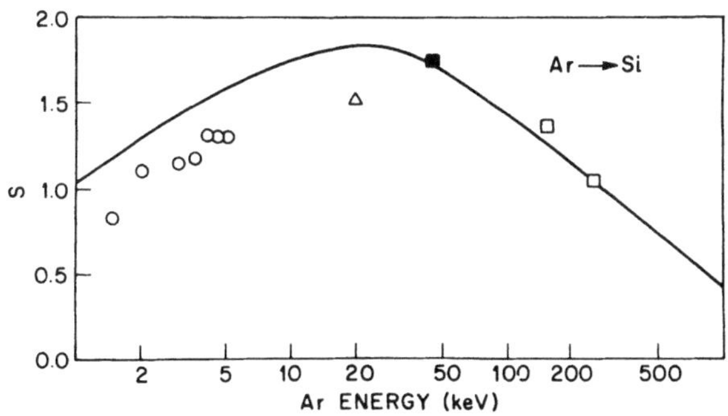

Figure 5. Sputtering of Si by Ar ions. The theoretical curve is from Sigmund [17] and data compiled by Andersen [16]

4. Damage and Formation of Amorphous Layers

When an energetic ion penetrates a solid target, sufficient kinetic energy may be imparted to lattice atoms during nuclear collisions to cause atomic displacements. Typical displacement energies are 20 eV. The recoiling lattice atom may itself possess sufficient kinetic energy to displace many other lattice atoms. As a result, a cascade of displaced atoms may originate from a single collision between the implanted ion and a lattice atom. During its path, the implanted ion may initiate many such displacement cascades within a volume surrounding the ion track. The time scale over which the cascade occurs will typically be 10^{-12} sec. The effects of these cascades have been captured by Narayan in some beautiful high-resolution transmission electron microscopy (HRTEM). Figure 6 shows a lattice image micrograph of an individual amorphous Si zone produced by 100 keV Bi bombardment of (100) Si. The ion traverses the solid until it undergoes a violent scattering event which, in this case, resulted in a collision cascade and amorphous zone formation. Significantly, each Bi^+ ion has directly amorphized a volume containing several thousand silicon atoms. This is a very efficient damaging event which cannot be modeled using simple collision theory. In contrast to the observations in Figure 6, bombardment of silicon with light ions such as B^+ does not result in large amorphous zones about individual ion tracks. Indeed, light ion bombardment produces isolated damage clusters which contain few displaced atoms and are separated by several atom spacings along the ion track. Thus, there is a clear distinction between damage produced by individual light and heavy ions in silicon. For light ions, isolated defect clusters are created in essentially crystalline silicon, whereas for heavy ions, distinct amorphous zones are directly produced by individual ion impact.

A simple qualitative picture can be used to describe amorphization by heavy ion bombardment. Typical heavy ion damage (as in Figure 6) builds up with ion dose, via an increase in the density of amorphous zones, until zone overlap eventually leads to the formation of a continuous amorphous layer. From the zone dimensions for the Bi implants of Figure 6, the minimum dose required to form a continuous amorphous layer would be $\sim 5\times 10^{12}/cm^2$. This number is close to those observed experimentally, which are typically $>10^{13}/cm^2$. For light ions, where amorphous layers are produced by the accumulation and overlap of regions of discrete defects, amorphous threshold doses are much higher than for heavy ions (typically $>10^{15}$ ions/cm^2) and the amorphization process

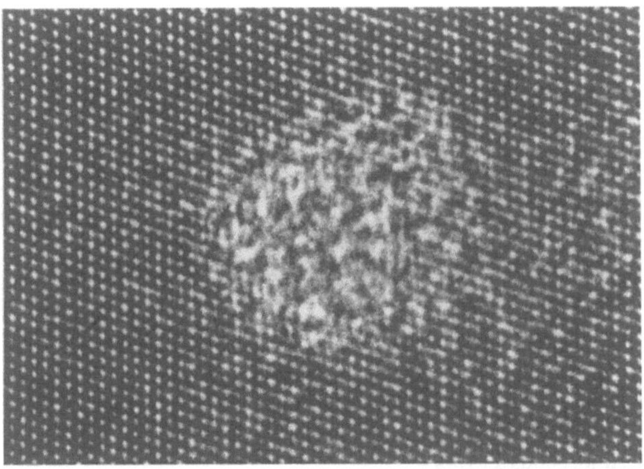

Figure 6. HRTEM lattice image of an amorphous Si zone produced by a collision cascade from a single 100 keV Bi ion. From J. Narayan (private communication)

is considerably more complex. The fact that annealing can take place during bombardment severely complicates an assessment of the nature and degree of damage arising from ion bombardment. At temperatures above absolute zero, the observed damage will be the result of competing disordering and annealing processes. In semiconductors, precise details of the defects involved and the annealing mechanisms which give rise to the final disorder structure are complicated and difficult to characterize.

The ability to form clean, amorphous layers by implantation has contributed significantly to our understanding of solid phase recrystallization processes. Silicon, germanium and gallium arsenide have well-defined temperatures at which crystallization is observed to take place, and these are determined by the activation energy for recrystallization and the growth kinetics. For ion-implanted layers, the recrystallization process usually proceeds epitaxially on the underlying crystalline substrate. This epitaxial growth process is clearly illustrated [18] by the series of cross-section TEM micrographs in Figure 7. Annealing for various times at 525°C results in regrowth of the amorphous layer produced by the implantation of 200 keV Sb^+ at $4.4\times10^{15}/cm^2$ into (100) silicon. The epitaxial regrowth rates are about 1.5Å min^{-1} for these annealing conditions. Note that the dark band of defects at ~1600Å below the surface corresponds to incompletely annealed end-of-range damage resulting from beam heating. The velocity of the phase boundary as a function of temperature is shown in Figure 8 from Olson [19]. These data are quite remarkable as they show a single activation energy of 2.7 eV for regrowth over a velocity range spanning ten orders of magnitude. There is still not yet a complete atomistic picture of the crystal growth process at the amorphous-crystalline interface.

This solid-phase epitaxial process occurs because of the peculiar features of the implantation process. The interfaces are atomically clean and the amorphous Si layers are impurity and void, or microstructure, free. It is quite difficult experimentally to achieve the solid-phase epitaxy of deposited amorphous Si films on Si. The slightest interfacial impurity or internal oxidation of microstructures will retard or completely halt the interfacial recrystallization and polycrystalline films will result. Deposited films have been produced that recrystallize epitaxially, but the highest vacuum (MBE) techniques are necessary to maintain interface and film purity.

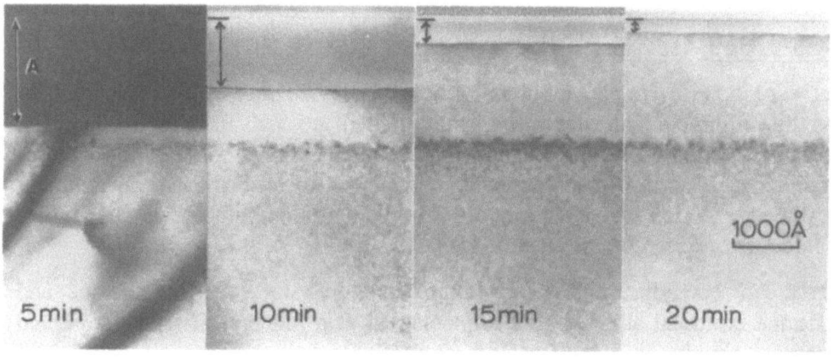

Figure 7. Transmission electron microscope cross-sectional micrographs for 200 keV Sb^+ ($6\times10^{15}/cm^2$) implants in Si annealed at 525°C. From Fletcher *et al* [18]

Fig. 8

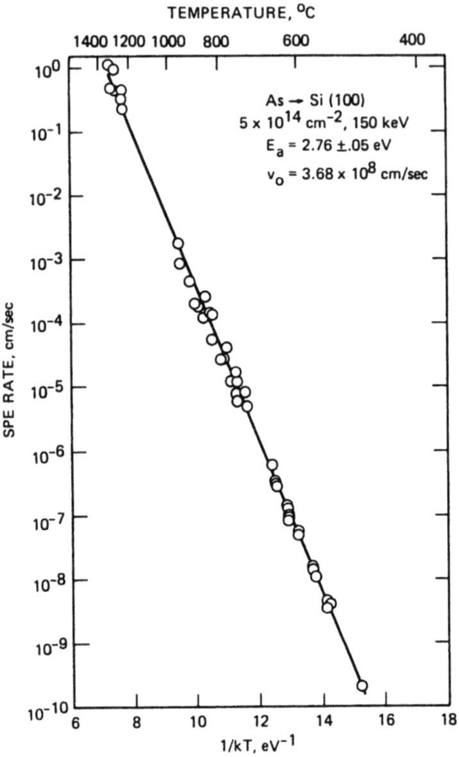

Fig. 9

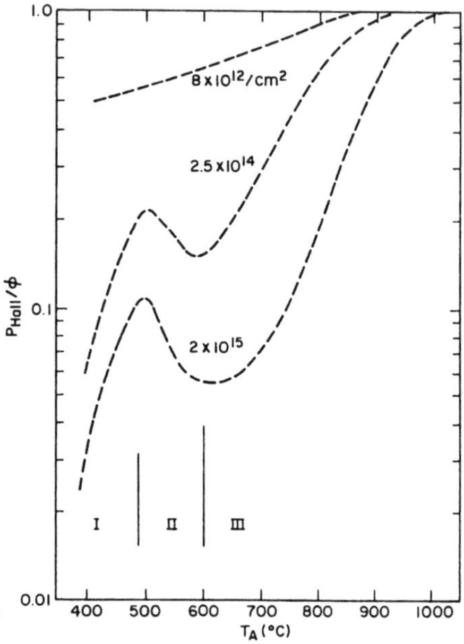

Figure 8. Dependence of solid phase epitaxy rate on temperature in As-implanted, Si(100) samples. From Olson [19]

Figure 9. Isochronal annealing of 150 keV B implanted into Si, at room temperature, at the doses shown for annealing times of 30 min. The ratio of free-carrier content (P_{HALL}) to dose (ϕ) is plotted against annealing temperature T_A for the three doses. Complete activation occurs for $T_a \gtrsim 900°C$. From Seidel and MacRae [21]

5. Annealing and Dopant Activation

The successful evolution of implantation as a tool for device fabrication has crucially depended upon the ability to anneal the implantation damage and produce the requisite electrical activity of the dopants [20]. The previous section outlined some of the features of damage production. The aim of the annealing process is to restore the semiconductor lattice to crystalline perfection and incorporate the implanted atoms on lattice sites where they can act as donors or acceptors of electrons.

The process can be quite complex. Figure 9 shows the data of Seidel and MacRae [21] for the electrical activity of B implanted Si as a function of temperature and dose. The electrical activity is plotted as the ratio of free-carrier concentration to total dose. A ratio of unity means that all the dopant atoms are electrically active. None of the B implantation doses will make Si amorphous at room temperature. In fact the low dose implant ($8\times10^{12}/cm^2$) introduces so little damage that the dopants are 50% activated at low annealing temperatures. The behavior is quite different for the higher dose implants where three annealing regions are observed: I) is characterized by point-defect disorder which is annihilated with annealing; II) shows a reverse annealing characteristic which has been identified with the formation of a dislocation structure and a reduction in substitutional B concentration; [22] III) causes complete removal of damage and full electrical activation. The annealing characteristics are different for heavier ions which create amorphous layers. Nevertheless annealing at temperatures greater than 1000°C will produce 100% electrical activation in Si as long as the concentration of dopant does not exceed solubility limits.

The annealing temperatures and times employed in current device production to remove implantation damage and activate the dopants are typically ~1000°C for half an hour. This is overkill in terms of annealing times and results in substantial diffusion of the dopants. The annealing times are determined, not by defect and dopant activation constraints, but by the cycle times of conventional tube furnaces. There is considerable activity at present in constructing rapid annealing furnaces which will heat wafers in seconds. The main advantage of the rapid thermal annealing approach is that dopant diffusion will be minimized. This concept is important for the next generation of devices where dimensions are shrinking to the submicron level.

6. Thermodynamic and Kinetic Properties of Amorphous Si

The ability to fabricate clean amorphous (a) Si layers by implantation has allowed us to investigate some of the fundamental properties of the Si condensed phases. Amorphous Si recrystallizes because it is in a higher free energy state than the crystalline (c) phase. We [23] were able to measure the free energy of a-Si by use of novel implantation and differential scanning calorimetry (DSC) techniques. Typical implants with energies of 100 keV result in a-Si layer thicknesses of approximately 1000Å. The amount of heat that would be released on recrystallization would be only a hundredth of a joule. This amount of heat is barely detectable in scanning calorimeters. Therefore to utilize the advances of implantation, we fabricated a-Si layers greater than a micron in thickness by bombardment with MeV ions. Figure 10 shows the DSC trace from a (100) wafer amorphized with Xe ions in the energy range 0.5 - 3.0 MeV. This implantation schedule produced a 2μm thick a-Si layer with an average Xe impurity concentration of about 1 part in 10^4. The calorimeter scan rate was 40 K/min. Heat is released exponentially until 970 K when solid phase epitaxy has ceased i.e. the c-a front has reached the surface. The power was integrated with time to obtain ΔH_{ac}, the heat of crystallization of a-Si. Calorimetric measurements on 20 samples amorphized under different conditions gave a value of ΔH_{ac} of 11.9 ± 0.7 kJ/mole. The rate of heat release by the advancing c-a Si interface depends only upon the velocity of the interface. The velocity can therefore be determined as a function of temperature and its absolute Arrhenius dependence agrees well with the data of Figure 8.

The enthalpy of crystallization was used by Donovan et al [23] to calculate the Gibbs free energy of amorphous and liquid (ℓ) Si referenced to c-Si. Figure 11 shows these free energy calculations. The two curves of the amorphous free energies give upper and lower limits. The hatched band represents the error which comes from lack of knowledge of the heat capacity differences. Bagley and Chen [24] and Spaepen and Turnbull [25] first estimated the free energy curves using data on a-Ge. They proposed that the amorphous phase melts at a discrete temperature $T_{a\ell}$ which is lower than the crystalline melting temperature $T_{c\ell}$. The intersection of the amorphous and liquid free energy curves gives,

Fig. 10

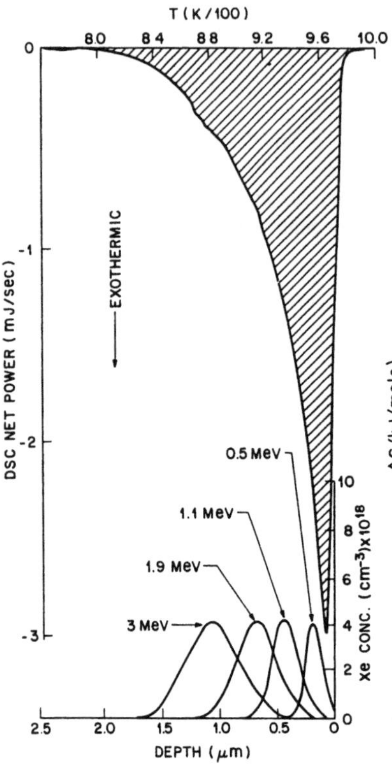

Fig. 11

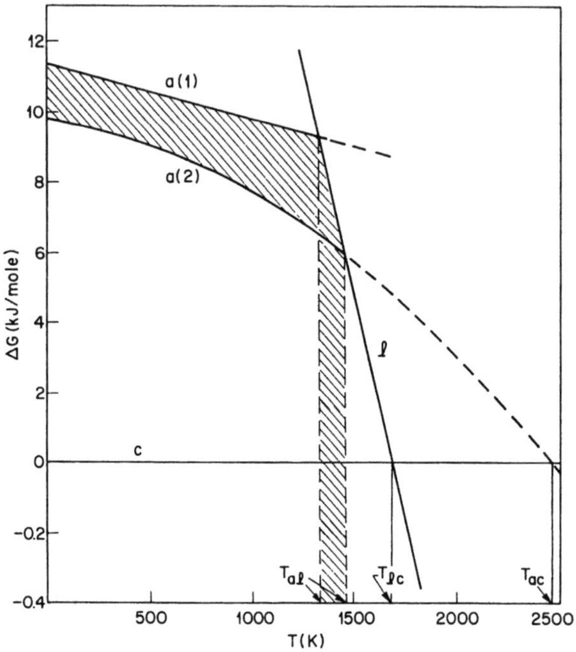

Figure 10. Net DSC power as a function of temperature (top) schematically compared (bottom) with the Xe implantation profiles. The low dose Xe implantations are tailored to give a uniform Xe concentration. From Donovan *et al* [23]

Figure 11. Gibbs free energy differences for Si - crystal (c), metallic liquid (ℓ), and amorphous (a). The different estimates for the amorphous melting temperature, $T_{a\ell}$, correspond to the two estimates of the heat capacity. From Donovan *et al* [23]

by definition, the amorphous melting temperature. The hatched region gives an idea of the uncertainty in $T_{a\ell}$. We see that $T_{a\ell} \approx 0.8\ T_{c\ell}$ or $T_{c\ell} - T_{a\ell} \approx 250^{\circ}$K which is a remarkable difference in melting temperatures for two solid phases of the same element.

The argument regarding this large depression in melting temperature is based on the premise that the phase transition from the amorphous to liquid phase is discontinuous and first order. The physical reasoning behind this premise is plausible when it is realized that the transition involves a fundamental change in bonding from the covalent, amorphous phase with four-fold coordination to the metallic liquid with (11-12) fold coordination. This transition will only occur in very rapid heating measurements otherwise the amorphous phase will recrystallize directly in the solid phase. We [26] have directly observed this transition using pulsed laser melting in the nsec regime and the transient conduction technique. The a-Si melts at a temperature of 225 ± 50 K beneath $T_{c\ell}$ of 1685 K. It should therefore be possible to quench a-Si from the melt if the melt is undercooled by more than 225°K. Such undercoolings can be achieved if c-Si is melted and solidified using ultra fast laser heating techniques. We [27] have determined that liquid Si will be undercooled by more than 225°K when the solidification velocities exceed 15m/sec. At these velocities a-Si can be formed on (100) c-Si. These solidification velocities correspond to quench rates of 10^{12} K/sec.

7. Materials Synthesis: Heterostructures

Ion implantation has traditionally been used in device fabrication for doping Si at quite low concentration levels. In the previous section we showed how the formation of a-Si layers, a metastable phase of Si, has lead to fundamental studies of phase changes. Recently there has been much activity in the use of implantation to directly synthesize binary compound phases. The first studies involved the formation of amorphous SiO_2 by oxygen implantation. Very high oxygen doses, $\sim 1\times10^{18}/cm^2$, were used to form layers of a 1000Å or so in thickness. The motivation behind this work is to form a Si-on-insulator structure where the Si layer will be dielectrically isolated thus permitting the fabrication of faster, radiation hard circuits. The early work in oxygen implantation for dielectric isolation was performed by Izumi [28].

The oxygen must be implanted at energies of at least 100-200 keV so that the SiO_2 layers will be buried under 1000-3000Å of crystalline Si. Moreover the implantation must be carried out hot, in the 450-600°C range, to preserve the crystallinity of the Si surface layer. The evolution of the implantation depth profiles with dose demonstrates some of the physical and chemical processes. Nuclei of SiO_2 are formed at the lowest doses but the implantation profile has the overall Gaussian shape. At higher doses, the nuclei coalesce and a flat-topped or layered structure forms at the SiO_2 stoichiometric level. There is a considerable chemical driving force for the formation of SiO_2 causing diffusion of oxygen within the SiO_2 layer to the wings of the distribution where more SiO_2 is formed. The critical dose, for example, for forming a continuous SiO_2 layer at 200 keV is $1.4\times10^{18}/cm^2$. The wafers must be annealed following the implantation to anneal damage in the overlying Si and clean up the Si/SiO_2 interfaces. Celler and his colleagues

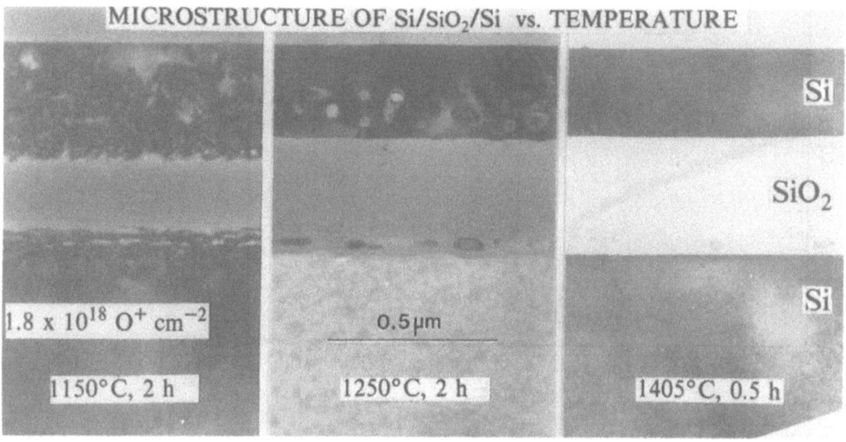

Figure 12. TEM cross-section of three samples implanted with the same oxygen dose. From Marsh *et al* [30].

[29] have shown that very high temperature anneals of 1405°C are particularly effective in enhancing the interfacial and stoichiometric perfection of the heterostructures. Figure 12 shows a TEM cross-section [30] of samples annealed at different temperatures. The 1405°C anneal results in a well-defined structure. This whole implantation and annealing process is remarkable and holds considerable promise as a device technology. At present, however, there is a serious disadvantage with the technology. The dislocation densities in the Si are in the range of $10^6/cm^2$. This is a very high density of defects but the success in fabricating devices in the structures demonstrates that the defects are largely electrically inactive.

The success of the SiO_2 synthesis lead White and her colleagues [31] to explore other compound formation by high dose implantation in Si. They have demonstrated that silicides can be formed by implantation. Silicides are used in devices as metallic conductors not only because of their good conductivity but also because of their high chemical and thermal stability in contact with Si. Since the conduction is metallic it is not necessary that the film be single crystal i.e. grain boundaries do not affect the conductivity. The silicides therefore do not have to be epitaxial with Si. Indeed, virtually all silicides currently employed in devices do not have structural matches with Si and are polycrystalline. There are, however, silicides that can be grown epitaxially on Si. In particular $CoSi_2$, which has the CaF_2 structure and a small lattice mismatch with Si, has been grown epitaxially [32] on Si by Co deposition and reaction techniques. By implanting Co at elevated temperatures and subsequent annealing, White *et al* [31,33] have grown single-crystal silicide heterostructures. Figure 13 shows a Rutherford

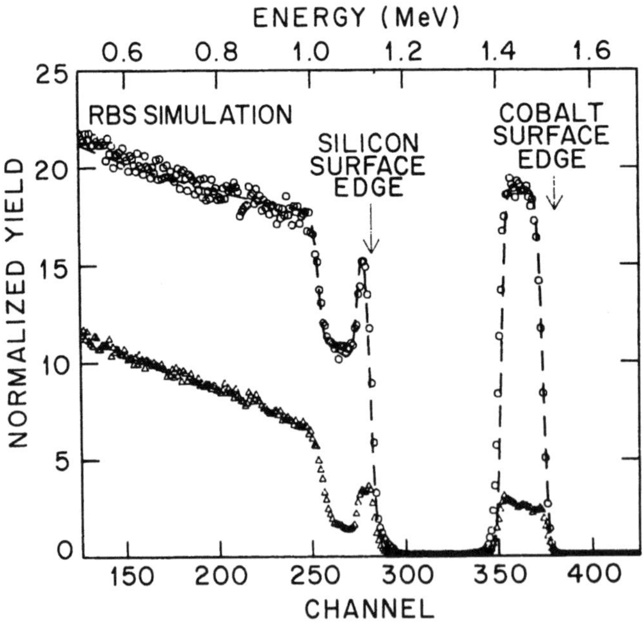

Figure 13. Rutherford backscattering and channeling spectra of an implanted $CoSi_2$ heterostructure. The dashed line is a fit for 1100Å $CoSi_2$ buried under 600Å of Si. From Alice White *et al* [33]

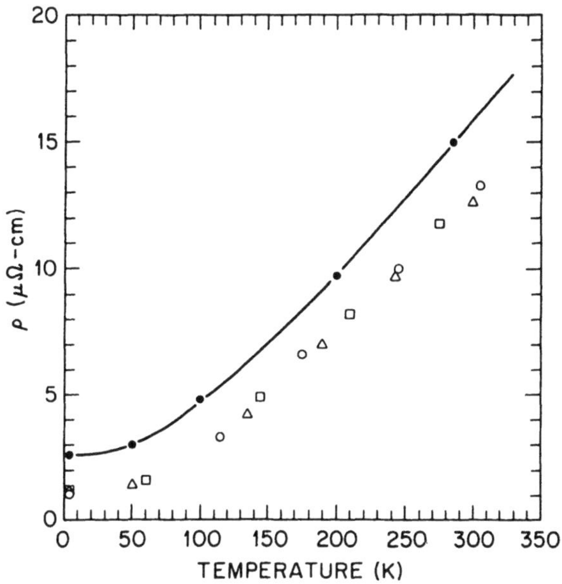

Figure 14. Resistivity vs. temperature characteristics for several of the implanted $CoSi_2$ layers (open symbols). Data of Hensel *et al* for MBE-deposited films (closed circles) is shown for comparison. From Alice White *et al* [34]

backscattering and channeling spectra [33] of a 1100Å $CoSi_2$ film buried under 600Å of Si. The 200 keV Co implant of $3\times10^{17}/cm^2$ was carried out at 350°C. The sample was then annealed at 600°C for 1 hr and 1000°C for 1/2 hr. The driving force for stoichiometric $CoSi_2$ formation is strong as evidenced by the sharp interfaces and low levels of Co in the overlying Si.

The structural perfection of these $CoSi_2$ layers is demonstrated by their electrical transport properties. Figure 14 shows [34] the temperature dependence of the resistivity. The data from implanted heterostructures is compared with results from surface silicide films grown by MBE deposition and reaction techniques [35]. The similarity in the shape of the curves shows that the temperature dependent phonon contribution to the resistivity is the same. However, the displacement of the curves indicates that the implanted silicides have a lower residual resistivity. This is a significant result and may be due to several factors. The implantation process is inherently clean because the Co beam is mass selected and contains none of the contaminants that may be present in evaporation sources. In addition, the buried films are capped very effectively and can be annealed at higher temperatures than conventional surface films. Moreover it is believed that the MBE deposited films are slightly metal rich. The implanted films which nucleate and grow in a Si-rich environment with relatively little mass transport may be more stoichiometric.

8. What Next?

There has been a resurgence of interest in ion implantation science and technology in the past five years or so. Some of the reasons have been outlined in this chapter. The ability to investigate basic phase transitions and materials properties has given us different perspectives on some basic problems. The field is also moving in some new directions. For example, high energy ion beams are being used to stimulate solid phase crystal growth at low temperatures. Atomic displacements at the Si a-c interface, produced by the MeV irradiation, have produced epitaxial crystal growth at temperatures as low as 200°C. This is an unexplored regime of crystal growth and unusual dopant segregation and diffusion phenomena have been observed [36].

Machine developments in recent years have also pushed the field in new directions. High current implanters, producing beam currents of mAs of heavy ions, have facilitated the development of the materials synthesis studies such as the formation of buried SiO_2 and $CoSi_2$ layers. High energy implanters, producing ion beams with energies in the MeV range, are allowing the device engineer to replace some of the conventional CVD steps with implants which are several microns deep. There is another machine development which will probably have a much greater impact on compound semiconductor technology than on Si technology. Implantation is already an essential part of compound semiconductor processing technology where large, swept area beams are used in a self-aligned fashion. Machines are now coming on line that can focus the ion beams to micron dimensions. This direct writing and doping technology is slow but may find use in optoelectronic circuits.

REFERENCES

1. R. S. Ohl, Bell Syst. Tech. J. *31,* 104 (1952).
2. W. Shockley, U.S. Patent 2787564.
3. J. W. Mayer, L. Eriksson and J. A. Davies in *Ion Implantation in Semiconductors,* [Academic Press, New York 1970].
4. R. G. Wilson and G. R. Brewer in *Ion Beams with Applications to Ion Implantation,* [John Wiley & Sons, New York 1973].
5. G. Dearnaley, J. H. Freeman, R. S. Nelson, and J. Stephens in *Ion Implantation,* [North Holland, Amsterdam 1973].
6. B. Smith in *Ion Implantation Range Data for Si and Ge Device Technologies,* [Research Studies Press, Forest Grove, Oregon 1977].
7. N. Bohr, Phil. Mag. *25,* 10 (1913) and N. Bohr, Dan. Vid. Selsk., Mat. Fys. Medd *18,* No. 8 (1948).
8. U. Fano, Am. Rev. Nucl. Sci. *13,* 1 (1963).
9. F. Fermi and E. Teller, Phys. Rev. *72,* 399 (1947).
10. J. Lindhard, M. E. Scharff, and H. E. Schiott, Kgl. Danske Vid Selsk. Mat. Fys. Medd., *13,* No. 14 (1963).
11. W. K. Hofker, Philips Res. Repts. Suppl., No. 8 (1975).
12. D. K. Brice in *Ion Implantation Range and Energy Deposition Distribution,* [Plenum Press, New York 1975].
13. J. F. Gibbons, W. S. Johnson, and S. W. Mylroic in *Projected Range Statistics in Semiconductors,* [John Wiley & Sons, New York 1975].
14. V. Littmark and J. F. Ziegler in *Handbook of Range Distributions for Energetic Ions in All Elements,* Vol. 6 [Pergamon Press, New York 1980].
15. D. S. Gemmell, Reviews of Modern Physics *46,* No. 1 (1974).
16. H. H. Andersen in *Physics of Ionized Gases 1974,* ed. V. Vajnovic [Institute of Physics, Univ. Zagreb 1974].
17. P. Sigmund, Appl. Phys. Lett. *14,* 114 (1969).

18. J. Fletcher, J. Narayan and O. N. Holland in *Microscopy of Semiconducting Materials 1981,* Inst. Phys. Conf. Ser. *60,* eds. A. G. Cullis and D. C. Joy (Inst. Phys., London-Bristol, 1981), p. 295.
19. G. L. Olson, Materials Research Society Proceedings *35,* 25 (1985).
20. T. E. Seidel, Chapter 6 in *VLSI Technology,* ed. S. M. Sze [John Wiley & Sons, New York 1983].
21. T. E. Seidel and A. U. MacRae in *First International Conference on Ion Implantation,* eds F. Eisen and L. Chadderton [Gordon and Breach, New York 1971], p. 149.
22. J. C. North and W. M. Gibson, ibid, (1971) p. 143.
23. E. P. Donovan, F. Spaepen, D. Turnbull, J. M. Poate, and D. C. Jacobson, J. Appl. Phys. *57,* 1795 (1985).
24. B. G. Bagley and H. S. Chen, AIP Conf. Proc. *50,* 97 (1979).
25. F. Spaepen and D. Turnbull, AIP Conf. Proc. *50,* 73 (1979).
26. M. O. Thompson, G. J. Galvin, J. W. Mayer, P. S. Peercy, J. M. Poate, D. C. Jacobson, A. G. Cullis, and N. G. Chew, Phys. Rev. Lett. *52,* 2360 (1984).
27. M. O. Thompson, J. W. Mayer, A. G. Cullis, H. C. Webber, N. G. Chew, J. M. Poate, and D. C. Jacobson, Phys. Rev. Lett. *50,* 896 (1983).
28. K. Izumi, M. Doken, and H. Ariyoshi, Electron Lett. *14,* 593 (1978).
29. G. K. Celler, P. L. F. Hemment, K. W. West and J. M. Gibson, Appl. Phys. Lett. *48,* 532 (1986).
30. C. D. Marsh, G. R. Booker, K. J. Reason, P. L. F. Hemment, R. J. Chater, J. A. Kilner, J. A. Alderman, G. K. Celler, in *Proc. European MRS, Vol. XII,* eds. G. Bentini, E. Fogarassy, and A. Golanski, [Les editions de physique, 1986], p. 137.
31. Alice E. White, K. T. Short, R. C. Dynes, J. P. Garno, and J. M. Gibson, Appl. Phys. Lett. *50,* 95 (1987).
32. R. T. Tung, J. M. Gibson, and J. M. Poate, Phys. Rev. Lett. *50,* 429 (1983).
33. Alice E. White, K. T. Short, R. C. Dynes, J. P. Garno, and J. M. Gibson, MRS Symp. Proc. *74,* 481 (1987).
34. Alice E. White, K. T. Short, R. C. Dynes, J. M. Gibson, and R. Hull, MRS Symp. Proc. *100,* 3 (1988).
35. J. C. Hensel, R. T. Tung, J. M. Poate, and F. C. Unterwald, Appl. Phys. Lett. *44,* 913 (1984).
36. J. M. Poate, J. Linnros, F. Priolo, D. C. Jacobson, and J. L. Batstone, Phys. Rev. Lett. *60,* 1322 (1988).

Low Temperature Epitaxial Crystallization of Amorphous Si by Ion-Beam Irradiation

F. Priolo, A. La Ferla, and E. Rimini

Dipartimento di Fisica, Corso Italia 57, I-95129 Catania, Italy

1. Introduction

Amorphous Si (a-Si) is a metastable phase and it will transform into the stable crystalline phase (c-Si) upon thermal annealing. On <100> oriented Si substrates and at 600 °C the amorphous-crystal (a-c) interface velocity is 10~Å/sec and it reduces to just 0.006 Å/sec at 450 °C[1]. This a-c interface velocity is characterized by an Arrhenius-type behavior with a single activation energy of 2.7 eV over more than eight orders of magnitude[2]. Moreover the interface motion is remarkably planar over growth distances of several thousands of angstroms.

Recently it has been shown that the epitaxial crystallization of amorphous Si layers can be obtained also at temperatures as low as 200 °C during ion-beam irradiation [3-7]. In this paper we will briefly review this process presenting some of the experimental results. Moreover we will demonstrate that amorphous Si layers deposited onto single crystal substrates without any cleaning procedure prior to the deposition can be epitaxially regrown by means of this novel technique. A more detailed description of the ion-beam induced epitaxial crystallization can be found in the refs.[3-9].

2. Experimental Procedure

Amorphous Si layers, 170nm thick, were produced by Ge or Si implantation on <100> Si substrates at 77K. The crystallization was induced by a 600 keV Kr beam electrostatically scanned over a 2.5 cm in diameter circular area at an ion flux of $1x10^{12}$ ions/cm^2sec. This Kr beam has a projected range of 350 nm, i.e. greater than the thickness of the amorphous layer to be regrown. In fact this is a necessary condition to obtain a crystallization induced by the ion beam.

During irradiations the samples were mounted on a resistively heated copper block whose temperature, monitored by a thermocouple, was maintained in the range 200-450 °C. The epitaxial recrystallization was measured "in situ" during the irradiation monitoring the reflectivity of a 5mW He-Ne laser beam focussed onto the irradiated area. The experimental apparatus is schematically shown in Fig.1. A small mirror placed near the sample reflects the laser light back to the wafer surface and then to the detector. The reflected light is detected by a photodiode connected to an X-Y recorder whose X-

Springer Series in Materials Science, Vol. 13
Semiconductor Silicon Editors: G.C. Harbeke and M.J. Schulz

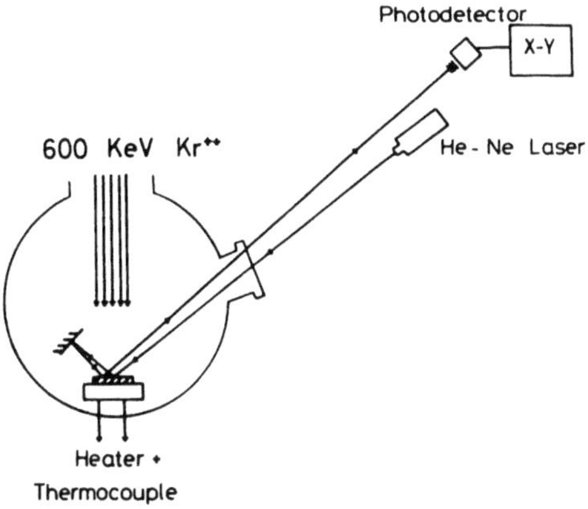

Fig.1-Experimental set-up (from ref.6)

axis is driven by a signal proportional to the Kr dose. The reflectivity of the sample will oscillate due to successive constructive and destructive interferences occurring between the light reflected from the surface and from the advancing a-c interface. In the adopted set-up the change in reflectivity from a peak to a contiguous valley corresponds to an advance of the a-c interface of 33 nm. This value was also checked analyzing the samples after the irradiations by Rutherford backscattering and channeling. It should be noted that due to the presence of the mirror and the consequent double reflection the squared reflectivity is actually measured.

3. Experimental Results

The ion-induced epitaxial recrystallization of a <100> Si crystal preamorphized to a depth of ~165 nm by Ge implantation is illustrated in Fig.2 by the transient reflectivity signal (upper part) The oscillation in the reflectivity signal indicates the occurrence of an interference between the light reflected from the surface and from the moving a-c interface. Decreasing the distance from the surface the amplitude of the oscillations increases due to a reduced absorption in the

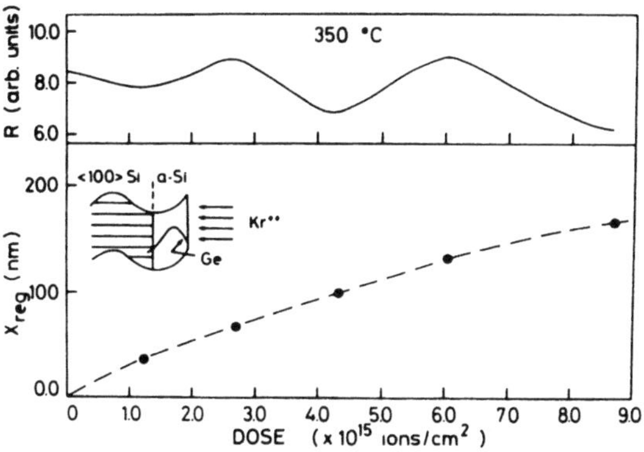

Fig.2-Reflectivity signal (up) and regrowth thickness (down) vs. Kr dose (from ref.6)

amorphous layer. The lower curve reports the regrown thickness as a function of the Kr^{++} dose.

The growth rate, i.e. the slope of the regrown thickness vs. dose curve, is not constant and decreases as the a-c interface approaches the surface. This depth dependence can be correlated with the energy deposited by the impinging Kr ions into those nuclear collisions able to displace an atom from its lattice site. This energy has been calculated by a Monte Carlo simulation program, TRIM, assuming a threshold energy for displacements of 13eV. The results are shown in Fig.3 (dashed line) as a function of depth. In the same plot is reported the depth dependence of the growth rate as deduced from Fig.2 (full dots). The experimental rates closely follow the calculated distribution of the deposited energy. This result is quite remarkable and confirms unambiguously that the ion-beam induced growth is associated with the generation of point defects in the vicinity of the a-c interface. No correlation has been observed between the data and the electronic energy loss.

Experiments similar to those so far reported were carried out varying the irradiation temperature. The experimental regrowth rates, normalized for the energy deposited into nuclear collisions, have an Arrhenius-type temperature dependence with an activation energy of 0.32±0.05eV.The actual meaning of this activation energy is not yet understood. It should be noted that the growth rate, however, depends on the impurities dissolved in the a-layer. For instance, the presence of electrically active dopants can increase the ion-induced growth rate by a factor of 2-3[5,6]. Also the substrate orientation influences the rate which is a factor of 2 faster on <100> with respect to <111> substrates[5]. Despite the huge amount of experimental data a unifying theory able to explain all the results is still lacking.

An interesting application of this novel technique is the possibility of regrowing epitaxially deposited Si layers[8,9]. The epitaxial growth of deposited layers is dominated by the presence of contaminants at the interface with the single crystal. Ultra high vacuum deposition or sophisticated cleaning procedures should be used in order to obtain the epitaxial regrowth. These constraints can be eliminated by the use of an ion beam. The crystallization was investigated in Si layers deposited at 540 ^{o}C by silane dissociation in a chemical vapor reactor onto <100> Si wafers. No cleaning procedure was performed prior to deposition. After deposition the samples were implanted at room temperature to make sure that the structure of the deposited layer was amorphous. The ion-beam induced crystallization was performed at 450 ^{o}C and monitored "in situ" by transient reflectivity. The extracted results are shown in Fig.4 where the regrown thickness is shown as a function of the irradiating Kr-dose. The plot shows an initial regrowth of the a-layer formed by the preimplantation beyond the deposited Si-substrate interface and a subsequent stop of the regrowth at this interface. After a critical dose (which has been observed to depend on the irradiation temperature) the a-c interface propagates in the deposited layer and epitaxial

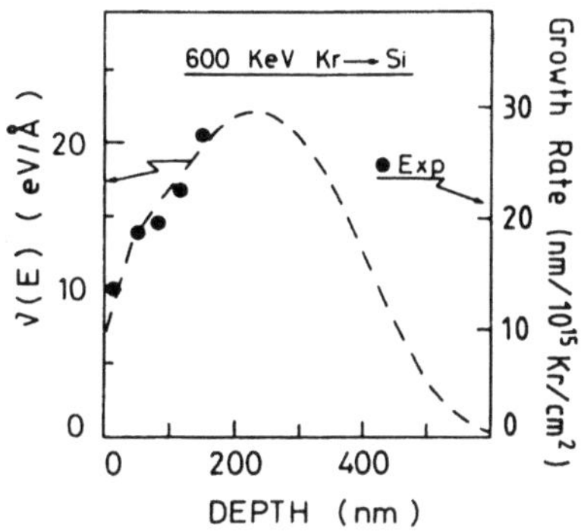

Fig.3- Calculated nuclear energy loss (dashed line) and experimental interface (dots), (from ref.6)

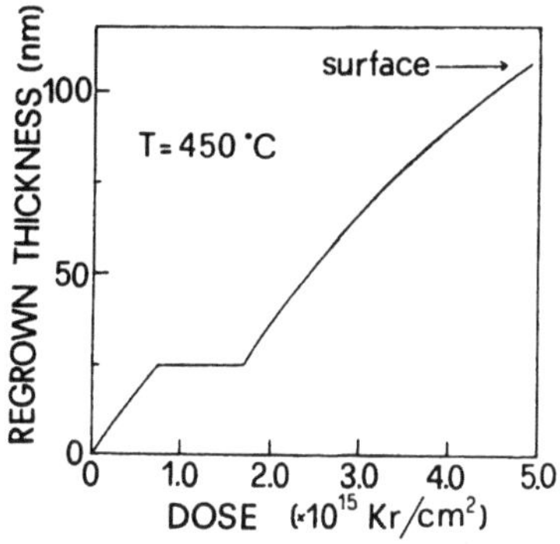

Fig.4- Regrowth thickness vs. dose for a deposited a-Si layer

crystallization occurs as also observed by Rutherford backscattering and Transmission Electron Microscopy. This critical irradiation dose is thought to be responsible of breaking the thin native interfacial oxide and of dissolving the O atoms by radiation-enhanced diffusion processes. When the interfacial O concentration is low enough the a-c interface can pass through and epitaxy occurs.

References

1) L. Csepregi, J.W. Mayer and T.W. Sigmon, Phys. Lett. 54A,157, (1975)
2) G.L. Olson, S.A. Kokoroski, J.A. Roth, and L.D. Hess, Mat. Res; Soc. Symp. Proc., vol.13 pag.141
3) J. Linnros, G. Holmen, and B. Svensson, Phys. Rev. B 32, 2270, (1985)
4) J.S. Williams, R.G. Elliman, W.L. Brown and T.E. Seidel, Phys. Rev. Lett. 55, 1482, (1985)
5) A. La Ferla, S.Cannavò, G. Ferla, S.U. Campisano, E. Rimini, and M. Servidori, Nucl. Instrum. Meth. B19/20, 470, (1987)
6) F. Priolo, A. La Ferla and E. Rimini, J. Mat. Res., Nov./Dec. 1988 (in press)
7) J.M. Poate, J. Linnros, F. Priolo, D.C. Jacobson, J.L. Batstone and M.O. Thompson, Phys. Rev. Lett. 60, 1322 (1988)
8) A. La Ferla, E. Rimini, and G. Ferla, Appl. Phys. Lett. 52, 712, (1988)
9) F. Priolo, A. La Ferla, C. Spinella, E. Rimini, F. Baroetto, G. Ferla and A. Licciardello, Appl. Phys. Lett. 53, 2605 (1988)

On the Generation of Ripples on Silicon

D. Jost[1] *and G. Benedek*[2]

[1]Max-Planck-Institut für Quantenoptik, D-8046 Garching, Fed. Rep. of Germany
[2]Dipartimento di Fisica dell' Università, Via Celoria 16,
I-20133 Milano, Italy

1. Introduction

In this paper we discuss a possible generation mechanism for surface acoustic waves, in particular Rayleigh waves, on silicon which can be used to make submicron gratings. We will briefly present some of the experimental observations which have led us to think of such mechanisms as viable. In conclusion we will identify the problems associated with these mechanisms in the hope of stimulating further discussion on the subject of picosecond laser pulse interaction with semiconductors. The mechanism which we have proposed [1] and discuss here relies on the generation of an exponentially decaying electronic strain (into the depth of the crystal away from the surface) which launches a Rayleigh wave, a surface acoustic wave (SAW). The electronic strain is a direct consequence of the photogenerated electron-hole plasma gradient resulting from the absorption of the photons contained in a picosecond laser pulse.

2. Experimental Evidence

Scanned irradiation of a Si(111) crystal with a focused cw mode-locked Argon or Nd:YAG laser to its melting threshold has generated a type of surface grating with a periodicity which is dependent on the laser temporal pulse width [2]. These gratings or surface corrugations we designate as ripples and have attributed their origin to a frozen out surface acoustic wave. In order to try to understand how such phenomena come about we have looked in detail at the interaction of a single laser pulse with the semiconductor lattice. Under the experimental conditions which allow observation of this phenomena, an area with $1\mu m$ diameter will see about 10^6 laser pulses. We think that the surface acoustic wave generated by one pulse is amplified by each subsequent pulse and the resulting corrugation is frozen out after each pulse. The end result of this amplified SAW we call a ripple.

3. Rayleigh Wave Mechanism

3.1. Interaction of a Laser Pulse with Silicon

In the field of material processing the interaction of laser pulses with crystalline or amorphous Si has been and continues to be intensely studied. The phenomenology of laser pulse-semiconductor irradiation is rather rich since this usually leads to physical regimes not before accessible *i.e.* without lasers. Of particular interest is the generation of nonequilibrium high density electron-hole plasmas [3] by very short injection times corresponding to the temporal laser pulse width, which nowadays can be below picoseconds. The kinetics of high density plasmas in semiconductors have recently received some detailed attention [4, 5]. The mechanisms involved are rather complex because of the intrinsic nonlinearities occurring in the time domain which require stringent approximations in order to allow a solution to the problem.

Springer Series in Materials Science, Vol. 13
Semiconductor Silicon Editors: G. C. Harbeke and M. J. Schulz

3.2. Density of Photogenerated Carriers

Within say a 100 ps laser pulse with about 10^{27} photons , the energy deposited by the laser beam as photons generates a large number of electron-hole pairs per unit volume n. As a consequence of the presence of these the penetration depth of subsequent photons within the same pulse is rapidly shortened by the increased plasma density which leads to shorter plasma decay lengths with shorter pulses. This density falls off exponentially into the z-direction. We start by approximating the carrier density, n, in an equation which includes the laser pulse parameters as follows

$$n = \frac{F}{\varsigma h\nu}\frac{\tau_0}{\tau}\left(1 - e^{-\tau/\tau_0}\right) \tag{1}$$

where

$$\varsigma = \frac{c}{e}\left(\frac{m^*}{4\pi n}\right)^{1/2} \tag{2}$$

and $h\nu$ is the laser photon energy, F is the laser fluence, m^* is the carrier reduced effective mass, c is the speed of light in vacuum, e is the elementary charge, τ is the laser pulse duration,and τ_0 is a plasma overall relaxation time. The prefactor $F/\varsigma h\nu = n_\infty$ gives the asymptotic carrier density for $\tau << \tau_0$ corresponding to as many electron-hole pairs in the volume $\pi R_0^2\varsigma$ as there are photons absorbed.

The use of the plasma penetration depth in (2) is justified whenever the surface plasmon energy $\hbar\omega_p$, at some n becames larger than the exciting laser energy. That is, to avoid confusion we must emphasize that in (1) and (2) ς must be understood as a plasma decay length taken at the time just before n starts to decay from its peak value, thus at some n_{max}.

Eliminating n from (1) and (2) and solving for ς we obtain the following expression,

$$\varsigma = \frac{m^* c^2 h\nu}{4\pi e^2 F_{th}}\frac{\tau/\tau_o}{\left(1 - e^{-\tau/\tau_o}\right)} \tag{3}$$

which we designate as the electron-hole plasma decay length. In this fashion we get at ς and later at the ripple periodicity Λ without ever explicitly evaluating (1), the only problem being what to use for m^*, which we estimate from the literature [6].

3.3. Perturbing Force

By assuming that the electronic strain gradient has the same decay length as the photon field in the presence of the hot plasma, that is $\varsigma(z)$, we write the Fourier components of the force density induced by a gaussian laser pulse of temporal width τ, focused to a circular area of radius R_0 as

$$f_z(\mathbf{K}, z; \omega) = f_0 exp(-\frac{1}{4}R_0^2 K^2 - \frac{z}{\varsigma} - \frac{1}{4}\tau^2\omega^2) \tag{4}$$

where f_0 has the meaning of momentum transfer.

3.4. Response of the Rayleigh Waves

The response of the Rayleigh waves to the perturbing force is given by the elements of a Green's function [1,7]. We assume a linear dispersion relation and an isotropic velocity for the Rayleigh waves. Since the ripples are not observed with circular symmetry but focus along certain directions we take on a 1-dimensional integration in K space.

For an isotropic medium, the Rayleigh wave polarization vectors for propagation along x are given by [8, 7],

$$
\begin{aligned}
w_x(\mathbf{K}, z) &= iN(\gamma_T^{1/2} e^{-\gamma_T K z} - \gamma_L^{-1/2} e^{-\gamma_L K z}), \\
w_y(\mathbf{K}, z) &= 0, \\
w_z(\mathbf{K}, z) &= N(\gamma_T^{-1/2} e^{-\gamma_T K z} - \gamma_L^{1/2} e^{-\gamma_L K z}),
\end{aligned} \tag{5}
$$

where

$$
\gamma_{T,L} \equiv (1 - v_R^2 / v_{T,L}^2)^{1/2} \tag{6}
$$

$v_{T,L}$ stands for either the transverse of longitudinal sound velocities respectively, K is the absolute value of $\mathbf{K}$, and N is a normalization constant. Each component is the sum of two exponential terms, a slow decaying one proportional to $exp(-\gamma_T K z)$ and a fast decaying one proportional to $exp(-\gamma_L K z)$. When v_R is very close to v_T and $\gamma_T << \gamma_L$ the first term in w_z is dominant. In this limit the Rayleigh wave can be approximated by a single slow decaying component linearly polarized along z, given by

$$
w_z(\mathbf{K}, z) = (2\gamma_T K)^{1/2} exp(-\gamma_T K z), \tag{7}
$$

where the prefactor ensures normalization. For silicon $\gamma_T \sim 0.63$ is not sufficiently small for the longitudinal component to be neglected, however since in our case the force density is a consequence of the electronic strain gradient, and this, at large laser irradiance, *i.e.* large electron-hole plasma density at the surface, is much larger along the z-direction than parallel to the surface. In this particular physical situation the response of Rayleigh waves occurs mostly via their vertical component, whereas the longitudinal one might be much less effective.

3.5. Shape and Period of Ripples

If the integration of the z component of the displacement vector at the surface $u_z(\mathbf{R}, t)$ is done in one dimension as determined by the experimentally observed focusing effect [1], the ripple shape frozen out after the laser pulse is turned off ($t = \tau$) as

$$
\begin{aligned}
u_z(R, \tau) &= f_0 (2\rho v_R \varsigma)^{-1} exp(-\tau^2 v_R^2 + R_0^2/(2\gamma_T \varsigma)^2) \\
&\times [I_+(R/R_0, \tau)\, exp\Big(\frac{-i(R - \tau v_R)}{\gamma_T \varsigma}\Big) \\
&+ I_-(R/R_0, \tau)\, exp\Big(\frac{-i(R + \tau v_R)}{\gamma_T \varsigma}\Big)]
\end{aligned} \tag{8}
$$

where the integrals

$$
\begin{aligned}
I_\pm(R/R_0, \tau) &= \int_{-\infty}^{+\infty} dp\, exp(-ip\frac{\tau}{R_0} \pm p\frac{v_R \tau^2}{2R_0 \gamma_T \varsigma} - p^2 \frac{\tau^2}{4R_0^2}) \\
&\times \int_{-\infty}^{+\infty} \frac{dq}{q} \frac{exp(iq(R/R_0))}{v_R q \mp (p + i0^+)} exp(-\frac{1}{4}q^2 + \frac{qR_0}{2\gamma_T \varsigma}),
\end{aligned} \tag{9}
$$

are slowly varying functions of R decaying over a distance of the order of the focusing radius R_0. The ripple shape is shown in (9) as a combination of sinusoidal components of wavelenth Λ,

$$\Lambda = 2\pi\gamma_T\varsigma = \frac{\gamma_T m^* c^2 h\nu}{2e^2 F_{th}} \frac{\tau/\tau_0}{1-e^{-\tau/\tau_0}} \tag{10}$$

and period equal to $2\pi\gamma_T/v_R$, corresponding to a propagation velocity v_R. The integrals $I_\pm(R/R_0,\tau)$ modulate the intensity of the sinusoidal components producing a rapid decay beyond the radius R_0, namely outside of the illumination area. The prefactors give the overall ripple amplitude as a function of the relevant parameters of the incident laser pulse and of the crystal. We note, in particular, the dependence on the radius of the illuminated area. Since $I_\pm exp[-(\tau^2 v_R^2)/(2\gamma_T\varsigma)^2]$ have a maximum when R_O is of the order of $\tau \cdot v_R$.

4. Amplification

So far we have described the interaction of a single laser pulse with the silicon lattice; however it must be emphazised that besides being for a single distinct pulse the above description is for the first pulse. Under the experimental conditions [2,9], the second pulse will *see* a lattice which has recorded the effects of the first pulse. This occurs through modulation of the optical properties by the SAW. The initial deformation modulates the band gap at the sample surface, which in turn modulates the absorption coefficient. With subsequent incoming pulses this process gets repeated thus reinforcing the initial deformation. Although the details of this amplification have not yet been worked out, it is necessary to involve it in order to account for the large amplitude of the observed ripples.

5. Recording

That these ripples are observed is due to the fact that the same high fluence pulse needed to create the large n is also used to cause transient melting at the surface of the illuminated area. Transient melting implies a resolidification of the material before the next pulse arrives. In the absence of transient melting, although the physics of this particular situation would still be taking place, the surface would remain undisturbed because the lattice would relax to its equilibrium state leaving no trace of the deformation. A picosecond pulse causing transient melting also cools it very fast below the melting temperature, thus literally freezing out the displacement of the lattice.

6. Discussion

In the Figure we plot Λ as a function of τ according to (9) for a fixed value of the prefactor $\gamma_t m^* c^2 h\nu/2e^2 F = 4.12 \times 10^{-6}\,cm$, where $m^* = 0.43 m_e$ (electron mass), this mass being extrapolated from reference [6]. The available experimental points for silicon agree well with the theoretical curve predicted by this Rayleigh wave mechanism. We note also that the observed deviation from linearity in the Λ *versus* τ is well reproduced.

We have not directly calculated n, rather we have exploited its temporal and spatial behavior as a function of τ (laser pulse width), which is much more relevant to the dynamics of this problem than the absolute value.

Recent experiments in Ge [9] differ in two results from those on Si [2]. First, the ripples in Ge do not show a strong Λ on τ dependence. Secondly their orientation is always perpendicular to that of the electric field vector of the incident laser beam, in marked contrast to those in silicon which orient parallel to one ofthe crystallographic axes. From the $\Lambda(\tau)$ behavior in Ge compared to that of Si we can only conclude that if the above mechanism is correct, then the plasma kinetics of these two materials must differ considerably. Furthermore the difference in the orientation of these ripples can be understood if two different mechanisms of photoacoustic amplification are involved.

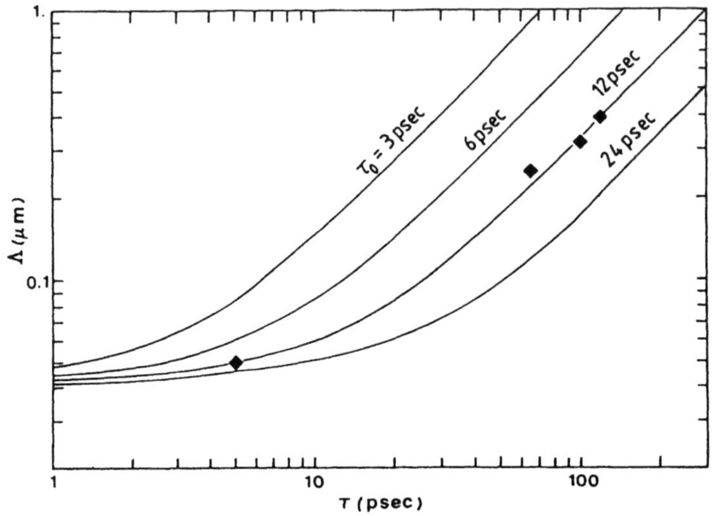

Fig. 1: Dependence of ripple periodicity on laser pulse in Si. The black symbols correspond to experimental values

In conclusion it seems that the generation of Rayleigh wave ripples at normal incidence of a pulsed MHz laser, whose periodicity is smaller than the wavelength of the incident light is not a peculiarity of silicon and can also be observed in germanium. Comforting as this thought may be the new Ge data also points out the deficiencies in the understanding of the phenomena involved, namely the plasma kinetics of the two materials and the amplification.

Acknowledgments. One of us (D. J.) is very grateful to the European Physical Society for a grant to participate in this school. We are both indebted to our collaborators in Toronto, Melanie Ledgerwood, Henry van Driel and John Sipe, for readily making their germanium data available to us prior to publication and for the continued discussions.

REFERENCES

[1] D. Jost, H. P. Weber and G. Benedek, *J. Electr. Spectr. Related Phenom.* **44** (1987) 345

[2] D. Jost, W. Lüthy, H. P. Weber and R. P. Salathe, *Appl. Phys. Lett.* **49** (1986) 625

[3] The word *plasma* is used in this paper to refer to a high density electron-hole plasma

[4] H. M. van Driel, *Phys. Rev.* **B 35** (1987) 8166

[5] E. J. Joffa, *Phys. Rev.* **B 21** (1980) 2415

[6] H. M. van Driel, *Appl. Phys. Lett.* **44** (1984) 617

[7] M. G. Cottam and A. A. Maradudin: In *Surface Excitations*, V. M. Agranovich and R. Loudon, eds., North-Holland, Amesterdam, 1984, pp 1-194

[8] G. W. Farnell: In *Physical Acoustics*, Vol. 6, Academic Press, New York, 1969, pp 109-166

[9] M. Ledgerwood, D. Jost, H. M. van Driel and J. E. Sipe, *Joint CAP/APS Congress*, Montreal, Quebec, Canada, June 20-22, 1988

Part III

Defects

Theory of Defects in Crystalline Silicon

M.A. Schlüter

AT&T Bell Laboratories, Murray Hill, NJ 07974, USA

ABSTRACT

Point defects in silicon are capable of localizing electrons whose energy depends strongly on the local lattice configuration. The strong coupling can lead to exotic features such as "negative U" behavior, electronically enhanced migration or doping dependent defect reactions. Examples of several detailed and quantitative theoretical studies are presented.

1. Introduction

These lecture notes describe the physics associated with some point-like defects in silicon. Because of the strong covalent bonding in silicon and the associated charge in-homogeneities intrinsic lattice defects and impurities show a strong position dependent coupling to the surrounding lattice. In many cases the defects also induce deep bound states in the gap acting as electron or hole traps. The energetic position of these traps varies itself strongly with lattice geometry and defect configuration. This enables the occurrence of electronically driven defect processes as they are coupled to trapping and detrapping of carriers. The characteristic energy scale of these processes is set by the energy gap and the associated lattice distortions are of order few tenths of Å.

In the following we will briefly introduce some of the theoretical tools used to obtain quantitative descriptions and then selectively discuss some defect systems in detail. We begin with the description of the isolated lattice vacancy and its negative U behavior, then continue with some analysis of electronically driven migration processes of impurities like, e.g. interstitial Al and Si and conclude with a description of hydrogen diffusion and passivation.

2. Theoretical Description

The key theoretical notion is a local defect energy which depends on the number of trapped electrons which are in contact with a reservoir of electrons characterized by a Fermi energy. The evaluation of this energy from first principles calculations has become possible in recent years and numerous useful results have appeared. These recent successes are based on a long history of systematic studies and developments with the goal to describe in a simple yet accurate fashion the inherently complicated underlying many electron systems [1].

Springer Series in Materials Science, Vol. 13
Semiconductor Silicon Editors: G. C. Harbeke and M. J. Schulz

To describe a localized defect we first consider a perfect, translationally invariant, infinite semiconductor crystal. A point defect is created by e.g., removing one crystal atom, or replacing it by an impurity atom or by similar operations. We define a small region of action around the defect in which *strong* chemical effects occur. In general, atoms in this region will distort significantly from their original perfect crystal position. In semiconductors, this region is typically a few screening lengths (≈5-10Å) in diameter and contains of the order of 20-50 atoms. The region of *action* is surrounded by a much larger region of *reaction* over which the crystal's response to the defect can best be described by elastic and dielectric means. The task to deal with these problems can be separated into two categories: first, how to describe the atomic arrangement and second how to account for the strong chemical changes around the defect. The number of approaches available to handle these problems is large.

We report here results primarily obtained with two types of methods. One is the Greens function method which describes an isolated defect in an otherwise perfect crystal. It is aesthetically the method of choice, but computationally rather involved. Detailed accounts of the method are given elsewhere [1,2]. In brief, the method requires initially the calculation of the perfect crystal Greens function. Defect related quantities, such as bound state energies, phase shifts and defect induced electronic charge densities can then be obtained by solving Dyson's equation for the defect Green's function. It is customary in this context to expand the electronic wavefunctions in a set of localized orbitals. Dyson's equation then becomes a matrix equation in the basis of local orbitals centered at and near the defect. As mentioned above this method is attractive since it explicitly only calculates defect-induced changes. In particular, bound state energies can be evaluated accurately for a given defect potential. The complexity of the method has, however, somewhat impeded its widespread use.

As an alternative, many studies on defects are done in the so-called supercell geometry [3]. Here a cell is defined containing a sufficiently large number of atoms (typically 20-50). The defect is placed into the cell and the configuration is periodically repeated creating a period "defect crystal". The advantage is obvious; with the help of Bloch's theorem standard bandstructure techniques can be applied like e.g., an expansion of the wavefunction in plane waves. As the cell size is increased the distance between defects becomes larger and isolated defect properties can in principle be simulated. In practice, the method is particularly suited for the calculation of configurational analyses which converge more rapidly as a function of cell size than do e.g., bound state energies of deep traps. The simplicity of the method often outweighs the increased (but straightforward) numerical effort associated with large cell sizes containing many atoms.

The most central theoretical issue is the choice of the Hamiltonian for the description of the many electron systems. Most, but not all defect calculations are today done within the Density Functional (DF) framework and using the Local Density Approximation (LDA) [4,5]. Several reviews on this subject have recently appeared [6-8].

The Density Functional theory represents an attempt to integrate the lattice ion potential V_{ext} with the mutual interaction of electrons. The whole theory is based on the remarkable theorem by Hohenberg and Kohn which states that all properties of the many-particle ground state are given by the ground-state electron density distribution. The idea is to replace the system of interacting electrons by a system of *noninteracting* particles which move in a fictitious additional field, the so-called exchange correlation potential, such that their density remains the same and then solve the noninteracting system. The solution of the many-body problem is thus reduced to the solution of a set of one-particle equations, not more difficult to solve than the Hartree problem. The theoretical difficulty lies in determining the unknown fictitious potential $V_{xc}[\rho] = \delta E_{xc}[\rho] / \delta\rho(r)$. Here, a second important concept yields the framework for practical approximations. In analogy to the Thomas-Fermi method, an approximation to V_{xc} is given by regarding a small neighborhood of the electron system as behaving like jellium, i.e., $\rho = \text{const}$, at the value of the local density. Then well-known approximations for the homogeneous electron gas exchange correlation energies E^h_{xc} can be used [6].

While there is a large body of successful applications of local-density or local-spin-density theory, there are also failures, notably in two areas. First, some ground-state properties of open-shell systems, such as 3*d* row transition metal atoms, are rather sensitive to the approximate form of exchange and correlation functionals. The presently used functionals sometimes yield wrong ground states. Similarly, some surface properties such as the existence of a long-range image potential are not given by simple LDA theories. The inability of the LDA to describe the stability of certain negative ions falls into this category as well. Improved functionals, going beyond the local approximation, seem to be necessary and in some of these cases a correct description of the ground state can be obtained by going beyond LDA.

The second point refers to one electron or quasiparticle energies. It is general practice, although there is no formal justification, to interpret the eigenvalues of Kohn-Sham density functional equations as quasiparticle energies, especially for extended systems where relaxation effects are negligible. While the overall features of experimentally measured quasiparticle spectra (photoemission or inverse photoemission) are quite well reproduced sizable quantitative errors do occur. Although the origin of this has been formally recognized since long, only recently have quantitative quasiparticle calculations been done for both semiconductors and metals [9,10]. These calculations are outside the framework of density functional theory.

An interesting situation occurs for semiconductors. Here the eigenvalue difference $\varepsilon_c - \varepsilon_v$ between the first unoccupied conduction band state and the last occupied valence band state of the N-electron system are customarily interpreted as the energy gap. Large errors ranging from 0.5 to 1 eV occur in comparison to experiment. Of course the quasiparticle argument could be used to discount the appropriateness of a ground state theory. However, the semiconductor gap can also be viewed as a ground state property (by neglecting excitonic effects), because it can be expressed as the difference between ionization energy and electron affinity, each of which is a difference of total energies of ground states with different particle numbers. The difference can be shown to be equivalent to the difference of the highest occupied eigenvalues of the N and N + 1 electron systems. A subtle feature in density functional theory explains the apparent paradox. The Kohn-Sham potential $V_{xc}[\rho]$ needs not to be the same for N and N + 1 particle systems, but can differ by a constant Δ in space which does not change the density. In other words, the functional derivative $\partial E_{xc}[\rho]/\partial\rho$ can be discontinuous across the semiconducting gap. A series of recent calculations shows that this discontinuity can be linked to the quasiparticle spectrum and can thus be quantitatively evaluated [10]. The results show that the LDA is a rather good approximation for semiconductors, provided the discontinuity Δ is explicitly taken into account. The quantitative evaluation of Δ is, however, laborious and simplified approaches are clearly needed. For most defect calculations, either uncorrected LDA values or empirically adjusted gaps (by use of the "scissor operator") [11] have been used as basis to derive deep levels. Empirically, it has also been found that defect energies based on (incorrect) LDA gaps agree better with experiment than those derived from adjusted gaps. However, the adjustment is sometimes necessary to obtain the correct ground state occupation. This seemingly paradoxical situation needs further investigation.

In the description of the valence electron system is often advantageous (but not necessary) to eliminate the degrees of freedom associated with core electrons explicitly. This is done by the use of pseudopotentials. Originally considerable success was achieved in describing band structure and optical properties of semiconductors and simple metals with use of empirical pseudopotentials [12]. In a later, more sophisticated approach only the unscreened portion of the pseudopotential originating from the core electrons was parameterized thus retaining explicitly the screening part due to the (pseudo) valence electrons [13]. Certain, nonunique empirical dependences remained. These were all but eliminated by the introduction of "norm-conserving" pseudopotentials [14,15]. The essence of these types of potentials is to guarantee for certain reference systems (atoms) that the pseudo wavefunction is identical to the true valence wavefunction outside the core region *and* that it is correctly normalized. While containing no nodes in the core region the normalized pseudowavefunction yields correct electrostatic potentials outside the core. The norm-conservation feature furthermore assures that the pseudopotential not only correctly simulates the scattering properties of the reference system but that for systems that differ from the reference configuration there is *no* first-order transferability error. This is the main advantage of norm-conserving pseudopotentials and makes them very useful for the description of molecules and solids. For details, e.g. the inclusion of relativistic effects the reader is referred to Ref. [15].

3. The Simple Lattice Vacancy

Although not of direct technological importance, the vacancy in silicon is one of the most studied deep-level defects. Both theorists and experimentalist consider its physics as prototypical. Here we will first outline the basic physical picture of the vacancy and then describe the results of early Green's function studies which partly confirmed that picture and partly extended or modified the previously accepted interpretations.

The originally accepted picture of the vacancy is due to the work of Watkins [16]. The underlying model is the Coulson-Kearsley [17] defect molecule where attention is focused on those orbitals in the crystal which are most perturbed by removal of the single atom to form the vacancy. There are four such orbitals, one on each of the four atoms nearest the vacancy, each associated with one of the four valence bonds broken by removal of the central atom. One-electron molecular orbitals are constructed as linear combinations of the four broken bond orbitals. Four such independent molecular orbitals can be constructed. The completely symmetric combination (A_1) is expected to be lowest in energy because it is nodeless. The other three are partner functions for a three-dimensional representation (T_2) of the T_d symmetry group of the undistorted vacancy.

Most of the recent one-electron calculations of the electronic states of the undistorted silicon vacancy find that the threefold degenerate T_2 level lies deep in the gap and that the A_1 state, or the resonance to which it gives rise, is located about 1 eV below the top of the valence band.

The total charge disturbance at the vacancy, a rather short-range charge density has been calculated [18] (see Fig. 1, lower panel). It is essentially confined to a cavity terminated by the nearest-neighbor atoms. This is in contrast to the behavior of individual states which can be quite extended (see Fig. 2). The fast healing of a perturbance has also been found for semiconductor surface systems. It is of the order of a bond length, coinciding with the range of the spatially dependent static dielectric

Fig. 1

Fig. 2

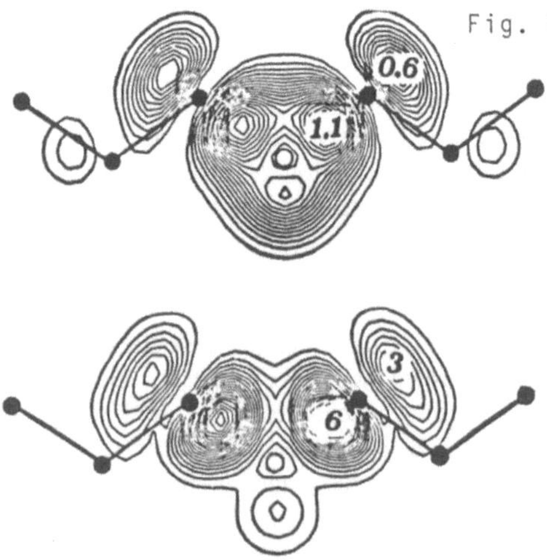

Fig. 1 Contour plots of the unperturbed (top), perturbed (middle), and total change (bottom) of charge densities induced by a silicon vacancy (from Ref. 18)

Fig. 2 Contour plots of charge distributions of the A_1 resonance (top) and the T_2 bound state (bottom) induced by the silicon vacancy (from Ref. 18)

function [19]. Also shown in Fig. 1 are contour plots of the total unperturbed silicon charge (top) and the total charge in the presence of a vacancy (middle). The total valence charge in the presence of the vacancy (Fig. 1, middle panel) clearly shows the disappearance of bonding charge with the removal of an atom. This situation is very reminiscent for that of the (111) surface which also showed a smeared-out and rounded charge distribution which decays rapidly into the vacuum.

We now return to the question of occupying the vacancy states in the defect molecule picture. For the vacancy to be, e.g., in the singly positively charged state V^+, three electrons must be distributed among these four states. Two of them go into the A_1 state (or that resonance) with their spins paired. The third goes into the T_2 level, a situation which, because of the orbital degeneracy, is unstable with respect to Jahn-Teller distortions. The type of distortion which is tetragonal can e.g. be inferred from EPR experiments. It drops the energy of the T_x state and raises the T_y and T_z energy *without* splitting them. For the vacancy to be in its neutral state, V^0, an additional electron is needed. This too goes into the T_x state, further increasing the amount of tetragonal distortion and pairing the spins. An additional electron is needed to produce the single negatively charged state V^-. It goes into one of the unsplit T_2 levels, a situation which, because of the orbital degeneracy still remaining, is again unstable with respect to another Jahn-Teller distortion.

Experimental support for this picture was provided by Watkins' EPR experiments, out of which has also emerged a level scheme with the lowest E(0/+) level near the valence band maximum. The assignment and nature of this level was contradicted by early Green's function calculations [18]. The claim was that the lowest electrical level is not E(0/+), but is E(0/++). It was proposed that a *two-electron* transition level E(0/++) exists and lies near the valence band top, and that the states V^0, V^+ and V^{++} form an Anderson negative-effective-U system. This is unusual, because normally one would expect that electron-electron repulsion requires an extra energy, call it U, to add a second electron to a singly occupied state. Thus, as the Fermi energy is raised, the sequence of *expected* equilibrium states would be V^{++}, V^+ and V^0. The essence of Anderson's effective-negative-U, however, is that an energy-lowering structural distortion may be sufficiently enhanced by the presence of a second electron that the energy gain more than compensates the repulsive energy cost. In this case, a second electron will rapidly follow the first one on or off the center.

The meaning of this is illustrated in Fig. 3 for the simplifying example of a single configuration coordinate Q. The defect energy $E_{tot}(N,Q,\mu)$ vs. Q is plotted using one value of μ in the top panel and another value of μ in the middle panel. Each of the two curves separately has the appearance of a configuration coordinate diagram, and the equilibrium value of Q for each N is the minimum of that particular curve. The lowest of the three minima gives the ground state of the system. The state for which the system is lowest depends on μ, as is illustrated in Fig. 3 (top) and (middle). By plotting that minimum energy $E_{min}(N,\mu)$ vs. μ as done in Fig. 3 (bottom), one can determine which N is the ground state for any value of μ and thus determine the μ-dependence of the charge state of the vacancy. The value of μ at which the charge state changes is the level position. Figures like this have become rather useful and standard for discussing the charge state dependence and stability of defect systems. As seen from Fig. 3, $N = 1$ is never a ground state occupation, regardless of Fermi

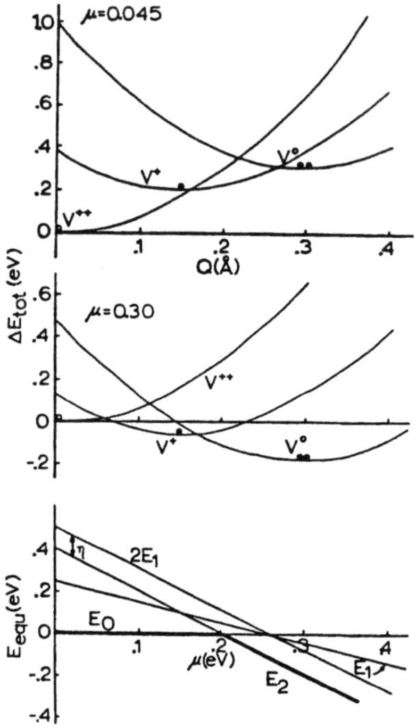

Fig. 3 Total energy functional E_{tot} (N,Q,μ) for the silicon vacancy as a function of distortions Q for states V^{++}, V^{+} and V^{0} with the Fermi level $\mu \approx 0$ (top) and $\mu \approx 0.3$ eV (middle). The bottom panel shows the respective energy minima as a function of μ. (from Ref. 18)

level position μ. This negative U situation occurs when the electron-electron repulsion energy $U < 2E_{JT}$, where $E_{JT} = V^2/2k$ with V the deformation potential of the gap level and k the associated spring constant.

It is interesting to ask whether the negative-U situation is rare and pathological or one to be met frequently. A simple scaling argument, familiar from polaron theory, gives some insight into this question. Let us regard the localization length of the defect as a characteristic coordinate. Then $U \sim 1/R$ scales like a typical Coulomb energy and $\hat{E}_{JT} \sim -1/R^3$ scales like a contact term which is directly proportional to the amount of charge in the defect level driving the Jahn-Teller distortion. It follows from his argument that for large R, i.e. for delocalized centers, $U_{eff} > 0$, while increasing localization tends to reverse the sign of U_{eff}. There is, however, a cut-off for small R values, roughly equal to the interatomic distance when the electron-lattice coupling becomes inoperative and the Coulomb repulsion becomes again dominant. Thus, if at all, U_{eff} could become negative for intermediate localization and then it depends on the material parameters U and $E_{JT} = V^2/2k$. A soft lattice (small k) favors a negative U_{eff}, so do a large coupling constant V and a small Coulomb repulsion U. The latter two are, however, counteracting each other in the sense that dielectric screening decreases both of them.

It can be regarded as a rather rare coincidence that the early calculations in 1980 predicted a negative U_{eff} (by ≈0.13 eV) for the silicon vacancy. In fact, due to the numerical uncertainties in these estimates, the situation remained unsettled until 1982 when Newton et al. [20] unambiguously measured the negative-U character of the silicon vacancy. By now several other point defect systems in semiconductors are known to be also negative-U systems.

4. Enhanced Migration of Interstitial Impurities

The migration of interstitial atoms in semiconductors is a long standing and fascinating problem. In particular, the phenomenon of capture or recombination-enhanced migration has been recognized as an important feature of defect reactions and device degradations. The mechanism for this enhancement is, generally speaking, that the energy stored in free electrons or holes (i.e. E_g/per pair) can be partially released upon carrier capture or recombination and can thus supply part of the energy necessary for the defect to make a diffusional jump. Various theoretical models have been proposed for this mechanism, but a basic understanding is only slowly emerging [21].

In analogy to the prototype silicon vacancy, it is natural to first consider the silicon self-interstitial which should occur together with vacancies as a first-generation radiation defect. Unfortunately, there is very little known experimentally about the Si self-interstitial. It has so far completely eluded direct experimental observation, however, several theoretical models have been proposed. We shall come back to this problem at the end of this section.

Detailed experimental information is, however, available for the migration of interstitial Al in silicon [22]. Interstitial Al is produced by electron irradiation of Al (substitutional)-doped p-type silicon at room temperature. It can be monitored by DLTS and EPR techniques. The results are compatible with an E(+/++) level at $\approx E_v$ = 0.17 eV, exhibiting tetrahedral symmetry. With no free carriers present the defect diffuses with an activation barrier of ~1.2 eV, while in the presence of carriers this barrier is lowered to ~0.27 eV. This remarkable result is shown in Fig. 4. The process is very energy efficient, considering a 0.9 eV barrier lowering in a material with a 1.1 eV band gap.

To obtain the quantities needed to evaluate the barrier lowering one considers Al at two different interstitial sites, the tetrahedral (T_d) site, which, according to the experimental results, is probably the energy minimum along the diffusion path, and the hexagonal (D_{3d}) site, which is likely to be the highest saddle point along the diffusion path [11,33]. For the tetrahedral position a deep A_1 bound state in the gap and a strong T_2 resonance at the edge of the conduction band (see Fig. 5) correspond to mainly p-like states of nonbonding character. The A_1 bound state gives rise to an electrical E(+/++) ≈ E_v + 0.1 eV in good agreement with experiment. Although there is no other deep level found in the gap, there will be a standard shallow donor level E(0/+) near the top of the gap. The natural charge state of the impurity will, therefore, be Al^+ for Fermi-level positions away from the band edges.

As one moves the interstitial atom from the high-symmetry tetrahedral site to the lower-symmetry hexagonal site, the threefold degenerate T_2 resonance in the

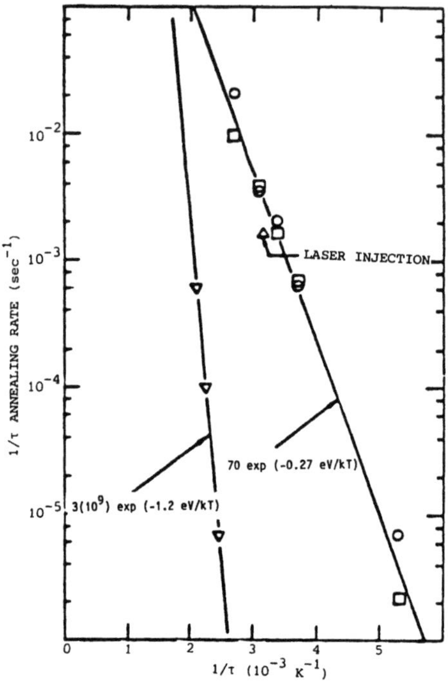

Fig. 4 Temperature dependence of the recovery rate of the interstitial Al defect in silicon. No free carriers (∇), with free carriers (□,0). (from Ref. 22)

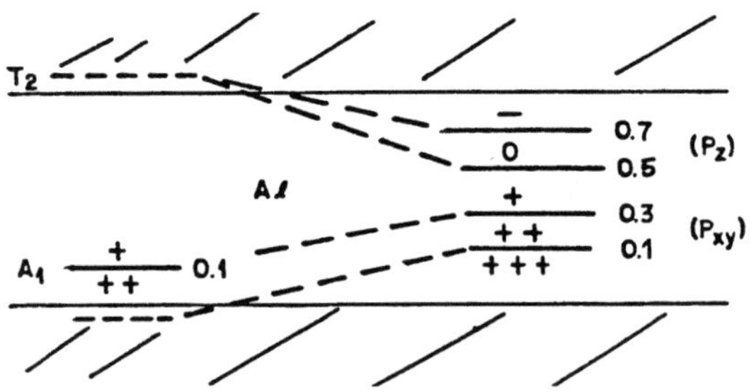

Fig. 5 Calculated electrical level structure for interstitial Al in silicon in the tetrahedral site (left) and the hexagonal site (right). (from Ref. 11)

conduction band splits into a twofold degenerate resonance which rises (not shown) and a onefold degenerate state which drops into the gap. The new deep state in the gap can be occupied by zero, one, or two electrons, creating two additional electrical levels, the lower of which was originally the shallow donor level (see Fig. 5). The calculated strong spectral changes between the T_2 and D_{3d} sites suggest their involvement in the migration enhancement process. There is a sizable energy barrier (measured at 1.27 eV and calculated at 1.3 ± 0.5 eV in p-type Si) associated with unassisted migration (see lowest curve in Fig. 6). The barrier results mainly from the

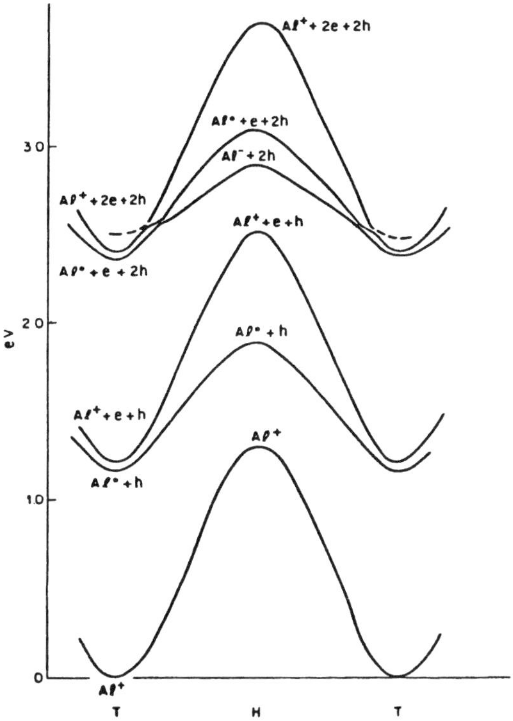

Fig. 6 Configuration coordinate diagram for the total energy variation of Al diffusion along the interstitial channel in silicon with no carriers trapped (bottom), one carrier trapped (middle) and two carriers trapped (top). (from Ref. 11)

drastic changes in bonding configuration and is also modified by different elastic distortions. Under electron injection, all the tetrahedral Al^{++} converts to Al^{+} at low currents. The positively charged Al^{+} can bind another electron in a shallow effective-mass-like state Al^{0}. The strong T_2 resonance near the conduction band minimum drops into the gap as the Al vibrates thermally from the tetrahedral position toward the hexagonal one. This allows a carrier to be localized in the vicinity of the Al interstitial leading temporarily to Al^{-}. The calculations show that the shallow-deep transition of these *two* electrons acts as a driving mechanism to help move the Al atom along the tetrahedral-hexagonal path (see Fig. 6). The barrier lowering can be deduced from Fig. 5, as $\Delta V_B = (0.8 \pm 0.5$ eV), which is in excellent agreement with the experimental value of 0.93 eV. The participation of *two* electrons in the enhancement process contradicts the commonly held belief that the maximum energy available for enhancement is equal to the gap E_G.

Apart from the remarkably high barrier lowering which according to the model above is due to the participation of two electrons, a drastic drop in the pre-exponential factor for the recovery rate of the defect has also been observed (see Fig. 4). The factor drops from $3 \cdot 10^9 s^{-1}$ to $70s^{-1}$, indicating very different mechanisms limiting the diffusion in the two cases. The value of $3 \cdot 10^9 s^{-1}$ is typical for a thermal process, i.e. the lattice vibrational frequency ($\sim 10^{13} s^{-1}$) divided by the number of migrational jumps required before the defect becomes trapped ($\sim 10^4$). The value of $70s^{-1}$ is compatible with a carrier capture limiting process. Another natural explanation of the small value is an extrinsic Coulomb effect. When the interstitial Al is in the Al^- state by virtue of having captured two electrons, Coulomb repulsion between it and the ionized acceptor will effectively raise the trapping barrier such that this trapping channel becomes unobservable. Only a small fraction of migrating atoms ($10^{-5} - 10^{-7}$) which happen to be in the Al^+ or Al^0 state while in the immediate vicinity of the acceptor would be trapped.

It is now believed that the shallow-deep alternation mechanism proposed here for the enhanced migration of interstitial Al is a quite common phenomenon in defect migration. The mechanism has some similarities with the electronic excitation mechanism invoked for the photostimulation F-center production in KCl, but it differs from the common models for enhanced migration in semiconductors, i.e. the charge state mechanism, the saddle-point mechanism or the energy release mechanism [21,22].

It is instructive to use the results for the Al diffusion to speculate on the silicon self-interstitial problem. If one repeats the Green's function calculations for Si, a level structure as shown in Fig. 7 is obtained. The calculations show *no* deep level and a (++) charge state for Si in the tetrahedral interstitial site. Tetrahedral interstitial silicon would, therefore, elude DLTS/EPR detection. The calculated level structure for the D_{3d} situation would in principle allow a two-electron shallow-deep alternation mechanism similar to that for Al migration. The barrier lowering would be $\Delta V_B \approx 2.2$ eV, close to the maximum possible in silicon. For an intrinsic calculated barrier of 0.4 ± 0.5 eV this ΔV_B would reverse the barrier and could produce *athermal* recombination-enhanced diffusion which would make Si interstitials very mobile even at low temperatures [23].

The problem of silicon self-diffusion at low and high temperatures is, however, more complex and has been subject to a longstanding debate. Car et al. presented extensive Green's function calculations [24] similar to those shown here, while Bar-Yam and Joannopoulos [25] treated the problem via a supercell technique. Both vacancy and interstitial mechanisms were considered. The calculated formation

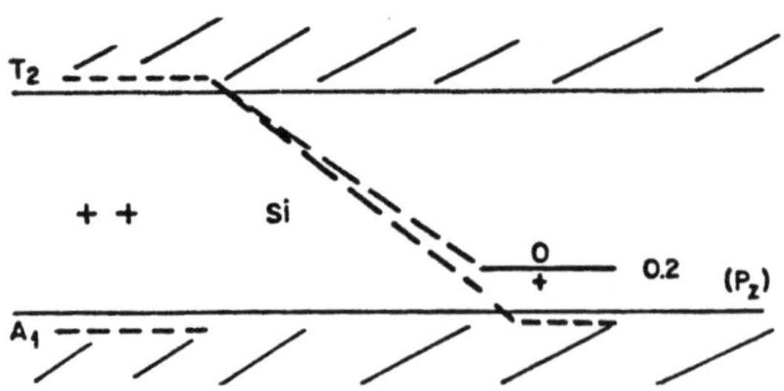

Fig. 7 Calculated electrical level structure for interstitial Si in silicon. Details as in Fig. 5

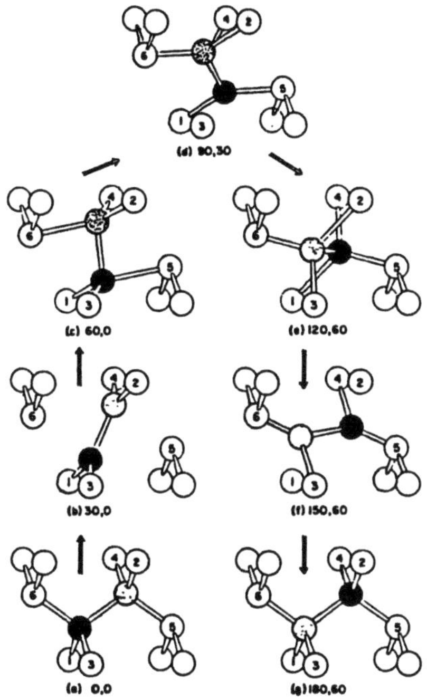

Fig. 8 Sketch of the atomic position change during the concerted exchange mechanism for two Si atoms as proposed in Ref. [26]

energies of about 4 eV and migration barriers varying between 0 and 1 eV depending on species and charge state are in rough overall agreement with experimental findings and may well explain *low* temperature diffusion. However, at least one experimental observation remained unexplained: the unusually high entropy of ~10k for *high* temperature diffusion. It is estimated that vacancy or interstitial diffusion mechanisms only yield entropies in the range of ~1k. To resolve this puzzle, Pandey [26] recently proposed a model of "concerted" direct atom exchange without vacancies or interstitials. This model introduces large distortions of many atoms and larger entropies are expected. Fig. 8 shows a sketch of the rather involved movements of atoms during this exchange. The essence is that during the breaking of one bond another new one is being formed thus lowering the overall energy barrier. Careful supercell calculations augmented with long range strain fields give a barrier estimate of ~4.3 eV which compares with experimental values of 4.1 eV at 1100K to 5.1 eV at 1500K. Although there are no calculated entropies yet, large values are expected since the exchange involves large displacements of more than 50 atoms and since it can be done in six independent ways.

5. Hydrogen Diffusion and Impurity Compensation

In the past paragraph we discussed possible modes of impurity and self-diffusion via electronically enhanced interstitial pathways and through concerted local exchange mechanisms. In this last example of hydrogen diffusion yet another novel mechanism appears to manifest itself. Moreover, many interesting phenomena like acceptor and donor passivation or defect generation which are associated with the presence of hydrogen in crystalline silicon seem to be related to each other.

Complete total energy surfaces $E_T(r_H,N)$ for a single H atom in a Si lattice with different charge states (H^+, H^0 and H^-) have recently been calculated by Van der Walle et al. [27] and by Chang and Chadi [28]. It was found that the energy surfaces for H in all its charge states could be divided into *two* distinct regions with qualitatively different behavior. In the region of high Si electron density, including the bond center site, nearby Si atoms show strong lattice relaxations and a H-induced donor level appears high in the gap. For H in the interstitial channels or regions of low electron density the surrounding Si atoms show little relaxation and a deep H related level appears near the bottom of the band gap. Figure 9 shows a contour plot of the calculated energy surface in the (110) plane for a positively charged H^+ [27]. The lowest energy for H^+ occurs in a roughly disc-shaped region (B-C) intersecting each bond half ways. In p-type material the Fermi level is near the top of the valence band and hydrogen is in its H^+ state. Thus electrons from neutral H-atoms annihilate the free holes in the valence band and directly compensate followed by (chemical) pairing between the H^+ and the negatively charged acceptors, forming an overall neutral complex. This picture is in good agreement with recent IR absorption measurements by Bergman et al. [29] which show vibrational fingerprints of a bond centered hydrogen atom, slightly off-center next to acceptors. The calculations also show that H^+ diffuses easily via the bond centered path with a very small barrier height of 0.1-0.2 eV.

In n-type material the calculations predict H^- to be stable in a low density interstitial site which would also imply direct compensation. However, numerical uncertainties associated with the donor level position do not allow to rule out H^0 as the stable species in n-type Si. IR measurements indicate hydrogen vibrational modes characteristic of interstitial positions as predicted. The issue of direct donor compensation is still under debate; it appears to be less efficient. The calculations [28]

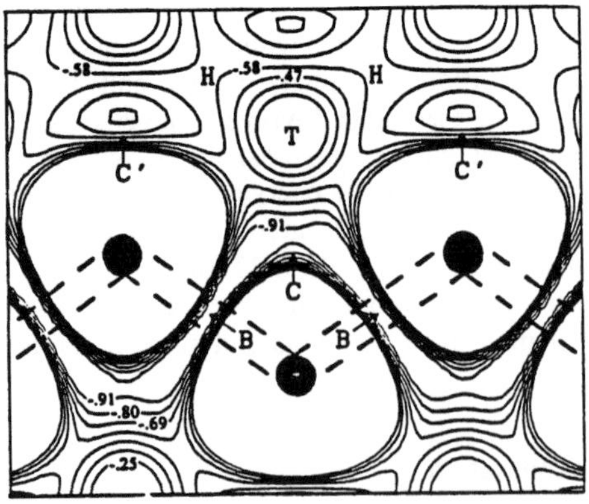

Fig. 9 Contour plot in the (110) plane, intersecting Si atoms, of the total energy surface (in eV) for the H^+ impurity. The reference E = 0 is chosen at the tetrahedral site. From Ref. [27]

indicate a ~0.5 eV decrease in binding energy between acceptor and donor complexes which could be an explanation for the difference in behavior.

Calculations for the interaction of several H atoms indicate that H_2 molecules are the preferred state in pure crystalline Si. The binding energy is only slight higher than binding to impurities. Frenkel pair formation with hydrogen tying off the dangling bonds of the generated Si vacancy is also energetically favorable but kinetically less likely. H_2 formation, on the other hand, is suppressed if the diffusing H atoms are charged while Si-H bond formation can still proceed. While more work is needed, these recent calculations elucidate considerably the interaction process of hydrogen with shallow impurities which has been rather unclear so far in contrast to the simple hydrogen passivation of damage induced dangling bonds.

6. Conclusions

The theoretical description of point defects in semiconductors has advanced considerably in recent years. This is mainly due to a combination of the development of computational algorithms with the more fundamental advances in Density Functional and Pseudopotential theory. The most remarkable physical findings are related to the strong coupling between electron states and lattice configurations. Quantitative pictures of electronically enhanced defect processes could thus be developed. Our theoretical capabilities are, however, still limited to rather simple defect systems. New concepts rather than improved existing schemes are needed for the next step in our understanding.

Acknowledgments

Most of my own work on defects has been done with Gene Baraff, to whom I am grateful for our long collaboration. I would also like to acknowledge the advice and help of E.. O. Kane, M. Lannoo, G. B. Bachelet, H. S. Greenside, G. Allan, D. R. Hamann, R. W. Godby, L. J. Sham, M. S. Hybertsen and M. Stavola.

REFERENCES

[1] M. Schlüter in "Semiconductor Physics", Ed. C. E. T. Goncalves da Silva, L. E. Olivera and J. R. Leite, Word Scientific, Singapore, 1987 p. 196.

[2] G. A. Baraff and M. Schlüter, Phys. Rev. B*30*, 1853 (1984).

[3] M. Schlüter, J. R. Chelikowsky, S. G. Louie and M. L. Cohen, Phys. Rev. Lett. *34*, 1385 (1975).

[4] P. Hohenberg and W. Kohn, Phys. Rev. *136*, B864 (1964).

[5] W. Kohn and L. J. Sham, Phys. Rev. *140*, A1133 (1965).

[6] "Theory of the Inhomogeneous Electron Gas", ed. S. Lundvist and N. H. March, Plenum Press, New York, 1983.

[7] "Local Density Approximations in Quantum Chemistry and Solid State Physics", ed. J. P. Dahl and J. Avery, Plenum Press, New York, 1984.

[8] "Density Functional Methods in Physics", ed. R. M. Dreizler and J. da Providencia, Plenum Press, New York, 1985.

[9] M. S. Hybertsen and S. G. Louie, Phys. Rev. B*34*, 5390 (1986) and references therein.

[10] R. W. Godby, M. Schlüter and L. J. Sham, Phys. Rev. Lett. *56*, 2415 (1986).

[11] G. A. Baraff and M. Schlüter, Phys. Rev. B*30*, 3460 (1984).

[12] For a review, see M. L. Cohen and V. Heine, Solid State Phys. *24*, 38 (1970); V. Heine and D. Weaire, Solid State Phys. *24*, 249 (1970).

[13] J. A. Appelbaum and D. R. Hamann, Phys. Rev. B, *8*, 777 (1973).

[14] D. R. Hamann, M. Schlüter and C. Chiang, Phys. Rev. Lett. *43*, 1494 (1979).

[15] G. B. Bachelet, D. R. Hamann and M. Schlüter, Phys. Rev. B, *26*, 4199 (1982).

[16] G. D. Watkins, in Point Defects in Solids, vol. 2, p. 333, ed. J. H. Crawford and L. M. Slifkin, New York, 1975.

[17] C. A. Coulson and M. J. Kearsley, Proc. R. Soc. London, Ser. A *241*, 433 (1957).

[18] G. A. Baraff and M. Schlüter, Phys. Rev. B *19*, 4965 (1979) and G. A. Baraff, E. O. Kane and M. Schlüter, Phys. Rev. B2*1*, 5662 (1980).

[19] R. Resta, Phys. Rev. B*16*, 2717 (1977).

[20] J. L. Newton, A. P. Chatterjee, R. D. Harris and G. D. Watkins: in Proceedings of the XII International Conferences on Defects in Semiconductors, ed. by C. A. J. Ammerlaan (Amsterdam, 1983) p. 219.

[21] M. Lannoo and J. Bourgoin, in "Point Defects in Semiconductors," I., Springer Series in Solid State Sciences No. 22 (New York, N.Y., 1981) p. 219.

[22] J. R. Troxell, A. P. Chatterjee, G. D. Watkins and L. C. Kimerling, Phys. Rev. B, *19*, 5336 (1979).

[23] G. A. Baraff, M. Schlüter and G. Allan, Phys. Rev. Lett. *50*, 739 (1983).

[24] R. Car, P. J. Kelly, A. Oshiyama and S. T. Pantelides, Phys. Rev. B*30*, 2260 (1984).

[25] Y. Bar-Yam and J. D. Joannopoulos, Phys. Rev. B*30*, 2216 (1984).

[26] K. C. Pandey, Phys. Rev. Lett. *57*, 2287 (1986).

[27] C. G. van de Walle, Y. Bar-Yam and S. T. Pantelides, Proc. Mat. Res. Soc., Ed. M. Stavola, S. J. Pearton and G. Davies, Vol. 104, p. 253.

[28] K. J. Chang and D. J. Chadi, Phys. Rev. Lett. *60*, 1422 (1988).

[29] K. Bergman, M. Stavola, S. J. Pearton and T. Hayes, preprint.

Defects in CZ Silicon

W. Zulehner

Wacker-Chemitronic GmbH, D-8263 Burghausen, Fed. Rep. of Germany

1. Introduction

Silicon single crystals used today for the production of electronic devices are grown completely dislocation-free. With the large crystal diameters of the current crystals and the relatively high growth velocity it is not possible to grow single crystals with only a few dislocations. If, by a disturbance, one dislocation is formed in a large dilocation-free CZ or FZ silicon crystal then multiplication and spreading of dislocations occurs driven by the high cooling strains and the crystal grows further with a high dislocation density and, shortly afterward, also the formation of twins and polycrystalline growth is initiated.

If there are no dislocations in a crystal, the self-point defect supersaturation which develops during cooling cannot be reduced by absorption of the self-point defects at dislocations. Therefore strong supersaturation of self-point defects occurs in the crystals. Especially in FZ silicon crystals, this supersaturation may lead to the formation of the so called "swirl defects" which are supposed to be agglomerates of self-point defects. By suitable pulling technique, however, the formation of the swirl defects can be totally pevented. Due to the high oxygen content, the situation in CZ silicon is more complex than in FZ silicon. In CZ silicon we have in every case two types of point defects which come into super= saturation during cooling: self-point defects and oxygen. In perfect crystals the supersaturation of point defects can only be reduced by homogeneous nucleation and subsequent precipitation which leads to the formation of microdefects: Precipitates, perfect dislocation loops and stacking faults.

2. Point Defects

2.1 Vacancies and Interstitials

Before discussing aggregation phenomena and microdefects it is necessary to inspect the most important point defects in Si. Silicon crystallizes in the diamond lattice type which has a low packing density, only 34 % in the hard sphere model. The geometrical conditions in the diamond lattice naturally favor the formation and incorporation of interstitial atoms and not so that of vacancies. The silicon lattice contains much less

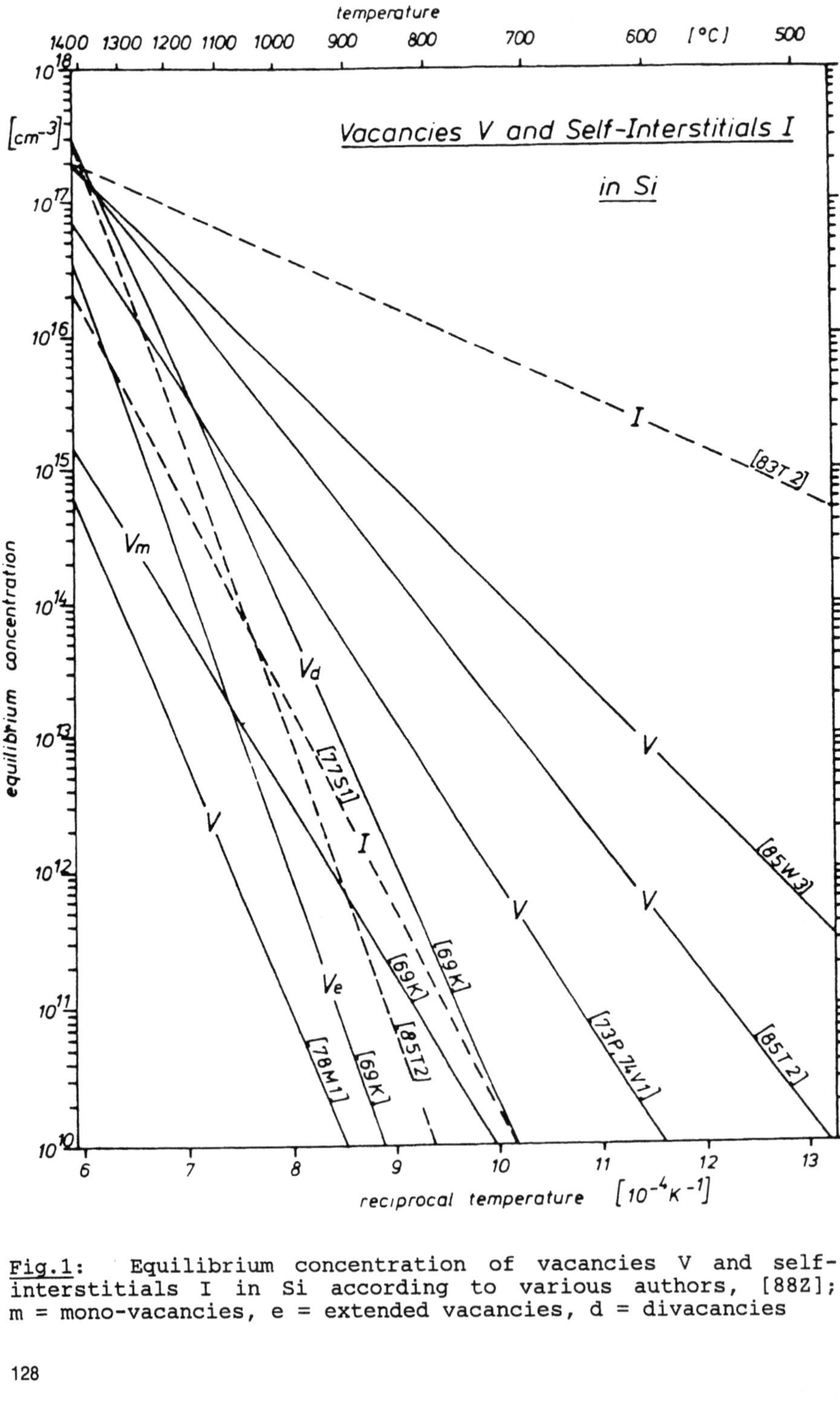

Fig.1: Equilibrium concentration of vacancies V and self-interstitials I in Si according to various authors, [88Z]; m = mono-vacancies, e = extended vacancies, d = divacancies

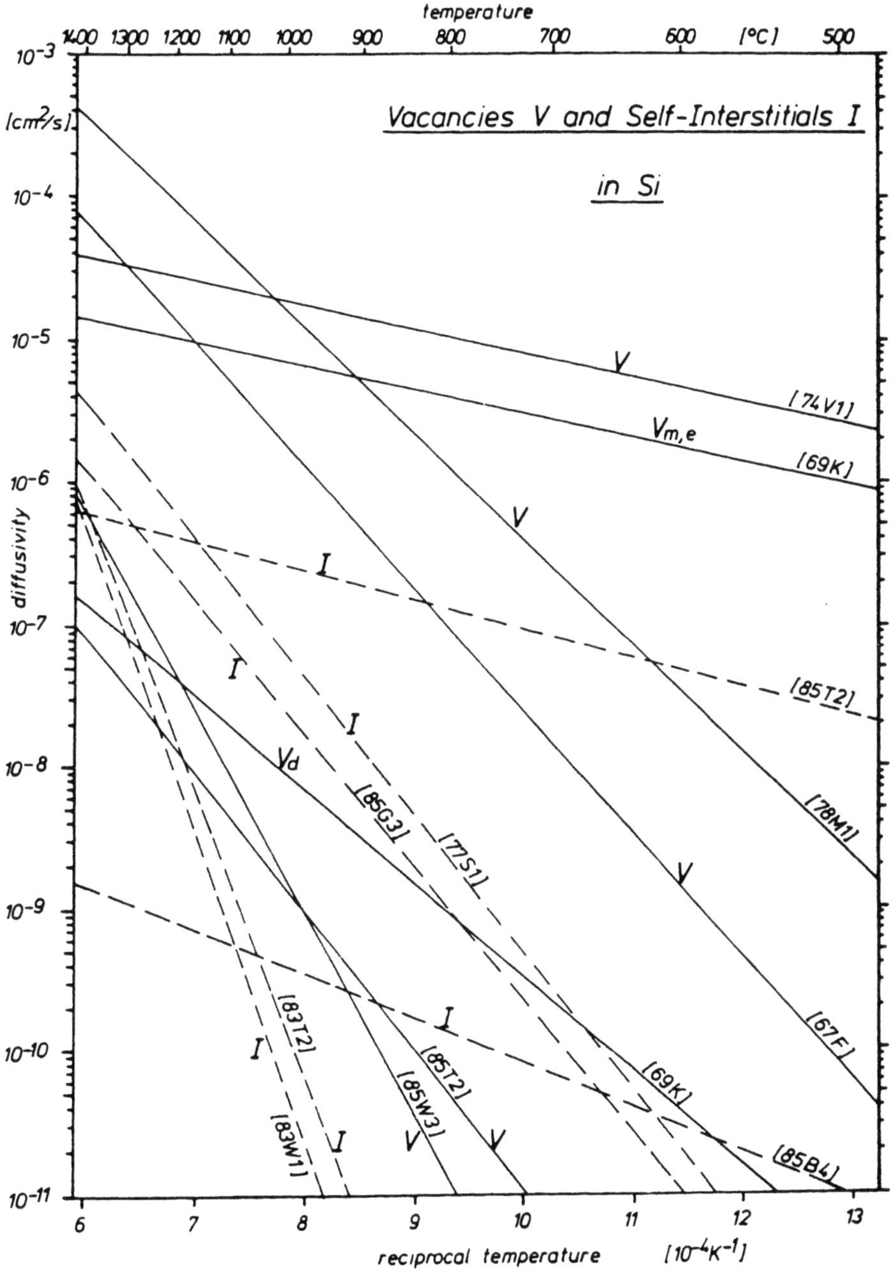

Fig. 2: Diffusivity of vacancies V and self-interstitials I in Si according to various authors, [88Z]; m = mono-vacancies, e = extended vacancies, d = divacancies

vacancies than a typical metal lattice in thermal equilibrium which contains about 10^{-3} vacancies at the melting point after freezing, [74S], whereas silicon only has between 2×10^{-9} and 6×10^{-6} under the same circumstances, [69K, 73P, 74V, 78M, 82V, 83T, 85G, 85W]. These values can be taken from Fig. 1, where the equilibrium concentrations of vacancies and self-interstitials in silicon for temperatures between melting point and 500°C are shown. The concentration values differ by several orders of magnitude from author to author. The situation is analogous for the diffusivity of the self-point defects, see Fig. 2. The values also differ by several orders of magnitude.

Up to now, there is no consensus about the role, the concentration and the diffusivity of vacancies and self-interstitials in intrinsic silicon. But most of the effects in silicon suggest a dominance of the self-interstitials at high temperatures (> 900°C). The self-point defect concentrations in **real crystals** are not only related to cooling rate and temperature. Their concentrations also depend on type and concentration of impurities, on processes like oxidation, oxygen precipitation, nitridation. The self-point defect concentrations can be largely altered in silicon and we have the case of **non-intrinsic silicon.**

2.2 Oxidation Effects

With regard to technical applications the increase of self-interstitial concentration by oxidation and oxygen precipi= tation is most important. It leads to the formation of extrinsic stacking faults and to enhanced or retarded diffusion of impurities depending wether these impurities diffuse via an interstitialcy or a vacancy mechanism.

Oxidation-enhanced diffusion (= OED) was observed with C, B, Al, Ga, In, P and As, whereas oxidation-retarded diffusion(= ORD) was observed with Sb; for references see [84F, 84K]. These observations correspond to the rule: The smaller atoms diffuse via an interstitialcy mechanism, also called kick-out mechanism, [80G], and the bigger Sb-atoms via a vacancy mechanism, also called dissociative mechanism, [56F].

The injection of self-interstitials during oxidation results from the effect that the diffusivity of oxygen in the SiO_2-layer is much higher than that of silicon, and that the reaction between oxygen and silicon, i.e. the formation of SiO_2 takes place at or nearby the SiO_2-Si interface, [85T]. This is shown in Fig. 3. Amorphous or crystalline (cristobalite) SiO_2 consumes about 2.3 times the volume as Si in the silicon lattice. This increase in volume causes compressive strain both in the silicon and in the oxide. In the oxide, the stress can be released by viscous flow and in silicon by the generation and emission of self-interstitials. This emission or injection of self-interstitials causes supersaturation of self-interstitials.

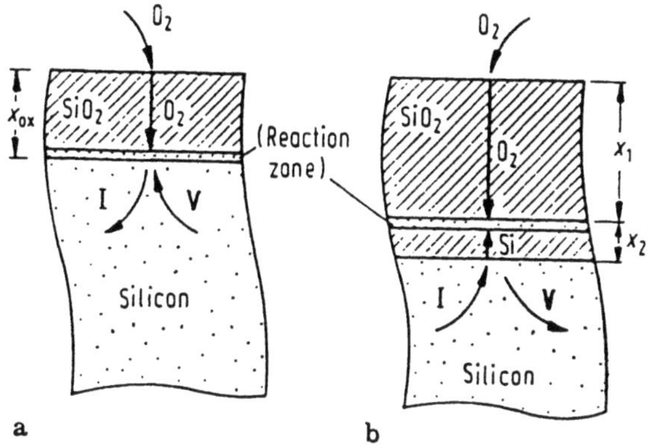

Fig. 3: Schematic of the oxidation of silicon: a = oxidation with thin and medium oxide layers leads to generation of self-interstitials and elimination of vacancies; b = oxidation with thick oxide layers leads to elimination of self-interstitials and generation of vacancies

Self-interstitials in supersaturation tend to form extrinsic stacking faults (= precipitates of self-interstitials) if there are precipitation sites or nuclei. In a lattice without crystal defects, however, no precipitation of self-interstials takes place in spite of strong supersaturation.

With longer oxidation and the formation of thick oxide layers, the oxygen and silicon atoms recombine to SiO_2 more and more in the oxide layer, [79F, 81T1, 83G]. In this case Si-atoms are removed from the Si-lattice with the consequence that:
a) vacancies are formed and emitted into the lattice and/or
b) self-interstitials are absorbed at the Si-SiO_2 interface.
Long oxidation therefore leads to undersaturation of self-interstitials and to supersaturation of vacancies and causes the shrinking of existing extrinsic stacking faults.

2.3 Carbon and Oxygen

Next to the self-point defects carbon and oxygen are the most important point defects in silicon crystals because they are always present in the highest concentrations of all impurities exept the intentional dopants. In the purest silicon, that is in FZ-silicon, always C- and O-concentrations of some 10^{15} cm^{-3} remain. The carbon contents in CZ- and FZ-silicon crystals are in average the same because they depend primarily on the carbon content of the polycrystalline silicon which is used for the pulling of the crystals. In typical crucible pulled silicon of today (1988) the oxygen concentrations are ranging from 5 x 10^{17} to 1 x 10^{18} cm^{-3}, that is by a factor of 100 higher than in typical FZ-silicon.

2.4 Other Impurities

The nitrogen content in silicon crystals is far lower than the carbon and oxygen content. In CZ-Si the nitrogen content is very low and can normally not be detected (detection limit $\approx 2 \times 10^{14}$ cm^{-3}). Next to the light elements C, N and O the fast diffusing transition elements like Cu, Ni and Fe are of great importance. The typical features of these elements are
a) very strong decrease in solubility with decreasing temperature, that is, high solubility at high temperatures (maximum near 1250°C) and very low solubility at lower temperatures, b) slow decrease in diffusivity with decreasing temperature, that is, very high diffusivity at high temperatures and still high diffusivity at lower temperatures. They all diffuse only or mainly via interstitial sites. These properties result in a strong capability for entering and spreading into the silicon at elevated temperatures and in precipitation with decreasing temperatures. As a result, a large fraction of the microdefects in silicon are precipitates of fast diffusing transition metals or contain these elements to some extend.

3. Aggregation Phenomena

Aggregation of impurities may lead to the formation of impurity complexes or to a new phase. Complexes consist of a few atoms and remain isolated. New phases are large aggregates of new compounds. Typical examples for complexes in silicon are the thermal donors, the C-O complexes and the transition metal-boron complexes. Typical examples for new phases are silicon carbide (SiC) and silicon oxide (SiO_2).

There are two models for the nucleation of a new phase in a supersaturated solution: homogeneous nucleation and heterogeneous nucleation [65B, 82I]: The **homogeneous nucleation** takes place randomly at any point by the condensation of highly supersaturated atoms. The **heterogeneous nucleation** takes place only at structural irregularities or impurity sites. A specific characteristic to distinguish the two types of nucleation is the degree of supercooling required for nucleation:
Large supercooling is typical for homogeneous nucleation, small supercooling is typical for heterogeneous nucleation.

3.1 Complex Formation

The smallest agglomerates of defect atoms in Si are complexes of two or a few impurity atoms which stick together due to elastic and/or electrical resp. chemical forces. The elastic forces are generated by atomic mismatch in the lattice. Two impurity atoms with opposite mismatch attract each other. It is not generally possible to differentiate between the elastic and the electrical/chemical forces. An example for mostly elastically bound complexes in Si are the carbon-oxygen complexes, first reported by [64N,68N]. In this case the expanding strain of the interstitial oxygen atoms combines with the contractive strain of the substitutional carbon atoms which are much smaller than the silicon atoms. Examples for mostly electrically bound complexes are the transition metal-boron complexes, [62L], which are formed only by Coulomb attraction. The most prominent representative of them is the iron-boron complex $Fe^{+}B^{-}$, a pair of an interstitial iron atom and a substitutional boron atom. The interstitial iron atom is a donor and the substitutional boron atom is an acceptor.

3.2 Homogeneous Nucleation

The presence of an embryo size distribution - a Boltzmann distribution - is assumed for describing homogeneous nucleation. The free energy $\Delta G(r)$ for formation of spherical embryos with radius r is a function of the negative volume free energy ΔG_V and the positive interface energy σ:

$$\Delta G(r) = \frac{4}{3}\pi r^3 \Delta G_V + 4\pi r^2 \sigma \qquad (1)$$

There is a critical nucleus size, at which the formation energy is largest. Its radius is called critical radius r_c:

$$r_c = -2\sigma/\Delta G_V \qquad (2)$$

The meaning of the critical radius r_c is:
a) $r > r_c$: Aggregates whose radius is larger than the critical radius at the annealing temperature are stable and grow.
b) $r = r_c$: Half of the critical nuclei grow and the rest shrinks.
c) $r < r_c$: The aggregates tend to shrink.

Further tendencies of the critical nuclei are:
d) They are the larger the higher the temperature, and
e) they are the smaller the higher the concentration.

Figure 4 shows the temperature dependence of the critical radius for different oxygen contents, [82T, 84L]. In the region above the respective curve, the nuclei tend to grow. In the region below the curve, the nuclei tend to shrink. The critical radius reaches atomic dimensions at temperatures around 450°C. At these temperatures, the stable nuclei have sizes of a single atom or of low complexes. The thermal or oxygen donors can be regarded as the smallest oxygen nuclei which are formed by homogeneous nucleation, [85G, 86W]. The thermal donor series almost entirely cover the critical radius range of oxide precipitates at temperatures where the nucleation frequently occurs, [86W]. At higher temperatures, the formation of stable nuclei needs a large supercooling. At 1000°C, for example, a supercooling of about 300°C is necessary to cause homogeneous nucleation in silicon with an oxygen content of 10^{18} cm^{-3}. A supersaturation of silicon self-interstitials (caused, for example, by oxidation) retards the formation of SiO_2 nuclei, [80H].

In general it is difficult or not possible to investigate **pure** homogeneous nucleation of oxygen in CZ Si crystals, it is

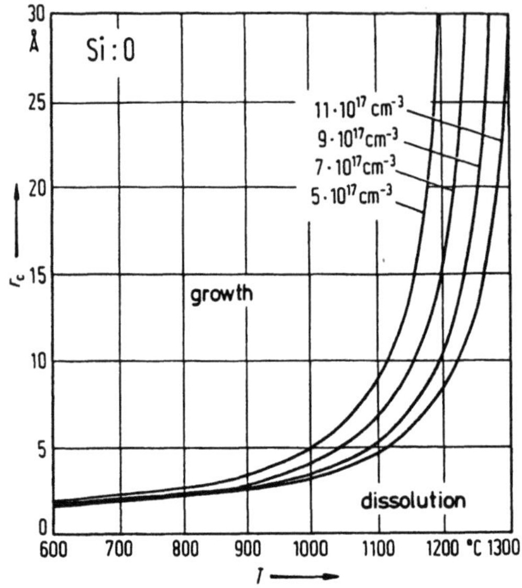

Fig. 4: Temperature dependence of the critical radius for O in Si according to [82T, 84L]. The saturation temperatures are T_S = 1223, 1271, 1309 and 1340°C for oxygen contents of 5, 7, 9 and 11 x $10^{17}/cm^3$, respectively

always mixed with heterogeneous nucleation, because the large industrial CZ-silicon crystals stay for many hours at temperatures between 1200°C and room temperature in the crystal puller during growth. Thus, the first annealing and the first precipitation, a homogeneous one, takes place during the first cooling.

3.3 Heterogeneous Nucleation

Heterogeneous nucleation is possible if there already exist precipitates or structural defects like dislocations, stacking faults, grain boundaries or interfaces. In this case, the supersaturated impurity or self-point defect preferably condenses or precipitates at these crystal defects because little supercooling is necessary. However, all the crystals used in electronics are grown dislocation free and no precipitation sites like pre-existing precipitates, dislocations, grain boundaries and interfaces exist after freezing. During the first cooling of the large crystals, no heterogeneous precipitation can therefore take place for most of the point defects, because the only heterogeneous precipitation site - the crystal surface - is too far away to be reached by most of the point defects. In any case, the first precipitation occurs with either homogeneous nucleation or with precipitation at complexes. The latter case is poorly investigated up today.

3.4 Influence of Inhomogeneities

An example for microscopic impurity variations is given in Fig. 5, [76M1]. It shows the variation of the antimony

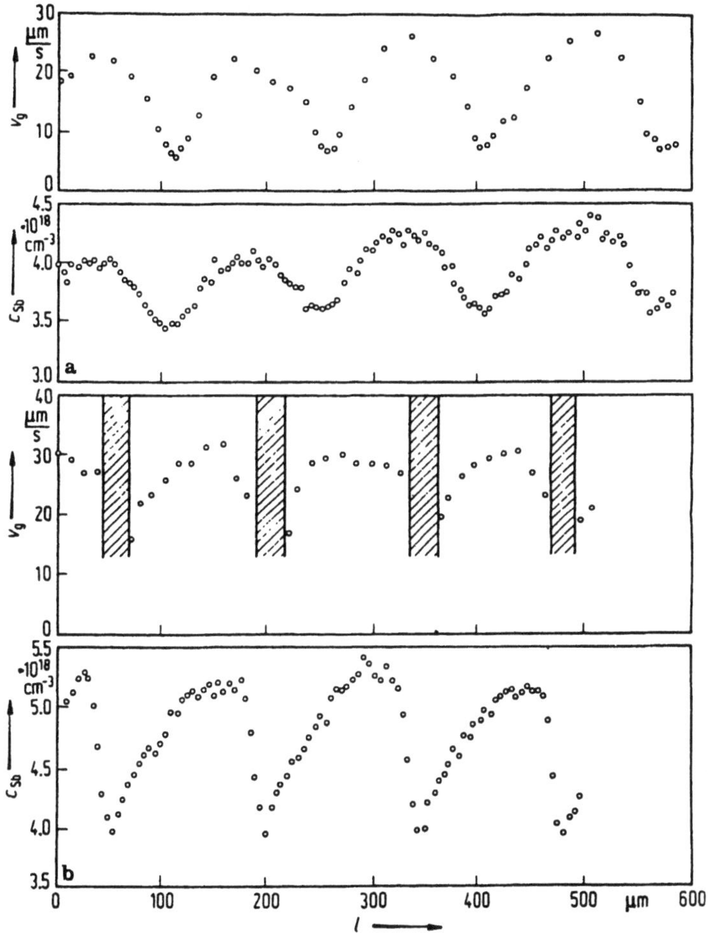

Fig. 5: Microsegregation analysis of an antimony-doped CZ-Si crystal, <111>, 18 mm Φ, pulled at 0.57 mm/min and with a seed rotation rate of 5 rpm; V_g = microscopic growth rate, C_{Sb} = antimony concentration, l = distance in pulling direction. The stronger concentration variations in the lower diagram are caused by backmelting effects (hatched region), [76M]

concentration at CZ Si pulling in a microscopic scale. The antimony concentration varies synchronically with the growth velocitiy. In the lower part of the diagram, the variations of the antimony concentration are larger because of a backmelting effect.

In Fig. 6 impurity striations in a CZ Si crystal, caused by the segregation of phosphorous, are shown. It is a micrograph of an axial section at the edge of an crystal. The phosphorous striations were made visible with etching and form waves in the sample surface. The facettes at the edge show, that it is a <111> crystal.

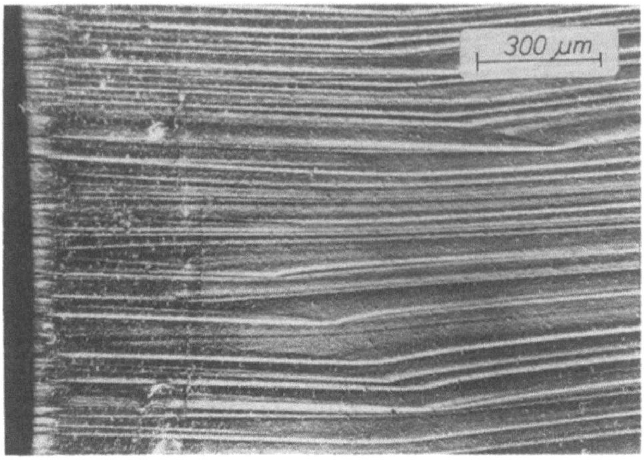

Fig. 6: Phosphorus striations in a <111> CZ-Si crystal made visible by etching, axial section near the edge (left)

Fig. 7: Axial section through the seed end of a CZ-Si single crystal showing convex and concave layers of oxygen precipitates, generated by annealing at 1000°C and revealed by etching. The microscopic oxygen distribution of this crystal is rather inhomogeneous

With respect to homogeneous nuclation and precipitation the microscopic impuritiy variations are most important. This becomes well perceptible in the case of oxygen precipitation:

Fig. 7 shows thin layers of SiO_2 precipitates in an axial section through the seed end of a CZ-Si crystal which was annealed for several hours at 1000°C. The precipitates were revealed by preferential etching. The oxygen precipitates mark the growth layers and make also visible the bending of the growth interface. With increasing diameter the growth interface is convex, as can be seen in the seed cone region. In the cylindrical part of the crystal, the growth interface

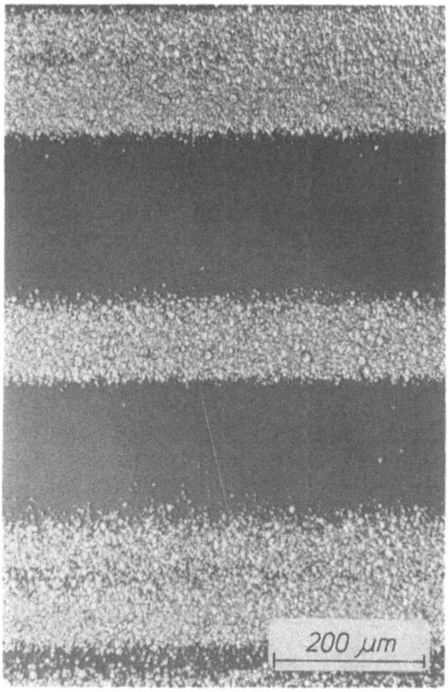

Fig. 8:
Micrograph of the sample of Fig. 7 showing layers of oxygen precipitates in very high density adjacent to layers without precipitates

becomes concave. The sample shows, that the oxygen precipitation and thus the oxygen distribution was not homogeneous. This inhomogeneity is even better seen microscopically. Fig. 8 shows a micrograph of the sample of Fig. 7. There are layers without precipitates side by side of layers with high or very high precipitate density. The differing oxygen concentrations and possibly additional influences result in very high concentrations of precipitates in the one crystal layer and often in zero precipitates in the adjacent layers. This phenomenon is most pronounced in the outer areas of the crystal. The nucleation formation takes place first in the layers of higher impurity concentration whereas in the adjacent layers of lower impurity concentration no nuclei are formed. By further annealing at constant temperature the nuclei grow due to diffusion and incorporation of impurity atoms. The absence of precipitates in the layers of lower oxygen content can be explained by

a) loss of oxygen due to diffusion of oxygen atoms to the precipitates in the layer of higher oxygen content, thus preventing homogeneous nucleation and/or
b) supersaturation with self-interstitials, caused by the precipitation, which also prevents homogeneous nucleation.

3.5 Precipitation

If precipitation nuclei are present, either homogeneous or heterogeneous ones, then precipitation can take place and depends primarily on the diffusivity of the impurity resp.

point defect. With advancing precipitation the precipitate particles become larger and, in most cases, exert compressive strain on the surrounding host lattice. This strain reduces the precipitation to a stop after some time if there is no mechanism which releases the strain. In the important case of oxygen in silicon the final precipitate composition is SiO_2, either amorphous or crystalline. Both aggregate states occupy about 2.3 times the volume of silicon, i.e. a SiO_2 particle exerts an enormous strain on the silicon lattice. The volume of the oxygen atoms is practically completely additional at the place of the SiO_2 particle because it was not present there prior to precipitation. Moreover, it takes only a small effective lattice volume in the dissolved, i.e. in the interstitial state.

The possible strain releasing mechanisms are: a) plastic deformation of the surrounding lattice, b) formation and emission of silicon interstitials and c) absorption of vacancies. Case a) has been confirmed several times by TEM investigations [76M2, 76T, 78Y, 79P, 79T, 80S], whereas up to now the decision between case b) and c) could not be made conclusively. However, case b) is favoured by the great majority of investigators today because of its consistency with the diffusion and precipitation phenomena in silicon. Like in the case of interstitial oxygen, interstitial silicon atoms also occupy only a low effective volume in the silicon lattice so that nearly the whole volume of a lattice atom is gained if a silicon self-interstitial is formed.

The number and distribution of precipitation nuclei is decisive for the process and the result of precipitation. If there are only a few precipitation nuclei then for each nucleus resp. each precipitate a large impurity supplying volume is available and the precipitates can grow large. If there are numerous precipitation nuclei then the impurity supplying volume for each precipitate is small and the precipitates remain small. Nevertheless, after longer or very long (depending on the diffusivity of the impurity) annealing times also larger precipitates can grow in the areas of initial high precipitate density. In this case the smaller precipitates are redissolved gradually and the larger ones grow further, a well known phenomenon in nature.

A concentrated overview over the precipitation behaviour of oxygen in dependence of temperature and concentration is given in Fig. 9. It shows the result of 20 h annealings at temperatures between 300 and 1300 °C. Two sets of samples were used for the annealings: The one set had an oxygen content of $8.8 \times 10^{17}/cm^3$ and the second set had an oxygen content of $13.5 \times 10^{17}/cm^3$. All the samples are preannealed at 600°C for 2 hours to eliminate the thermal donors which are formed during crystal growth. The diagram shows that the precipitation of oxygen was nearly the same at 300 and 600°C, both with medium and high oxygen content. At 450°C, however, the precipitation was much stronger. The amount of oxygen precipitated is given by the hatched region.

With higher oxygen contents, the precipitation is more complete, especially around 800°C. The residual concentration

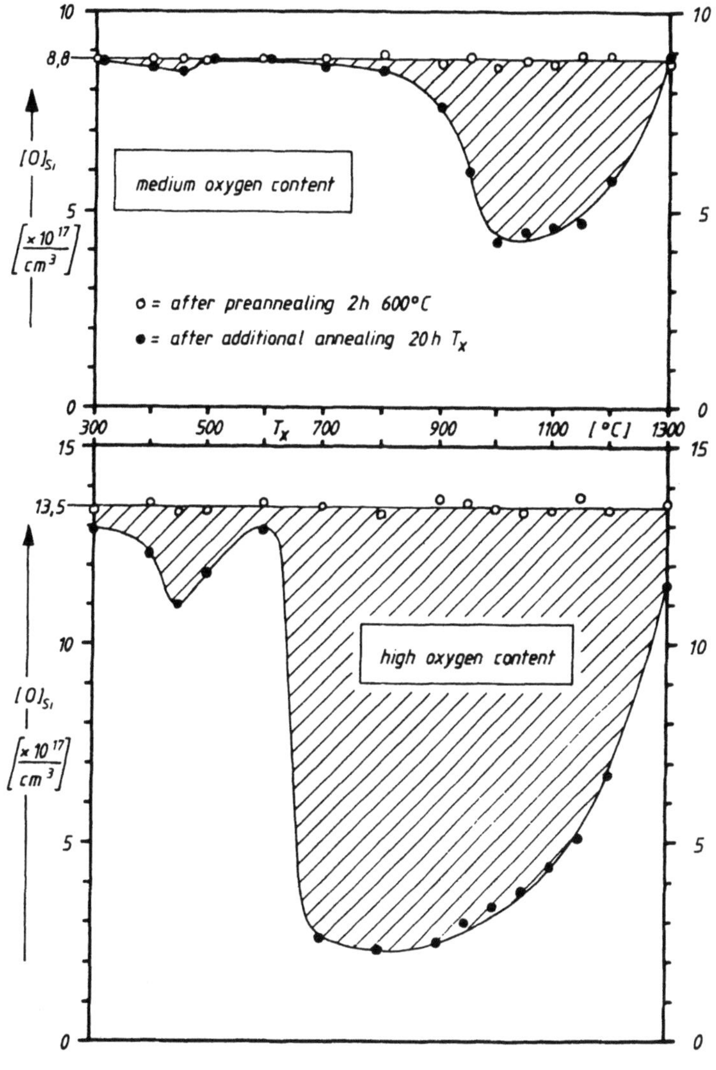

Fig. 9: Change in dissolved oxygen, i.e. precipitation of oxygen in CZ-Si by 20 h annealings at various temperatures in two different materials (medium oxygen content and high oxygen content); measured by IR absorption at 9 µm and 25°C

of dissolved oxygen in the high-oxygen samples is by a factor of 2 lower than in the medium oxygen samples which have the lowest residual dissolved concentration at 1000°C. This different precipitation behaviour can be explained by the different density of precipitation nuclei which is much higher in the case of high oxygen content. Due to this effect, the oxygen atoms in the high-oxygen samples had to diffuse only a short distance to the precipitates, so that the precipitation was faster and more complete especially at lower temperatures.

Above 1000°C, the precipitation was complete and the residual concentrations of dissolved oxygen correspond largely to the solubility of oxygen at the respective temperatures.

4. Oxygen Related Microdefects

4.1 Thermal Donors

As mentioned before, the thermal donors can be taken as the smallest oxygen precipitates. Their atomistic structure is still uncertain up today. According to IR-measurements, there exist at least 11 different thermal donors. They are mostly shallow double donors. The formation rate for the first thermal donors is proportional to about the fourth power of the oxygen concentration in Si. From this, Kaiser [57K] proposed in 1957 that the thermal donors contain 4 oxygen atoms and that they are SiO_4 complexes. Recent magnetic ENDOR measurements confirmed this assumption by finding 4 oxygen atoms in the lowest thermal donors. According to these measurements, the thermal donors contain 4, 6, 8, 10, 12 and so on oxygen atoms, that means, the next higher thermal donor is formed by the addition of two oxygen atoms. This result is in accordance with the theory of Gösele and Tan [82G], who proposed that below 500°C oxygen diffuses mainly in the form of O_2-molecules. Their theory explains the finding, that in this temperature range oxygen diffuses by several orders of magnitude faster than it is expected by the extrapolation from high temperatures.

In the more than 30 years since the discovery of the thermal donors, more than 20 models for their formation and structure have been published. Up today, there is no consensus about the exact nature of the thermal donors.

After prolonged annealing at 450°C, not only thermal donors are formed, but also electrically neutral complexes or precipitates of larger size than the thermal donors. After pulling, each CZ Si crystal contains thermal donors, and in most cases, also larger oxygen aggregates. The number of thermal donors depends the strongest on the oxygen content and secondly on the cooling conditions at crystal pulling. Figure 10 shows a typical axial thermal donor distribution in an as-grown phosphorous doped CZ Si crystal, which is given by the difference between the axial resistivity distribution before and after the elimination of the thermal donors. The thermal donors have been eliminated by an annealing of 2 h at 600°C and a following rapid cooling to room temperature. Thermal donors are not stable. Most of them can be eliminated within a few seconds at 650°C with rapid thermal annealing of a thin wafer. The diagramm gives a typical axial thermal donor distribution for a CZ Si crystal: The highest concentration at the seed end and the lowest concentration at the bottom end. This stems from the fact, that the seed end of the

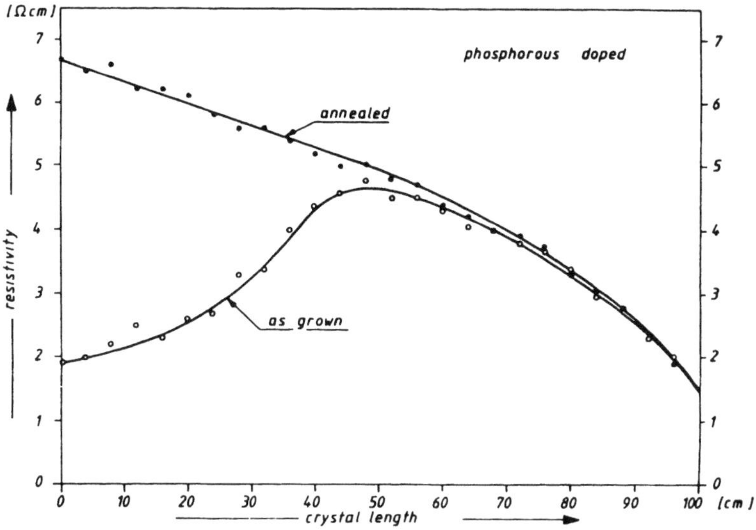

Fig. 10: Thermal donors: Axial resistivity variation in a phosphorus doped CZ-Si crystal with (= as- grown) and without (= after annealing) thermal donors. The thermal or oxygen donor concentration is given by the resistivity difference between the as-grown and the annealed state of the crystal

crystal contains the highest oxygen concentration and stays longest at temperatures below 500°C during pulling.

Besides the thermal donors which are formed below 500°C there are additional donors which are formed above 500°C up to about 800°C. They form more slowly, in much lower concentrations and also in strong dependence on the oxygen content. They are called "new donors", "high temperature donors" or "second thermal donors". They are large complexes or rather real SiO_2 precipitates, the interface of which shows electrical activity. Also the new donors resp. their electrical activity can be eliminated by annealing at higher temperatures.

4.2 Rod-like Defects

The smallest oxygen precipitates which could be detected by transmission electron microscopy are the "rod-like defects", [77T]. They are probably coesite needles. Coesite is a crystalline high-pressure modification of SiO_2. The needle-like shape and the coesite structure stem from the difficulties to build up larger SiO_2 particles in the hard and non-plastic silicon lattice at temperatures below 800°C. A needle-like precipitate causes less lattice strain than a spherical precipitate of the same volume. The coesite modification results from the high strength of the silicon lattice at low temperatures, [84O, 85B, 86T, 88Z]. The formation of rod-like precipitates is an alternative to plate-

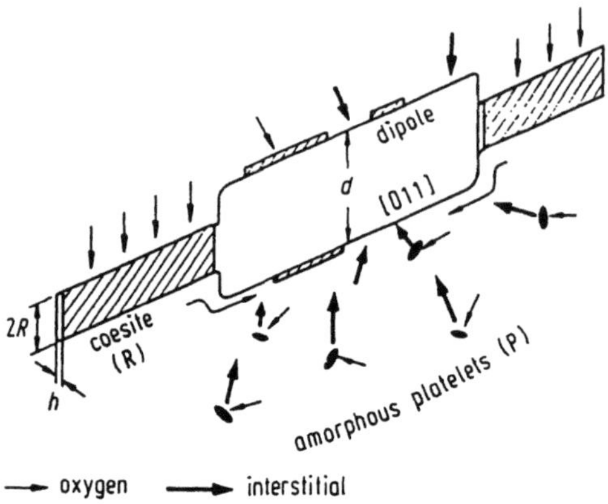

Fig. 11: O in Si: General scheme of the precipitation phenomenon at 650°C with the flux of oxygen atoms and silicon interstitial atoms indicated. Two types of precipitates are formed on different nuclei emitting interstitials either by pipe diffusion along the <110> ribbons (R) or in the bulk from the platelets (P), [84B]

like or spherical precipitates, in spite of the high surface energy of the needles.

Rod-like defects or coesite needles are observed up to 800°C, [81T2], they are typically 10 to 15 Å thick, 20 to 60 Å wide and up to 2 µm elongated in <110> directions and are lying in {100} planes. Figure 11 schematically shows the crystal defects and processes which are observed and assumed, respectively, for the formation of the rod-like or ribbon-like defects. Both coesite ribbons and amorphous SiO_2 platelets are formed by precipitation of oxygen, whereby they emit silicon interstitials. The silicon interstitials are in supersaturation and condense or precipitate in a dislocation dipole.

A different interpretation of the rod-like defects has been given by [87B] who explain it to be precipitates of self-interstitials.

4.3 Plate-like, Polyhedron and Spherical Precipitates

4.3.1 Plastic Deformation

Above 800°C, the silicon lattice becomes more and more plastic, i.e. perfect dislocations can be moved with moderate strain. The strain of the growing SiO_2 particles (platelets, polyhedrons, spheres) can be released by **generation and punching of prismatic dislocation loops.**

In Fig. 12, the process of strain release by plastic deformation is schematically shown. The SiO_2 particle consists of alternate layers of Si and O_2. For each layer of O_2 a layer of Si must be removed and driven away for strain release. The detached Si layers move into <110> directions and they have the shape and size of the SiO_2 particle. The edge of the detached Si layer is a perfect dislocation loop, and the space which is passed by the moving dislocation loop has the form of a prism. Such a loop is therefore called a prismatic dislocation loop. The dislocation loop moves in <110>

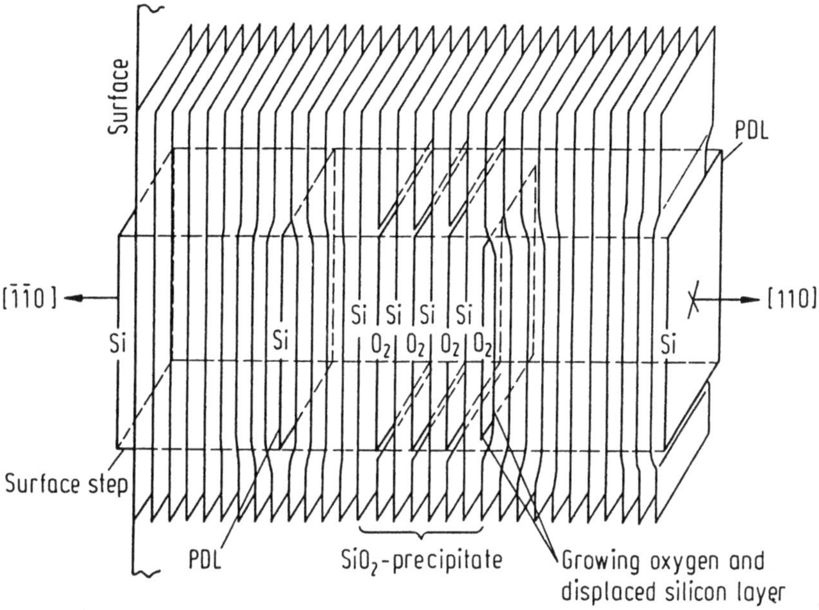

Fig. 12: Schematic drawing of the principle of strain release or volume compensation by punching of prismatic dislocation loops (PDL). The drawing shows {110} lattice planes vertical to the drawing plane. For simplification, the SiO_2 layers and the SiO_2 precipitate are lying parallel to {110} in contrast to normal SiO_2 precipitates. For each new layer of SiO_2, a layer of silicon atoms of the same size has to be detached from its original location and to be moved away from the precipitate as long as the strain is sufficiently high for movement. The PDLs have the contour of the precipitate which is mostly square or rectangular, and move always in <110> directions. The body enclosed by the moving dislocation is a prism. The PDLs repel each other. They remain nearly equidistant on this prism. If a PDL is pushed out to the surface then a surface step of (a/2)<110> is formed

direction by changing bond after bond with the neighboring atoms of the next lattice plane. This changing of bonds is a very fast process, so that a dislocation can move rather fast.

Figure 13 shows the first good TEM micrographs of this process published by Tan and Tice in 1976, [76T]. The generation and movement of the dislocation loops is also called prismatic punching and the dislocation loops are also called punched dislocations.

4.3.2 Emission and Precipitation of Silicon Interstitials

The strain release by prismatic punching is in competition with the strain release by the **generation, emission and precipitation of silicon interstitials.** This process is shown

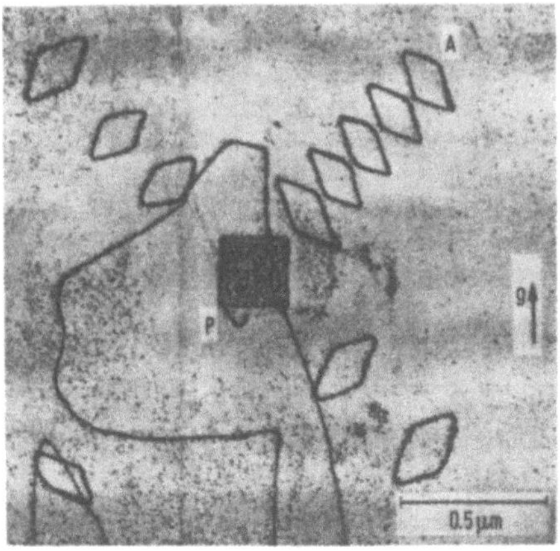

Fig. 13:
TEM micrograph of a square shaped SiO_2 particle in Si lying parallel {100} according to [76T]. The SiO_2 particle has pushed prismatic dislocation loops in four different <110> directions. The Burgers vectors of the dislocations are of the type (a/2)<110>. At P a generator dislocation is visible at the corner of the particle

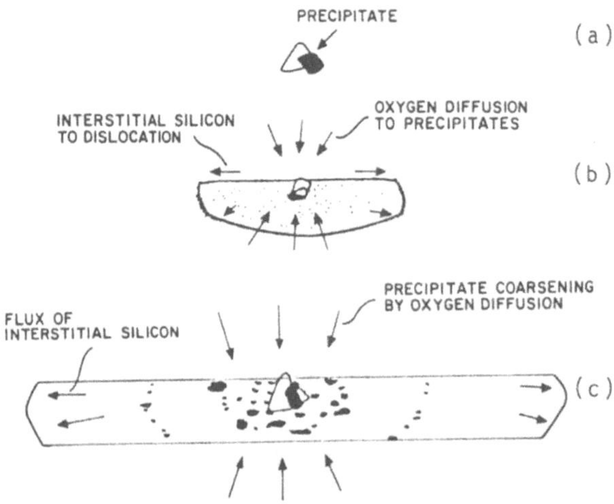

Fig. 14:
O in Si: Schematic of the stages in the development of an oxygen precipitate and its associated extrinsic stacking fault which is surrounded by a Frank partial dislocation loop

schematically in Fig. 14: Oxygen diffuses to the precipitate. For about each pair of oxygen atoms a silicon interstitial is formed and emitted. By the emission, a supersaturation of silicon interstitials occurs. Some of the oxygen precipitates or punched dislocation loops act as heterogeneous nuclei for the precipitation of silicon interstitials. Large extrinsic stacking faults grow around the nuclei thus reducing the supersaturation of silicon interstitials. Each dislocation - both perfect and partial dislocation - is a heterogeneous precipitation site for supersaturated impurities. Therefore, oxygen also precipitates at the partial dislocation loops of the stacking fault and additional SiO_2 particles are formed. At further growing of the stacking fault, the SiO_2 particles are left behind by the climbing partial dislocation loop.

Figure 15 shows 4 extrinsic stacking faults generated by oxygen precipitation and made visible by preferential etching. The surface of the sample is a {111} plane. Therefore, every

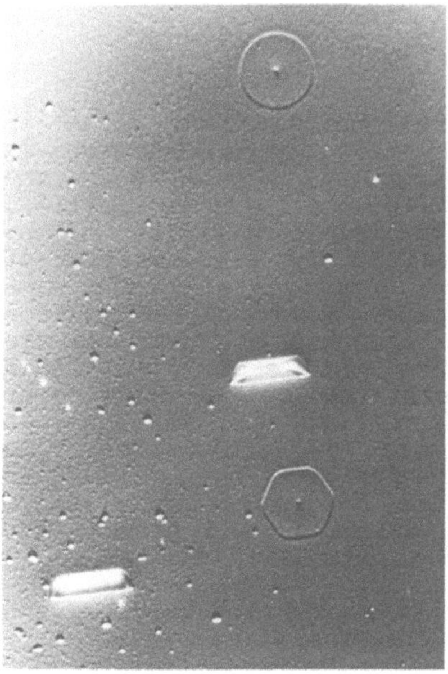

Fig. 15:
Extrinsic stacking faults in CZ-Si generated by oxygen precipitation for 20 h at 1200°C, {111} surface, Sirtl etched. Diameter of the stacking faults ≈ 20 µm

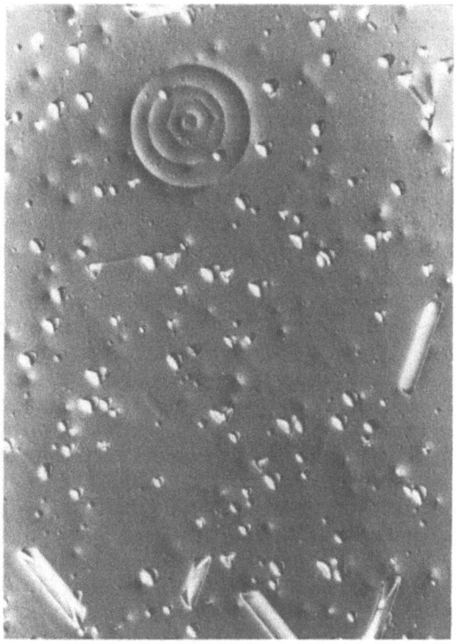

Fig. 16:
Fourfold stacking fault in a parallel sample to Fig. 15 with a higher defect density

fourth stacking fault is lying parallel to the surface. Some of them which are lying close enough below the surface are perceptible in their contours after etching. Most of them show the nucleating oxide particle or dislocation loop in the center. The diameter of the stacking faults is about 20 µm. The micrograph shows the two principal shapes of stacking faults of which the circular shape is by far the most frequent one. Two stacking faults are lying inclined to the surface and are dissected. At their sites the preferential etch has etched trapezoidal grooves into the surface. The other numerous etch pits in the surface are caused by SiO_2 particles without stacking faults, showing that at only a few precipitates stacking faults have grown. Figure 16 shows a fourfold stacking fault which is rather seldom. This sample contains a very high precipitate and stacking fault density.

If the oxygen precipitation takes place at high tempera= tures and for long annealing times, then stacking faults of more than 100 µm in diameter grow. At 1200°C, the silicon lattice is rather soft and dissociation of the extrinsic stacking faults occurs. The Frank partial dislocation loops of the extrinsic stacking fault partly dissociate into perfect dislocations and Shockley partial dislocations.

5. Other Impurity-Related Microdefects

5.1 Carbon and Nitrogen

Dissolved **carbon** occupies substitutional lattice sites in silicon and precipitates in the form of crystalline SiC. In crystalline SiC, a SiC molecule occupies only 52 % of the volume compared to 2 Si atoms in the silicon lattice. Precipitation of carbon causes a strong contraction in the silicon lattice in complete contrast to the precipitation of oxygen. If carbon and oxygen atoms precipitate in the same number, nearly no volume change takes place in the silicon lattice. Silicon interstitials are absorbed or vacancies are emitted if carbon precipitates form. For each precipitated carbon atom nearly one silicon interstitial can be absorbed or nearly one vacancy can be emitted. Due to this, carbon precipitates do not cause punching of dislocations and generation of extrinsic stacking faults, in contrast to oxygen, they prevent the formation of extrinsic stacking faults at oxygen precipitation or oxidation and promote the precipitation of oxygen.

Up today, **nitrogen** in silicon is poorly investigated and there is little kowledge about its lattice sites, solubility, diffusivity, precipitates and secondary defects. This is due mainly to its low solubility in silicon which is at most some $10^{15}/cm^3$ near the melting point, [74N]. Nitrogen impedes the formation of swirl defects (precipitates of silicon interstitials) in FZ silicon, what indicates, that nitrogen occupies at least partly substitutional lattice sites, comparable to carbon. If nitrogen precipitates into crystalline Si_3N_4 then it occupies only 48 % of the volume of pure silicon and the contraction of the lattice would be even stronger than in the case of carbon.

5.2 Transition Metals

The fast diffusing transition metals, e.g. Fe, Ni, Cu, occupy interstitial lattice sites and form metal silicides at precipitation. The volume change and defect situation at precipitation is comparable to the case of oxygen precipitation: Compressive strain, plastic deformation and/or emission of silicon interstitials. In wafers without gettering sites, the transition metals tend to precipitate at the surface or at the SiO_2-Si interface and form the so called **"haze defects"**.

6. Oxidation-Induced Stacking Faults (OSF)

6.1 OSF Nucleated by Surface Damage

As mentioned before, the oxidation of Si mostly leads to supersaturation of silicon interstitials. If there are no heterogeneous precipitation sites for the silicon interstitials, then they cannot precipitate. They are obviously not able to form homogeneous nuclei.

Soft **mechanical damage** at the surface of silicon, however, acts as a very efficient heterogeneous nucleation site for the precipitation of silicon interstitials. Upon heating up to oxidation temperature, the damage generates small perfect dislocation loops which dissociate into Frank and Shockley partial dislocation loops. The small Frank partial dislocation loops, which surround small extrinsic stacking faults, grow by absorbing silicon interstitials, thus forming large extrinsic stacking faults - the OSF. In this case, the OSF nucleating precipitation sites - the small Frank partial dislocation loops - all are situated immediately at the surface. All the **damage nucleated OSF** start their growth at the surface, at the same time and under the same conditions of self-inter- stitial supply and they all grow to **equal length.** Figure 17 shows oxidation induced stacking faults on a {100} surface nucleated by a scratch.

6.2 OSF Nucleated by Bulk Microdefects

In contrast to the surface nucleation sites caused by damage, the **bulk nucleation sites - the bulk microdefects** - lead to **OSF** of **various length.** Figure 18 shows OSF in a high density. In this case the stacking faults are nucleated by bulk microdefects which were present throughout the whole volume of the wafer. Depending on the distance of the bulk microdefects from the oxidizing surface, the stacking faults start the growth sooner or later and, additionally, they are supplied

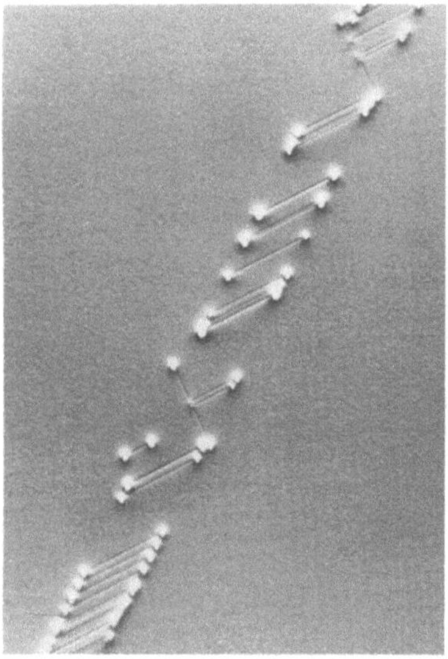

Fig. 17:
Oxidation-induced stacking faults (Sirtl etched) in a {100} surface nucleated by surface damage (a scratch)

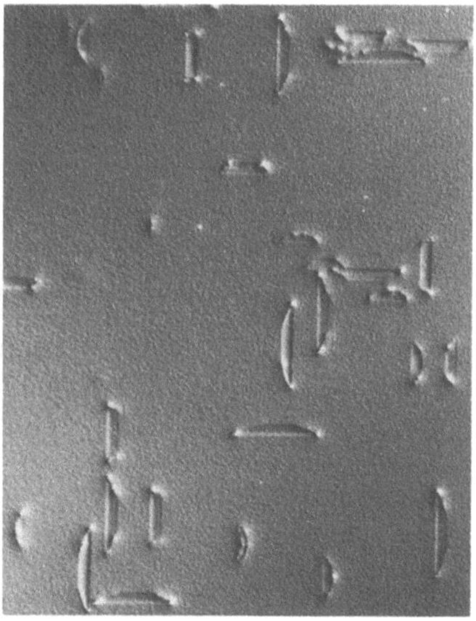

Fig. 18:
Oxidation-induced stacking faults (Sirtl etched) in a {100} surface nucleated by bulk microdefects

with silicon interstitials the less the greater their distance from the surface is. From this source, bulk nucleated OSF grow to various length.

By preferential etching, a silicon layer of about 8 µm was removed. The surface is a {100} plane, whereas all stacking faults are lying in {111} planes. Therefore, all the stacking faults are lying oblique to the {100} surface and the etching generates a vertical projection of the OSF at the surface. The straight lines are intersections of the stacking faults with the wafer surface before and after etching. The well developed straight lines are the intersections with the surface **before** etching, the less developed straight lines are the intersections with the surface **after** etching, i.e. about 8 µm deeper. The largest stacking faults started the growth first and were nucleated immediately at the original wafer surface. The bent lines are the partial dislocations which surround the stacking faults. The thicker end of the bent lines was the intersection of the partial dislocation with the wafer surface before etching. At many of the partial dislocations precipitates are seen which were probably formed by transition metals during cooling after the oxidation.

Figure 19 shows a TEM micrograph, [73R], of such a stacking fault. It is heavily decorated with impurity precipitates, in particular, at is edge which is a partial dislocation.

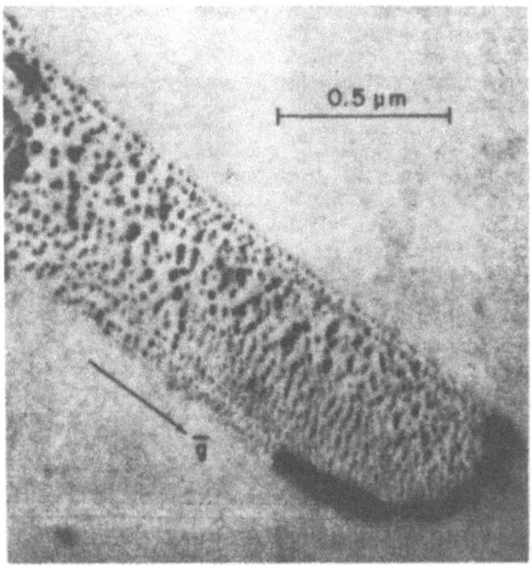

Fig. 19:
TEM micrograph of an oxidation induced stacking fault decorated with impurities both in the fault plane and larger at the surrounding partial dislocation (darkest area)

7. Intrinsic Gettering

"Intrinsic gettering" is the effect that crystal defects in the interior of a wafer gather the fast diffusing impurities during the cooling thus preventing that they precipitate at the surface or in the active areas of the electronic devices. The gettering sites are mainly the partial and perfect dislocation loops which are generated as secondary defects by the oxygen precipitates.

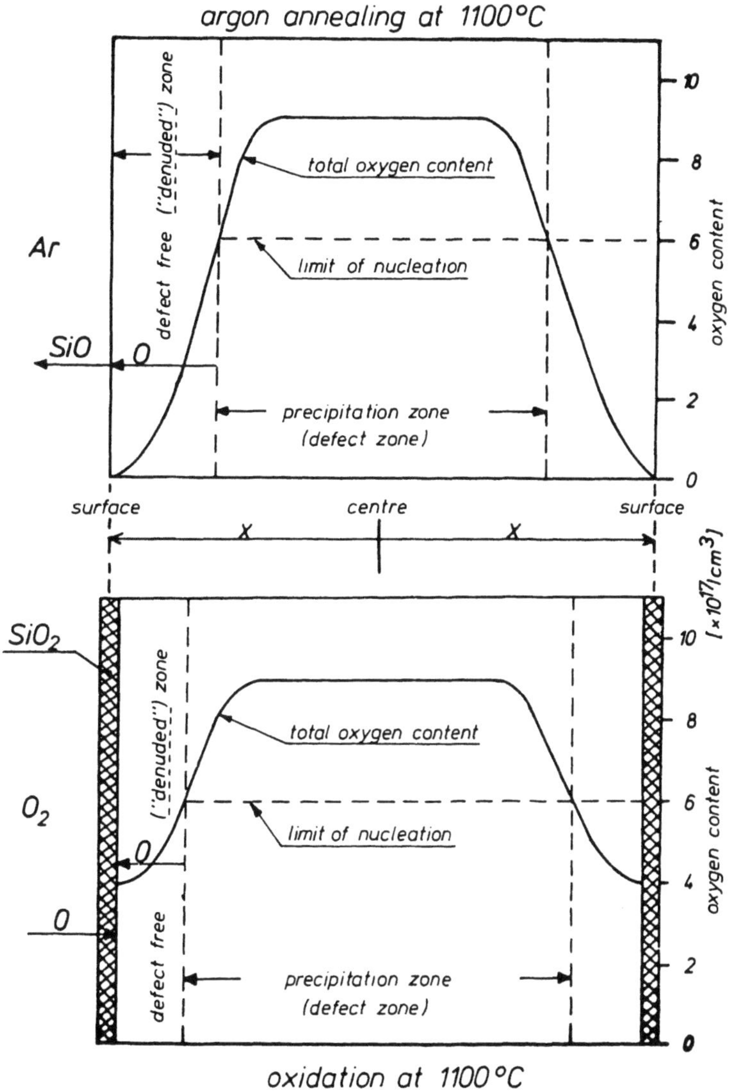

Fig. 20: Schematic of the formation of a denuded zone in an inert atmosphere and in an oxidizing atmosphere. In the case of an inert atmosphere the oxygen content at the surface of the Si is zero, in the case of oxidation the surface concentration assumes the value of the oxygen solubility at the oxidation temperature. In both cases, existing oxygen precipitates are redissolved in the denuded zone. The denuded zone is deeper in the case of an inert atmosphere

For the gettering effect it is also necessary that a defect free zone - a so called **"denuded zone"** (> 10 µm deep) - is formed before the cooling. This denuded zone must be free of oxygen precipitates and their secondary defects, otherwise they would deteriorate the electronic active areas at the surface. The formation of such a denuded zone is schematically shown in Fig. 20: If a CZ Si wafer is annealed both in an inert and in an oxidizing atmosphere, then oxygen diffuses out to the surface and an oxygen-depleted zone is formed.
In the case of an inert atmosphere, oxygen evaporates in the form of volatile silicon monoxide (SiO) and the oxygen concentration at the surface can be taken as zero. In the case of an oxidizing atmosphere a surface oxide layer is formed which acts towards the bulk silicon in the same manner as an oxygen precipitate (SiO_2 particle) in the interior of the wafer. The surface oxide thus acts as precipitation site for the oxygen of the bulk silicon and the oxygen concentration at the SiO_2-Si interface takes the value of the oxygen solubility at the oxidation temperature. For example, at 1100°C, the oxygen concentration at the interface reaches $4 \times 10^{17}/cm^3$. The depth of the denuded zone depends on annealing temperature and time. Because of the low oxygen concentration no oxygen precipitates and no secondary defects form in the denuded zone and pre-existing precipitates are redissolved.

In the interior of the wafer, however, the oxygen concentration remains high and oxygen precipitates and their secondary defects - the perfect and partial dislocation loops - form at the same time. At cooling of the wafer, the dislocation loops are stronger gettering sites for the fast diffusing elements than the surface, the surface oxide or the electronic active areas of the device. Thus, the active areas remain free from precipitates of impurities which would deteriorate the device.

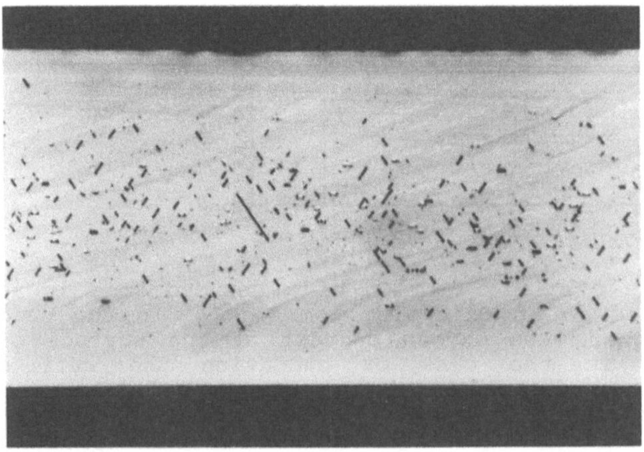

Fig. 21: Cleavage plane of a CZ-Si wafer with denuded zones (top and bottom) and internal gettering sites (interior). The polished wafer (polished side at the top) was cleaved and preferentially etched after the annealing procedure.

Figure 21 shows this situation in a real case. A CZ-Si wafer has been processed to form denuded zones (≥ 50 µm deep) and gettering sites. In the etched {110} cleavage plane, the large stacking faults are especially well perceptible, whereas the small oxygen precipitates and punched dislocation loops cannot differentiated from each other.

References:

[56F] Frank, F.C., Turnbull, D.: Phys.Rev. 104 (1956) 617.
[57K] Kaiser, W.: Phys.Rev. 105 (1957) 1751.
[62L] Ludwig, G.W. et al.: Solid State Phys. 13 (1962) 223.
[65B] Burke, J.: "The Kinetics of Phase Transformation in Metals", Chap. 6, Pergamon Press, London (1965).
[65N] Newman, R.C. et al.: J.Phys.Chem.Solids 26 (1965) 373.
[67F] Fairfield, J.M. et al.: J.Appl.Phys. 38 (1967) 3148.
[68N] Newman, R.C. et al.: J.Phys.Chem.Solids 30 (1969) 1493
[69K] Kendall, D.L. et al.: Semiconductor Silicon 1969, ed. R.R. Haberecht, E.L. Kern, The Electrochem. Soc., Inc., New York, N.Y. 10017, p. 358.
[73M] Mayer, K.R.: J.Electrochem.Soc. 120 (1973) 1780.
[73P] Patel, J.R.: J.Appl.Phys. 44 (1973) 3903.
[73R] Ravi, K.V. et al.: J.Electrochem.Soc. 120 (1973) 533.
[74N] Nozaki, T. et al.: J.Radioanal.Chem. 19 (1974) 109.
[74S] Schulze, E.R.: "Metallphysik", Springer Verlag, Wien, New York, 2.Aufl., 1974, 131.
[74V] Van Vechten, J.A.: Phys.Rev. B 10 (1974) 1482.
[76M1] Murgai, A. et al.: J.Electrochem.Soc. 123 (1976) 224.
[76M2] Maher, D.M. et al.: J.Appl.Phys. 47 (1976) 3813.
[76T] Tan, T.Y., Tice, W.K.: Phil.Mag. 34 (1976) 615.
[77T] Tempelhoff, K. et al.: Semiconductor Silicon 1977, ed. H.R. Huff, E. Sirtl, The Electrochem. Soc., Inc., Princeton, N.J. 08540, p. 585.
[78M] Masters, B.J. et al.: J.Appl.Phys. 49 (1978) 2717.
[78Y] Yang, K.H. et al.: Appl.Phys.Lett. 33 (1978) 225.
[79F] Francis, R., Dobson, P.S.: J.Appl.Phys. 50 (1979) 280.
[79P] Patrick, W. et al.: J.Appl.Phys. 50 (1979) 7156.
[79T] Tempelhoff, K. et al.: phys.stat.sol.(a) 56 (1979) 213
[80G] Gösele, U. et al.: Appl.Phys. 23 (1980) 361.
[80H] Hu, S.M.: Appl.Phys.Lett. 36 (1980) 561.
[80S] Shimura, F. et al.: J.Appl.Phys. 51 (1980) 269.
[81T1] Tan, T.Y., Gösele, U.: Appl.Phys.Lett. 39 (1981) 86.
[81T2] Tempelhoff, K. et al.: Semiconductor Silicon 1981, Huff, H.R., Kriegler, J., Takaishi, Y. (eds.), The Electrochem. Soc., Pennington, N.Y. 08534, p.244.
[82G] Gösele, U., Tan, T.Y.: Appl.Phys.A 28 (1982) 1.
[82I] Inoue, N. et al.: J.Electrochem.Soc. 129 (1982) 2780.
[82T] Tan, T.Y., Gösele, U.: Appl.Phys.Lett. 40 (1982) 616.
[82V] Voronkov, V.V.: J.Crystal Growth 59 (1982) 625.87B
[83G] Gösele, U., Tan, T.Y.: Defects in Semiconductors II, ed. S. Mahajan, J.W. Corbett, North-Holland, New York, 1983, p.45.
[83T] Tan, T.Y. et al.: Appl.Phys.A 31 (1983) 97.
[84B] Bourret, A. et al.: J.Appl.Phys. 55 (1984) 825.
[84F] Frank, W., Gösele, U., Mehrer, H., Seeger, A. in: Diffusion in Crystalline Solids, Murch, G., Nowick, A.S. (eds.), New York: Academic Press, 1984, p.63.

[84K] Kalejs, J.P. et al.: Appl.Phys.Lett. 45 (1984) 268.
[84L] Livingston, F.M. et al.: J.Phys.C 17 (1984) 6253.
[84O] Ourmazd, A. et al.: J.Appl.Phys. 56 (1984) 1670.
[85B] Babitskii, Y.M.et al.: Sov.Phys.Semicond. 19(1985)1221
[85G] Gaworzewski, P. et al.: phys.stat.sol.(a) 92(1985)129.
[85T] Tan, T.Y., Gösele, U.: Appl.Phys.A 37 (1985) 1.
[85W] Wada, K., Inoue, N.: J.Appl.Phys. 58 (1985) 1183.
[86T] Tiller, W.A. et al.: J.Appl.Phys. 59 (1986) 3255.
[86W] Wada, K., Inoue, N.: Semiconductor Silicon 1986, Huff, H.R., Abe, T., Kolbesen, B. (eds.), The Electrochem. Soc., Inc., Pennington, N.Y. 08534, p.778.
[87B] Bourret, A.: Inst.Phys.Conf.Ser. (1987), to be publ..

Tellurium Related Deep Traps in Silicon

I.D. Desnica and U.V. Desnica

R. Boskovic Institute, PO Box 1016, YU-41001 Zagreb, Yugoslavia

The electrical and optical properties of chalcogen impurities (S, Se and Te) in silicon have been extensively studied in last years, especially in regard of their potential use in the infrared detector devices [1] - [6]. If chalcogens replace Si atoms on regular sites in the silicon lattice, two extra electrons remain unpaired. Accordingly, they are expected to behave like double donors and give rise to two deep levels in the silicon forbidden energy gap. O, S and Se should be incorporated easily on a substitutional site, but due to pronounced differences in the tetrahedral radii between Si and Te, tellurium might have difficulties.

Survey of the literature data on the tellurium-doped silicon reveals many controversies. It seems that site selection, various complexes formation and thus the resulting electrical properties strongly depend on the doping methods and doping conditions as well as on the postdoping treatments, like thermal processing steps or laser annealing. So, in addition to the isolated impurities, usually a series of donor centers is observed. They are probably associated with complexes involving tellurium like Te_2, or $Te_{Si}+V_{Si}$ or Te+other impurities etc.

Table 1 presents ground state energy levels of Te centers in silicon as observed experimentally with different methods. Theoretical predictions following the "effective mass theory" are given as well. A fair agreement between the values for the electrical activation energies (Hall-effect, Deep Level Transient Spectroscopy, DLTS - when Pool-Frenkel effect is taken into account) and the optically (IR) determined ground state binding energies of the tellurium related donors is noticeable. This agreement holds well when the samples are tellurium-doped at the concentrations lower than the solid solubility limit, $<10^{17}$ cm^{-3}, and the mode of incorporation is either by diffusion or during the crystal growth process. Then the substitutional tellurium position is favored and it exhibits a pure double donor behavior with the ionization energies 0.2 eV for Te^{o} and 0.41 eV for Te^{+}.

On the other hand, the use of ion implantation as a method to introduce deep level impurities in silicon, circumvents the problem of low solid solubilities of tellurium and enables its incorporation in much higher doses. The implantation causes damage to the host lattice and complicated damage structures can be produced which have to be recovered by thermal or laser annealing. The electrical properties of implanted and thermally annealed silicon samples are complex and different energy levels are reported suggesting the possibility of tellurium positioned on some other sites or forming some complex defects.

However, it is interesting to compare the data on the microscopic properties of tellurium in silicon as obtained by Moessbauer spectroscopy

Springer Series in Materials Science, Vol. 13
Semiconductor Silicon Editors: G. C. Harbeke and M. J. Schulz

<u>Table 1:</u> Te in Si (electrical and optical properties)

ref.	sample preparation	DLTS E_1(eV)	Pool-Frenkel	IR E_1(eV)	HALL E_1(eV)
Grimmeiss et al.,[1]	diff. 1200°C, n-Si Te 10^{15} cm^{-2}	0.2 0.4		0.1988 0.4108	
Wagner et al.,[2]	doping during epitax. growth			0.1988 0.4108	
Lin et al. [3]	Te during growth $7x10^{16}$ cm^{-3}			0.1988	
Schaub et al.,[4]	Si:Te doped to max. solubility limit				0.200 0.440
Pensl et al. [5]	doping during epitax. growth	0.145 0.340	0.207 0.445	0.1987 0.4112	0.200 (0.42)
Hofmann et al.,[6]	Impl. Te $2x10^{10}$- $2x10^{11}$ cm^{-2}, 350keV, ann. in N_2,900°C	0.14 0.37	0.41		
Fischler [7]					0.14
Lyutovich et al.,[8]					0.13 0.5
Kaleyanaraman et al.,[9]	Impl. Te $10^{12}$$cm^{-2}$ 120keV, ann. in N_2 at 900°C	0.13 0.56			
Lee et al. [10]	Impl. Te $4x10^{12}$$cm^{-2}$				0.14
Ningh & Sah [11]	Effective mass theory prediction for Te in Si :			0.265 0.539	

and Rutherford Back Scattering (RBS). The results are systematized in <u>Table 2.</u> It can be seen that the substantial difference exists for the samples which had undergone thermal annealing and the samples which were laser annealed. For thermally annealed samples the results unanimously show that there is definitely not more than 30% of tellurium atoms on truly substitutional sites and probably much less. All the others are forming some kind of complexes what is reflected in a variety of observed ionization energy values in electrical measurements. The electrical activity of Te atoms derived from Hall-effect measurements in such samples is also low.

On the other hand, a considerable improvement is attained by laser annealing. Moessbauer and RBS data indicate that substitutional Te fractions of 80 - 90% in good quality single crystalline hosts can be realized. At the same time the electrical activity is increased to a value

Table 2: Te in Si ; microscopic properties (Moessbauer & RBS)

Larsen et al. [12]	impl. 60keV Te in 7° of <100>, anneal. thermally in N_2	dose/cm^2 $< 10^{13}$, $5x10^{14}$; T_{impl}/K 300, 300	Moessb. signals 2% 27% 63% 2% 4% 1% 27% 60% 3% 8% Te_i Te_{Si} Te+V Te^+
		laser anneal.	thermal anneal.
Kemerink et al. [13]	impl. 110keV, 20 cm strongly n- or p-Si, dose 10^{14}-$2x10^{15}$ cm^{-2}	85 - 90% Te in truly substitut. sites	< 30% Te atoms in substitutional sites
Dezsi et al. [14]	impl. 85keV in 10 cm n-Si, dose $2x10^{15}$ cm^{-2}	Te in pure substitutional sites	Te_{Si}+V_{Si}(?)>96% (no cubic symmetry)
Kemerink et al. [15]	impl. 120keV dose $2x10^{13}$ - $6x10^{14}$ cm^{-2}	Te substitutional (single component, no quadrupol int.)	
de Bruyn et al. [16]	impl. dose 4 - $7x10^{14}$ cm^{-2}	> 80% in pure substitutional sites, Te_{Si}	< 10% in pure substitutional sites

Also: Kemerink et al., [17], and [18]
de Waard et al., [19]
Foti et al., [20]
Dezsi et al., [21]

corresponding to 90% or more of the implanted atoms [19]. Obviously, these results were inspiring for more thorough investigation.

Single crystal samples of n-type silicon, 1.8 ohm cm resistivity were implanted at room temperature with the isotope ^{128}Te at ion energy 110 keV. The crystals with <100> perpendicular to the surface, were tilted by 7° relative to the beam axes to avoid implantations in channeled directions. The implanted dose was $5.4x10^{12}$, $1.8x10^{13}$ and $5.4x10^{13}$ cm^{-2}, with the average depth about 0.015 um. After the implantation, samples were subjected to different treatments: one was thermally annealed for 30 min at 800°C in vacuo $2x10^{-6}$ Torr, others were laser annealed with Q-switched ruby laser, pulse duration 25 ns (FWHM). The power density was 1.55 J cm^{-2}.

The characteristic DLTS spectra for variously treated samples are presented in Fig.1:

(a) is the sample which received the dose $1.8x10^{13}$ cm^{-2} and was thermally annealed for 30 min in vacuo $1x10^{-6}$Torr. The result is a single DLTS peak with an activation energy approximately 0.36eV.

(c) and (d) differ in implanted dose. They were laser annealed and in both samples two low energy peaks are prominent with the approximate activation energy 0.14 eV and 0.2 eV. The relation of the peak height to the implanted dose is evident, although it is not linear: while the implanted dose is increased by a factor 10, the peak height observed is increased by a factor 5 for the deeper energy level and a factor 4 for the shallower level. For the broad additional structure at the higher energies we have found it to be related to the surface layer.

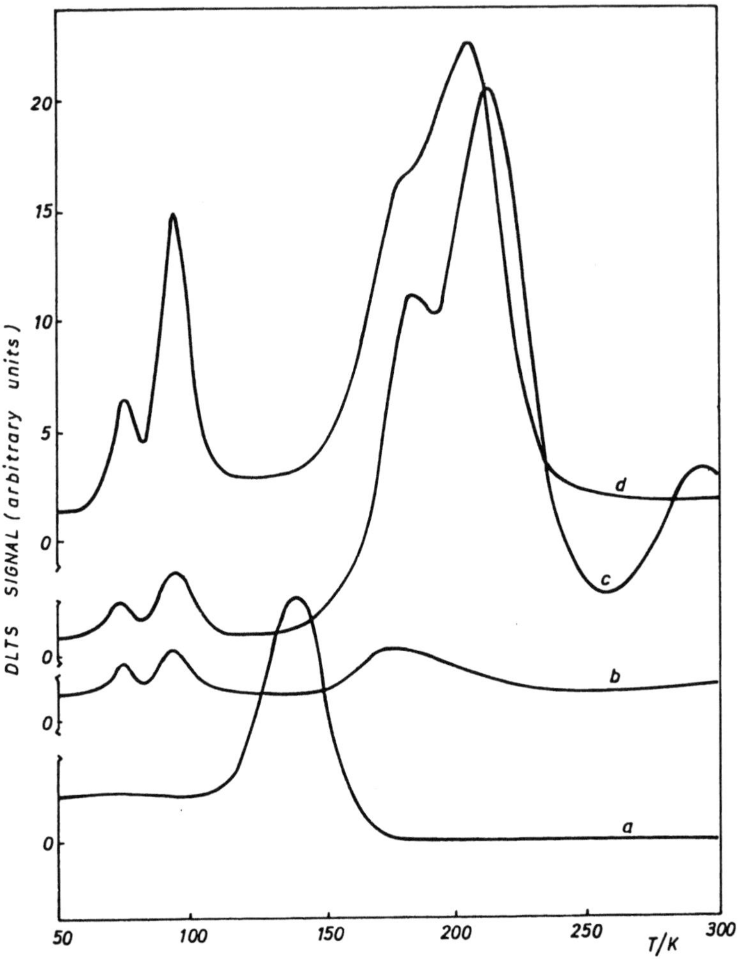

<u>Figure 1:</u> DLTS spectra obtained for n-type Si samples implanted with different doses of ^{128}Te: (a) $1.8x10^{13}cm^{-2}$, thermally annealed, (b) $1.8x10^{13}cm^{-2}$, laser annealed, 10nm surface layer removed, (c) $5.4x10^{12}cm^{-2}$, laser annealed, (d) $5.4x10^{13}cm^{-2}$, laser annealed

(b) 10nm thick surface layer was removed from the laser annealed sample by anodic oxidation. Now the broad, high energy peaks present in (c) and (d) are gone, so they were most likely due to the residual radiation damage close to the surface or Te complexes specific for the surface layer.

In conclusion we may summarize that for doping concentration of tellurium low enough to be below the solubility limit in silicon, Te atoms incorporate in substitutional sites giving rise to two deep levels E_c-0.2eV (for Te^o) and E_c-0.4eV (for Te^+).

For implanted samples with Te concentrations higher than $10^{17}cm^{-3}$, after thermal annealing it seems that up to 95% of Te atoms do not occupy truly substitutional sites. But if the implanted samples were laser annealed, Moessbauer data suggest that almost all Te is substitutional (Table 2) even at concentrations exceeding the solid solubility limit.

This knowledge led us to perform DLTS investigation on Te implanted and laser annealed samples to characterize the induced levels, hoping that this process (implantation followed by laser annealing) may circumvent the problem of low solubility of Te in Si, what is the main obstacle for its technological use. However, our results indicate much more complicated situation, with DLTS signals assigned to complex defects, quite the contrary to Moessbauer results, so obviously more investigations are needed to clarify this discrepancies.

Acknowledgments

The authors are grateful to Prof. H. de Waard and Dr. G.J. Kemerink, Laboratorium voor Algemene Natuurkunde, Rijksuniveriteit, Groningen, The Netherlands, for providing the implanted and annealed samples and to Prof. J. Lagowski, MIT, Boston, USA, for hospitality during performing the DLTS measurements.

References

1. H.G. Grimmeiss, E. Janzen, H. Ennen, O. Schirmer, J. Schneider, R. Worner, C. Holm, E. Sirtl and P. Wagner, Phys. Rev. B24, 4571 (1981)
2. P. Wagner, C. Holm, E. Sirtl, R. Oeder and W. Zulehner, Advances in Solid State Phys. XXIV, 191, P. Grosse ed., (Vieweg, Braunschweig 1984)
3. A.L. Lin, A.G. Crouse, J. Wendt, A.G. Campbell and R. Newman, Appl. Phys. Lett. 38, 683 (1981)
4. R. Schaub, G. Pensl, M. Schulz and C. Holm, Appl. Phys. A 34, 215 (1984)
5. G. Pensl, G. Roos, C. Holm and P. Wagner in Defects in Semiconductors, H.J. von Bardeleben, ed., Mat. Science Forum Vol. 10-12, 911 (Trans. Tech. Switzerland 1986)
6. K. Hofmann and M. Schulz, Appl. Phys. A 33, 19 (1984)
7. S.Fischler, Metallurgy of Advanced Electronic Materials Vol. 19 (Interscience, New York, 1963)
8. A.S.Lyutovich, V.P. Prutkin, V.M. Mikhaelyan, G.S. Gafitulina and Zh. V. Abramova, Sov. Phys. Semicond. 2, 728 (1968)
9. V. Kaleyanaraman, M. Mohan Chandra and V. Kumar, J. Appl. Phys. 54, 6417 (1983)
10. T.F. Lee, R.D. Pashley, T.C. McGill and J.W. Mayer, J. Appl. Phys. 46, 381 (1975)
11. T.H. Ningh and C.T. Sah, Phys. Rev. B4, 3482 (1971)
12. A.N. Larsen, G. Weyer, N.E. Holm and H.L. Nielsen, Phys. Rev. B12, 4939 (1980)
13. G.J. Kemerink, D.O. Boerma, H. de Waard and L. Niesen, Rad. Eff. 69, 83, 101 (1983)
14. I. Dezsi, R. Coussement, G. Langouche and M. Van Rossum, Appl.Phys. Lett. 44, 505 (1984)
15. G.J. Kemerink, Hyp. Int. 14, 37 and 53 (1983)
16. J.de Bruyn, R. Coussement, I. Dezsi, G. Langouche and M. van Rossum, Hyp. Int., 10, 973 (1981)
17. G.J. Kemerink, D.O. Boerma, H. de Waard, J.C. de Wit and S.A. Drentje, J. de Physique 41 C1, 435 (1979)

18. G.J. Kemerink, Hyp. Int. 10, 977 (1981)
19. H. de Waard and G.J. Kemerink, Physica 116B, 210 (1983)
20. G. Foti, S.U. Campisano, E. Rimini and G. Vitali, J. Appl. Phys. 49, 2569 (1978)
21. I. Dezsi, M. van Rossum, J. de Bruyn, R. Coussement and G. Langouche, Rad. Eff. Lett. 76, 221 (1983)

Part IV

Characterization Methods

High Resolution Electron Microscopy of Defects in Silicon

J. Thibault-Desseaux

Service de Physique/Département de Recherche Fondamentale,
Centre d'Etudes Nucléaires de Grenoble 85 X, F-38041 Grenoble, France

1. INTRODUCTION

Due to the more recent requirements, the silicon technology has become more and more complex. This leads to the necessity of an ever more detailed characterization of the produced devices. The defects induced during the processes have to be well-known, in order to be either avoided or neutralized. Thus a better knowledge of the structure and the behaviour of the intended or not-intented features is helpful to understand and to control the electronic properties as well as the mechanical properties of the material.

The high resolution transmission electron microscopy (HREM) is one of the powerful tools for the structure determination of the silicon devices on an atomic scale. This imaging technique gives a local structural information in a non-statistical way. The relative ease to produce "good" pictures with the recent microscopes has led to numerous results. However, the HREM has strong limitations which have to be kept in mind. Detailed informations on experimental and theoretical HR work are given in the reference book by J. Spence [1].

The course will be divided into two parts. In the first part the basic principles of the HREM will be recalled : the possibilities and the limitations will be emphasized. In the second part the images of isolated defects in silicon will be discussed : point defects, dislocations, precipitates. The possibility to get information on the mechanisms involving defects interaction will also be illustrated.

2. HREM IMAGING

The image formation in a microscope consists of two main steps.
i) The first step is the interaction of the incident electrons with the specimen. Generally this is strong and must be described with the aid of the dynamical N-beam theory of the diffraction.
ii) The second step is the "interaction" between the specimen out-going electron wave function and the microscope optics which introduces large aberrations. Fortunately the effect of these aberrations is mathematically known.

Springer Series in Materials Science, Vol. 13
Semiconductor Silicon Editors: G. C. Harbeke and M. J. Schulz

The final intensity distribution on the screen or on the film is the result of these complex interactions. Generally, the relationship between the image and the structure is not straightforward. Nevertheless in some well-defined conditions the image "looks like" the atomic structure. The silicon case is one of the simplest to interprete.

The interpretation of an image, i.e. the determination of the atomic positions requires a trial and error method : an atomic model of the defect is introduced in a computer, then the wave function resulting from the interaction between the incident electrons and the crystal containing the defect is calculated ; the phase and amplitude modifications of this wave function due to the aberrations are introduced. The intensity distribution is finally calculated. The computed image has then to be compared to the micrograph recorded in the same experimental conditions.

Sometimes, only geometrical informations are needed : Burgers vector determination, misorientation angle, periodicity, etc ... In these cases, the total determination of the atomic positions is not required. However, a perfect control of the experimental conditions is absolutely necessary (crystal orientation, microscope alignment).

In any case the best imaging conditions have to be checked on the image of the perfect crystal.

2.1 Interaction between the electrons and the crystal

The interaction between the electrons and the crystal is strong and has to be described by the dynamical theory of the diffraction. The propagation of the incident electrons through a potential $V(\vec{r})$ is governed by the Schrödinger equation. The potential is supposed to be time invariant. The incident electrons whose energy is 100 KeV or more, are relativistic and this has to be taken into account. The effect of the absorption by the crystal can be phenomenologically introduced by means of an imaginary part in the potential. The effect of the temperature on the lattice vibration can be introduced by the attenuation of the Fourier coefficients of the potential by an exponential term containing the Debye-Waller coefficient.

However, the major assumption is the forward scattering of the electrons : this is equivalent to say that the electrons are scattered at small angles and that is true for the accelerating voltages generally used (E>100 kV).

Two kinds of approaches have been developed.

The first one is based on the quantum mechanics solution of the Schrödinger equation. It can be expressed either as the superposition of Bloch waves in the crystal /2/ or as the superposition of plane waves /3,4/. It assumes that the crystal is infinitely periodic.

The second one is based on an optical description of the propagation of the electrons through a crystal cut into slices of infinitesimally small thickness. It is called "multi-slice" calculation /5,6,7/.

Whatever theory has been used, it has been established that the wave function resulting from the interaction between the incident electrons and a perfect crystal can be written as the superposition of plane waves :

$$\psi(\vec{r}) = \sum_g a_g \, e^{i\varphi_g} \exp i\vec{k}.\vec{r} \exp 2\pi i\vec{g}.\vec{r}$$

where the amplitudes a_g and the phases φ_g of the beams depend on the thickness and are calculated by any of the dynamical calculations. $\vec{k}$ is the incident wave and the $\vec{g}$ the diffraction vector.

In the case of the crystal defect imaging, a pseudocrystal is generally introduced, the cell of which is a supercell containing the defect. In this case, because of the presence of the defect, the reciprocal space is sampled N X N by the diffused scattered beams between the main diffracted beams. It has been shown /5/ that the computer time for the quantum mechanics approach is proportional to N^2 whereas the time for the optics approach is proportional to NLogN. It is clear that the "multi-slice" calculation is more suitable for the defects study.

<u>Multi-slice theory</u>

The specimen is cut into slices in which the phase grating approximation is valid, i.e. low angle diffraction and slow variation of the potential along the propagation direction of the electrons. Then the potential only changes the phase of the wave function ; the phase shift is given by the first Born approximation :

$$\chi_n(x,y,\Delta t) = \sigma \int_{z-\Delta t.}^{z} V(\vec{r}) \, dz = \sigma \, \bar{V}_n \, (x,y) \, \Delta t$$

where $\sigma = \frac{\pi e}{\lambda E}$. E, λ, e are respectively the total energy, the wavelength and the charge of the incident electrons. $\bar{V}_n(x,y)$ is the mean potential in the slice thickness. $\bar{V}_n.\Delta t$ is the projected potential (unit = volt.nm).

The wave function $\Psi_n(x,y)$ at the exit of the nth slice depends on the wave function $\Psi_{n-1}(x,y)$ via the combination of the slice transmission function $\exp i\chi_n$ and the propagation function along the slice. This leads to the formula

$$\Psi_n(x,y) = \exp i\chi_n(x,y,\Delta t).[\Psi_{n-1}(x,y) * P_n(x,y,\Delta t)] \qquad (1)$$

where * is the convolution product and P_n the propagation function given by the Kirchhoff formula of the propagation in the z direction.

$$P_n(x,y,z) = -\frac{i}{\lambda z} \exp - i\frac{\pi r^2}{\lambda z}$$

(1) allows a recurrence calculation of the wave function over M slices of thickness Δt, each slice could have different mean potential and thickness.

2.2 Influence of the microscope optics

The two dimensional wave function $\Psi_s(x,y)$ at the exist surface of the crystal is greatly affected by the microscope aberrations which modify the phase and the amplitude of the Fourier coefficients of the wave function. The main phase shift is introduced by the spherical aberration C_s and the focussing distance Δz

$$T(f) = \exp + i\gamma(f) \qquad \text{with } \gamma = -2\pi\left(Cs\lambda^3 \frac{f^4}{4} + \Delta z\lambda \frac{f^2}{2}\right)$$

The f represents the spatial frequency and is equal to 1/d.

The amplitude attenuation comes from the partial coherence of the incident beam and can be written in a first approximation as

$$G(f) = \exp - \left(\pi^2\lambda^2\delta z^2 \frac{f^4}{2}\right) . \exp - \left(\pi^2\lambda^2\alpha^2 \ (\Delta z + C_s\lambda^2 f^2) \frac{f^2}{2}\right) . D(f)$$

where δz is due to the chromatic aberrations and α is the half divergence of the incident beam, D(f) is the objective aperture cut-off.

The resulting wave function is

$$\Psi_i \ (x,y) = TF^{-1} \ ((TF \ \Psi_s).T(f).G(f))$$

The intensity distribution on the screen is the square modulus of $\Psi_i(x,y)$.

2.3 Perfect crystal image

In the case of a perfect crystal the intensity distribution can be written as

$$I(\vec{r}) = \sum_g \sum_{g'} a_g a_{g'} e^{i(\varphi_g - \varphi_{g'})} e^{+i(\gamma_g - \gamma_{g'})} \exp + 2\pi i(\vec{g} - \vec{g}').\vec{r}$$

where a_g and φ_g are the amplitude and phase of the g diffracted beam and γ_g is the phase shift introduced by the microscope. For simplicity the G function is not introduced.

If the specimen is oriented in such a way that a low index zone axis is parallel to the incident beam, and if the objective aperture only selects a limited number of diffracted beams, then the expression can be reduced. In the case of silicon which is a centrosymmetric crystal and if the crystal is in a Laue symmetric orientation along [011] or [001] the intensity distribution given by the interference between the four major equivalent beams ($\vec{g}_1$, $-\vec{g}_1$, $\vec{g}_2$, $-\vec{g}_2$ with $|\vec{g}_1| = |\vec{g}_2|$) and the transmitted beam can be written as :

$$I(\vec{r}) = a_o^2 + 4a_g^2 + 4a_o a_g \cos(\varphi_o - \varphi_g + \gamma_o - \gamma_g)(\cos 2\pi \vec{g}_1.\vec{r} + \cos 2\pi \vec{g}_2.\vec{r})$$

$$+ 2a_g^2 (\cos 4\pi \vec{g}_1.\vec{r} + \cos 4\pi \vec{g}_2.\vec{r})$$

$$+ 4a_g^2 (\cos(2\pi(\vec{g}_1 - \vec{g}_2).\vec{r}) + \cos(2\pi(\vec{g}_1 + \vec{g}_2).\vec{r})).$$

For [011]Si $g_1=1\bar{1}1$ $g_2=11\bar{1}$, and for [001]Si $g_1=220$ $g_2=2\bar{2}0$.

From this expression, it can be immediatly shown that the transfer of the fundamental spatial frequencies $\vec{g}_1$ and $\vec{g}_2$ depends on

$$E = \cos(\varphi_o - \varphi_g + \gamma_o - \gamma_g)$$

If $E = +1$ the intensity maxima represent the atomic columns, if $E = -1$ the maxima represent the tunnels in the structure. If $E = 0$ the image contains only the spatial frequencies $2\vec{g}_1$, $2\vec{g}_2$ and $(\vec{g}_1 +- \vec{g}_2)$ and is difficult to relate to the structure. E depends on two main parameters : the thickness of the specimen via $(\varphi_o - \varphi_g)$ and the defocus Δz via $(\gamma_o - \gamma_g)$. Consequently the contrast of the atomic columns depends on these two parameters.

However the intensity formula has been calculated for a complete centrosymmetrical case. If it is not the case (crystal misorientation, incident beam misorientation), the intensity distribution can vary and the image centrosymmetry can be destroyed.

Examples. Si[011] and Si[001] :

The main beams amplitude and phase variations with the crystal thickness in these two cases are presented in fig. 1. They oscillate when the thickness increases, giving rise to a pseudo extinction distance. As shown previously the main factor which governs the contrast is $E=\cos(\varphi_g - \varphi_o + \gamma_g - \gamma_o)$. It is more interesting to draw the curves $E(\Delta z, t) = \pm 1, 0$. The contrast variation is immediatly visualized which is extremely helpful to determine the best imaging conditions. It must be pointed out that this can be done because in these two cases only a few equivalent beams control the image intensity.

3. DEFECT IMAGING

HREM only allows to visualize atomic columns. The three dimensional information seems to be lost. However, the electrons have "felt" the three dimensional defect. Some authors have recently investigated the possibility

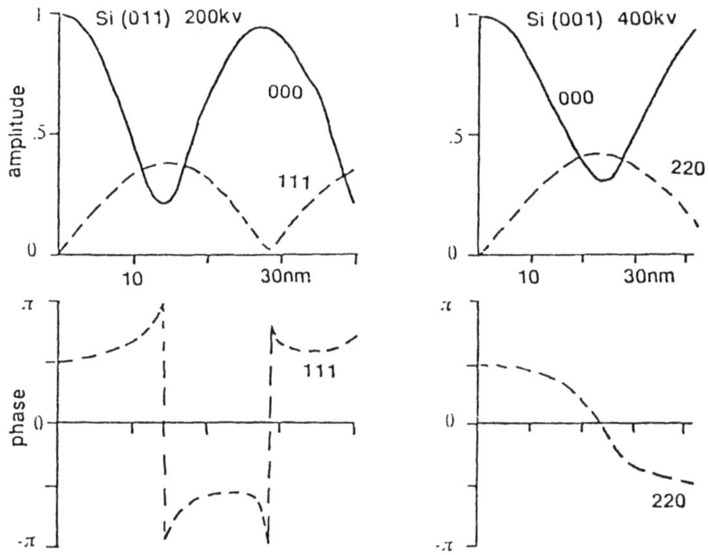

Fig.1 Amplitude and phase variations with the thickness of the transmitted and the diffracted beams in two cases : Si(011) at 200 kv and Si (001) at 400 kv. The phase of the 000 beam is taken as the reference

to extract this screened information : the interstitial or vacancy-type of the stacking faulted tetrahedron has been revealed to be unambiguously determined from a through-focus series /9/. Nevertheless the effect of chemical composition and distribution along a partially disordered atomic column seems to be extremely difficult to relate to the image features /10/. In some suitable cases the use of two different projection axes can lead to a reliable three-dimensional information /8/. In any case, these approaches need a perfect control of the experimental conditions and a lot of defect image simulations.

3.1 Point defect imaging

An image simulation of a perfect [011]Si with on average one half row of atoms missing among the 20 contained in a column is presented in Fig. 2a. The effect begins to be detectable. Only one vacancy leads to a non pertubated image. The lack of a complete column or a double column gives a recognizable signal (Fig. 2b). The lack of atoms might be considered as a thickness change between the faulted column and the neighbour one. This thickness change and the mean thickness must be appropriate in such a way that a small thickness variation induces a large phase shift giving rise to a large contrast variation.

As mentioned in /10/ the content of an atomic column is not straightforward to extract. Furthermore an atomic number Z and a filling variation are

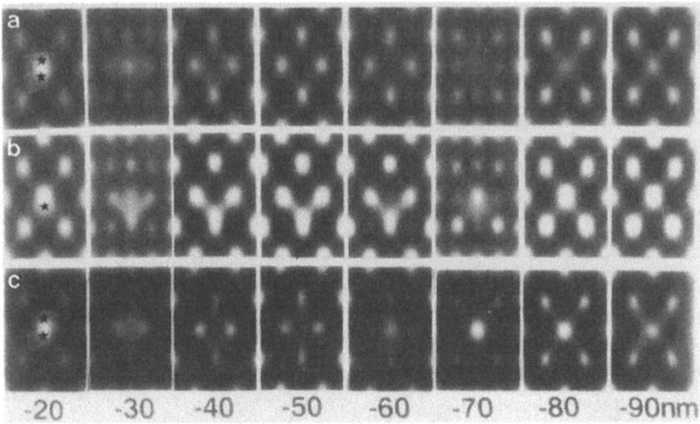

Fig. 2 : Through focus series of a Si crystal oriented in such a way that the (011) is parallel to the incident beam. a) one half of the columns (*) is missing on the average b) one complete row is missing c) two rows are missing. (200kV, C_s=1.05mm, α=.001rd, δ_z=10nm, crystal thickness 6nm)

extremely difficult to discriminate although, according to simulations /10/, it seems possible in well-defined conditions. The main problem remains to check if these conditions are experimentally fuffilled. One has to emphasize that the surface contamination modulates the columns contrast and can introduce artefacts.

3.2 Dislocation core imaging.

The role of the dislocations on the electrical properties of the silicon is well-known. However it is not trivial to relate the electrical behaviour with the differents contribution such as those of the dislocation core structure itself with bonds reconstruction defects along its line or/and with structural defect (kinks or jogs) or that of the point defects surrounding the dislocation. Atomistic calculations of the core energy and of the structure have been used to relate the defects with the electronic band structure /11, 12, 13, 14/. In the case of the silicon, the new potential introduced by Tersoff /15/ solves the problem of the dangling bonds which remained with the Keating potential /16/. However the resulting structures as obtained in /12, 14/ are not in exact agreement with the structures deduced from the HREM images /17, 18/.

3.2.1 60° Dislocation.

Except at high temperature, the dislocation line follows the <011> directions. Dislocations encountered in deformed or as-grown silicon are mainly 60° dislocations or screw dislocations. The 60° are dissociated into two 30° and 90° Shockley partials separated generally by an intrinsic stacking fault (Fig.3).

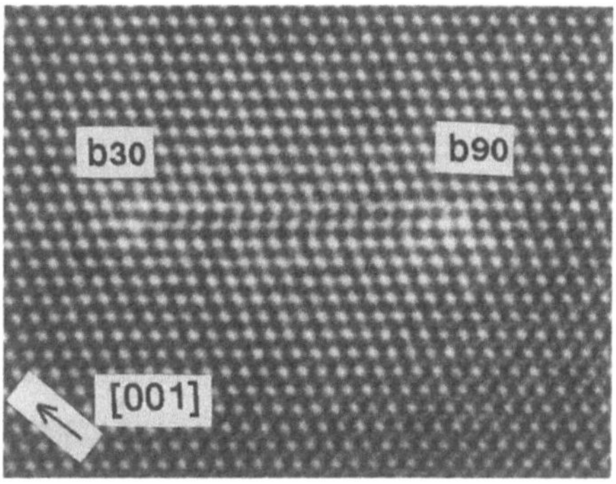

Fig.3 60° dislocation dissociated in two partial dislocations separated by an intrinsic fault

The core of these partials was extensively studied /17,18,19,20/. The main result is the difference feature between the 30° and 90° cores. According to the disparity of the images of the 30° core, this core is thought to be not well-defined /17,18/. The main part of the line seems to belong to the "glide" set (likely associated to reconstructed bonds as discussed in /14/), but the core certainly contains a lot of defects : "vacancy or interstitial shuffle "parts are, without any doubt, distributed inhomogeneously along the dislocation line. The 90° partial core is found to be more well-defined and the "HREM structure" corresponds fairly well with the calculated one.

The calculations predict, however, a completely reconstructed and purely glide core in the case of the 30° and 90° partials.

The discrepancy between the experiments and the calculations might be due to the fact that the experimental situation is much more complex than the ones described in the static atomistic calculations : the interaction between point-defects and dislocations during the deformation process will be illustrated in § 2.4.

3.2.2. Lomer edge dislocation.

This dislocation can be found in low angle boundaries, in epilayers or in deformed Si where it results from the interaction between two 60° dislocations gliding on two different 111 planes. As a matter of fact, its asymmetrical dissociation theoretically predicted /21, 22/ has not been observed (Fig.4) although this Lomer-Cottrell lock would have detectable stacking faults : d_1 = 0.9 nm, and d_2 = 3.4 nm. The B4 configuration (same global Burgers vector), with a symmetrical dissociation would be unlikely distinguished from a non dissociated Lomer configuration (d_1 = d_2 =.34nm). As discussed in /23/ this result indicates that the core energy governs the configurations : in the case of a Lomer-Cottrell lock the core of the stair-rod or/and of the 90° Shockley partials involved, certainly prohibits the dissociation. Furthermore the core structure deduced from the HREMicrographs turns out to be a configuration dissociated on one 111 plane and

Fig.4. Lomer edge dislocation. a) HREMicrograph b) predicted asymmetric dissociation of the Lomer-Cottrell lock

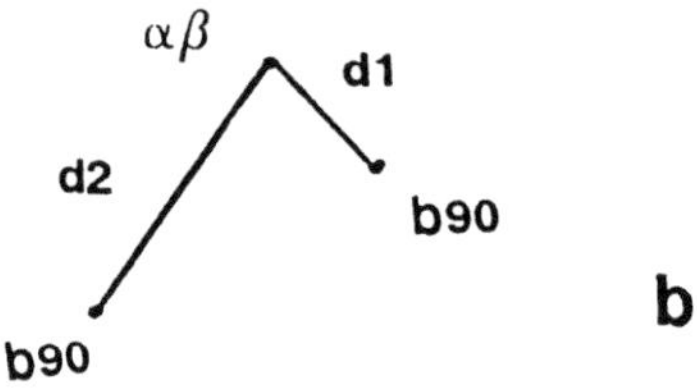

over only one atomic distance /17, 24/. This configuration was thought to be stabilized by impurities, mainly oxygen. This has to be reinterpreted in the light of new results /25/ which argue in favour of hexagonal silicon precipitation under point-defects supersaturation at low temperature (<700°C). The Lomer dislocation core with a stacking fault one atomic distance wide, as described in /17, 24/, contains essentially the seed of the hexagonal phase.

3.2.3 B2 lock.

It is also created by the reaction of two 60° dislocations and was also observed in deformed Si crystals /24/. Contrary to the previous case where the observed configuration does correspond to the theoretical one, the B2 lock is widely asymmetrically dissociated as predicted. The configuration is shown on Fig.4 where d_1 = 4.3 nm and d_2 = 9.6 nm.

As a matter of fact the partials involved in the dislocation of the B2 lock are not the ones concerned in the Lomer-Cottrell lock : the stair-rod

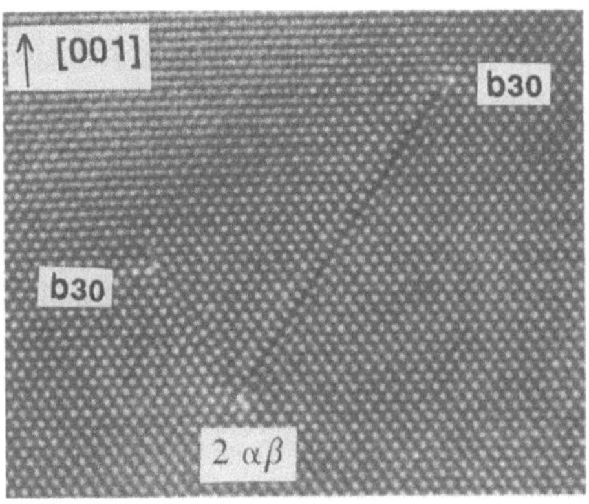

Fig.5 : B2 lock.
a) HREMicrograph
b) predicted dissociation.
(Micrograph M. Elkajbaji)

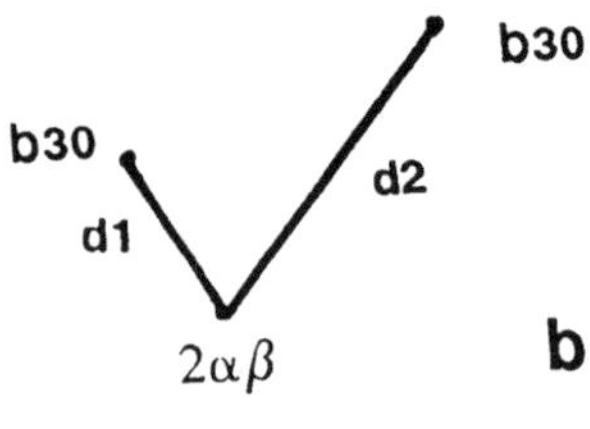

is $\alpha\delta$ in the LC whereas the stair-rod is $2\alpha\delta$ in the B2 lock, the partials are of the 90° type in the LC lock whereas they are of the 30° type in the B2 lock.

3.2.4 Climb of dissociated dislocations.

The interaction of point defect with the core of the dislocations has been clearly revealed by the HREM pictures obtained on 60° dislocations induced in a deformed Si crystal. Fig.6 shows the nucleation of perfect interstitial loop on the 90° partial. This happens mainly on dislocations stopped in dipoles or near obstacles such as grain-boundaries.

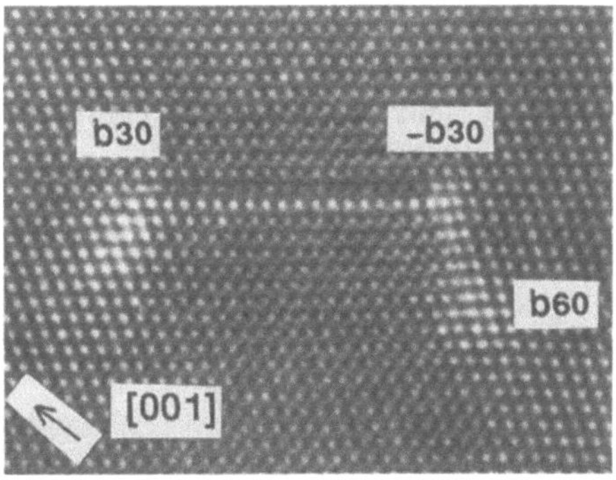

Fig.6. Nucleation of an interstitial loop on the 90° partial of a dissociated 60° dislocation

This phenomenon was first detected by weak-beam experiments /26/. A Franck loop nucleation would have been however energetically more favourable. As a matter of fact, the stair-rod created in this case would be of the $\alpha\delta$ type : this argues is favour of a high energetical structure of the $\alpha\delta$ stair-rod. In these deformation tests, the point defects mainly detected were interstitials and an estimate of the point defects supersaturation involved for the dislocations climbing leads to $C/Co = 10^6$ likely provided by the deformation.

This example shows that the point defects greatly affects the core structure and this will certainly have consequences on the electrical and mechanical behaviour of the material /27/.

3.3 Grain-boundary structure.

Boundaries are subject to extensives studies /review in 28/. More attention was paid however to low angle and special high angle tilt boundaries. The low angle boundaries which can be described with isolated dislocations are relevant to §2.2. the interaction between dislocations has however to be taken into account to describe more precisely the configurations /29/.

The structure of coincidence or near coincidence tilt boundaries have been investigated /30, 31, 32/. Pure tilt GB may be described with the aid

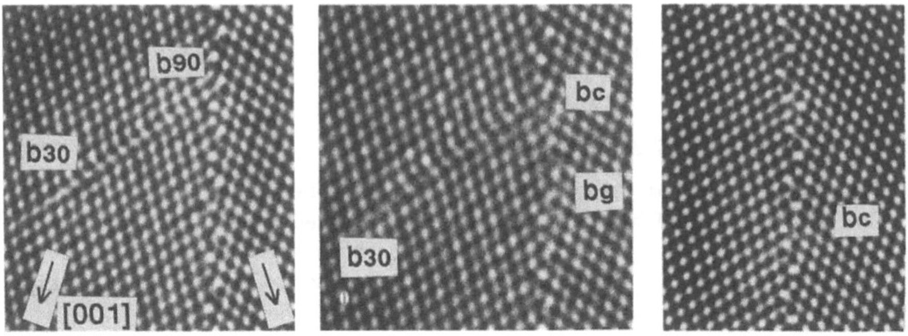

Fig. 7 : Entrance of a deformation induced dislocation into a Σ=9 GB. a). The 60° D remains dissociated close to the GB, b) the 90° leading partial touches the GB and composes into GB residual dislocations b_c and b_g ; b_g is easily glissile along the GB plane. c) residual dislocations remaining after b_g has moved away. (Micrograph M. Elkajbaji)

of the structural units concept /33, 34/. A number of coherent and incoherent twins were studied in as-grown Ge or Si bicrystals which reveal the same structural features. The use of the 400keV HREM allows to look at GB whose tilt axis is <001> /8/.

The electronic states associated with the Σ=3 (112) twin boundary were calculated /35/. It was concluded that the GB does not introduce deep levels in the band gap provided that the electronic rearrangement does not destroy the tetracoordination. From /32/ it was also concluded that the GB can generally be built with completely reconstructed models. Therefore the electrical activity of some as-grown or heat treated GB is thought to come from impurities or intrinsic point defects.

3.4 Interaction between dislocations and GB.

The interaction between dislocations and the tilt Σ=9 GB has been investigated in details /36,37,38,39,40/. It appears that the deformation induced dislocations are trapped by the GB where they decompose into dislocations whose Burgers vectors belong to the DCS lattice /41/. The entrance of 60° dislocation is shown on Fig.7a,b. The decomposition leads, in the case, to three DCS dislocations, one of them b_g is glissile along the interface. The main point is that the other residual dislocations are mobile along the GB by gliding and /or climbing /40/ (Fig.7c). The climb mechanism involves point defects /42/. The point defects help to release the stress at the dislocations pile-ups head created at the GB allowing the in-coming dislocations decomposition. The type of the point defects has not been determined yet, but it is known from §2.1 that interstitials are at least in supersaturation during deformation experiments.

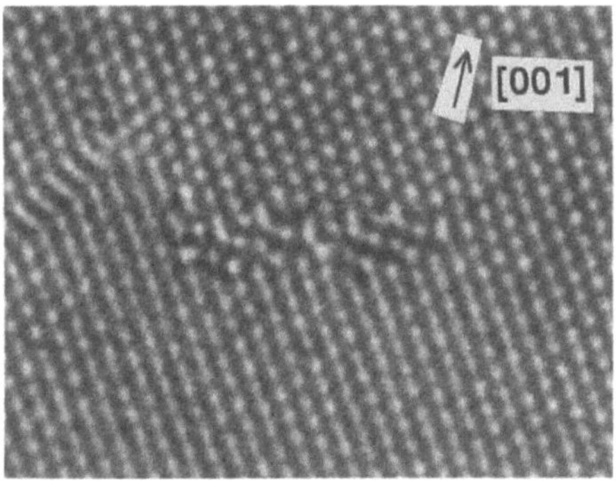

Fig.8 : Hexagonal silicon platelet detected in a sample deformed in tension at 800°C for a .3% strain. (micrograph J.L. Putaux)

3.5 Precipitates structure.

Electron microscopy is able to detect three dimensional defect created during the device processes and although strong limitations exit, HREM can be an adequate tool to determine the early stages of the growth of these defects. A typical example will be described here.

Point defect agglomeration.

A. Bourret /25/ has clearly demonstrated that the rod-like defects induced during Si heat treatment at T<800°C, are in fact made up of the Si hexagonal phase. At higher temperature the interstitials agglomerate into perfect or faulted dislocation loops /43/. He also showed that the "311" defects induced by electron or ion irradiation are relevant to the hexagonal Si.

As a matter of fact the same results were obtained in deformed Si. Large hexagonal Si platelets were observed /44/ in Si deformed at 700°C under hydrostatic pressure and at large strain. Furthermore, the nucleus of these platelets was also detected by HREM in Si deformed at T<800°C and for a .3 % strain (Fig.8). As mentioned in §2 the climb of the dislocations in the grains is a proof of the supersaturation of interstitials. These results are in agreement with the idea /25/ that interstitials at temperatures below 800°C induce a phase transformation from cubic to hexagonal Si. As mentioned in /25/ these defects can be built with structural units where the bonds are completely reconstructed.

4. CONCLUSION

HREM can help to determine the structure of defects in silicon. It appears that in most of the defects studied up to now the bonds can be reconstructed. It might be deduced that the electrical activity comes from defects such as impurities, point defects or peculiar structural defects such as jogs along the dislocation line.

References

1. J. Spence, (1981) Experimental High Resolution Electron Microscopy, Clarendon Press, Oxford.
2. H. Bethe, Ann. Phys. (1928) 87, 55.
3. C. Darwin, Phil. Mag. (1914) 27, 315.
4. A. Howie, M. Whelan, Proc. Roy. Soc. (1961) A263, 217.
5. J. Cowley, A. Moodie, Acta Cryst. (1957) 10 609.
6. P. Goodman, A. Moodie, Acta Cryst. (1974) A30, 280.
7. J. Cowley, Diffraction Physics (1975), North Holland, Amsterdam.
8. A. Bourret, J.L. Rouvière, J.M. Penisson, Acta Cryst. (in the press).
9. W. Coene, H. Bender, S. Amelinck, Phil. Mag. (1987) B52, 369.
10. W. Coene, D. Van Dick, J. Van Landuyt, S.AMELINCKX, Phil. Mag. (1987) B56, 415.
11. K.Lodge,S.Altmann,A.Lacipirella,N.Tomassini,(1984) Phil. Mag.B49,41
12. S.Marklund, Jour. Phys. (1983) C4-44, 25
13. H. Teichler, H. Veth, Jour. Phys. (1983) C4-44, 93
14. M. Heggie, R. Jones, Inst. Phys. Conf. Ser. (1987) 87, 367
15. J. Tersoff, Phys. Rev. Lett. (1986) 56, 632
16. G. Barraf, E. Kane, M. Shluter, Phys. Rev. (1980) B21, 5662
17. A. Bourret, J. Thibault-Desseaux, F. Lançon, Jour.Phys.(1983)C4-44, 15
18. F. Louchet, J. Thibault-Desseaux, Rev. Phys. Appl. (1987) 22, 207
19. A. Olsen, J. Spence, Phil. Mag. (1981) A43, 945
20. G. Anstis, P. Hirsch, C. Humphreys, J. Hutchinson, A. Ourmazd, Inst. Phys. Conf. Ser. (1981) 60, 15
21. A. Korner, H. Schmid, F. Prinz, Phys. Stat. Sol. (1979) a51, 613
22. A. Korner, M. Martinez-Hernandez, A. George, H. Kirchner, Phil. Mag. (1987) 55, 105
23. A. Bourret, J. Desseaux, A. Renault, Phil. Mag. (1982) A45, 1
24. M. Elkajbaji, H. Kirchner, J. Thibault-Desseaux, Phil. Mag. (1988) A57, 631
25. A. Bourret, Inst. Phys. Conf. (1987) 87, 39
26. D. Cherns, Dislocations 1984 ed. P. Veyssière, L. Kubin, J. Castaing (Paris-CNRS)
27. P. Hirsch, Defects in Semiconductors (1980) p257. MRS Ed. J. Narayan, T. Tan
28. Polycrystalline Semiconductors. Jour. Phys. (1982) C1-43
29. A. Bourret, J. Desseaux, Phil. Mag. (1979) A39, 405 & 415
30. C. D'Anterroches, A. Bourret, Phil. Mag. (1984) 49, 191
31. A. Bourret, J.J. Bacmann, Surf. Sci. (1985) 165, 495
32. A. Bourret, J.J. Bacmann, Rev. Phys. Appl. (1987) 22, 563
33. G.H. Bishop, B. Chalmers, Scripta Met. (1968) 2, 133
34. A. Mauger, J.C. Bougoin, G. Allan, M. Lannoo, A. Bourret, L. Billard, Phys. Rev. (1987) B35, 1267
35. R.C. Poind, V. Vitek, D.A. Smith, Acta Met. (1979) 27, 235
36. X. Baillin, J. Pelissier, J.J. Bacmann, A. Jacques, A. George, Phil. Mag. (1987) A55, 143

37. A. Jacques, A. Georges, X. Baillin, J.J. Bacmann, Phil. Mag. (1987) A55, 165
38. M. Elkajbaji, J. Thibault-Desseaux, M. Martinez-Hernandez, A. Jacques, A. Georges, Rev. Phys. Appl. (1987) 22, 569
39. J. Thibault-Desseaux, M. Elkajbaji, Jour. Phys. (1988) in the press
40. M. Elkajbaji, J. Thibault-Desseaux, Phil. Mag. (1988) 58, 325
41. W. Bollmann, Crystal defects and crystalline interfaces. Springer Berlin (1970).
42. J. Thibault-Desseaux, J.L. Putaux, A. Jacques, M. Elkajbaji, MRS (1988) Interfacial Structure, Properties and Design. Ed. Yoo, Clark, Briant
43. A. Bourret, J. Thibault-Desseaux, D. Seidmann, Jour. Appl. Phys. (1984) 55, 825
44. J.L. Demenet, J. Rabier, H. Garem, Inst. Phys. Conf. (1987) 87, 355

Tunneling Microscopy and Surface Analysis

H.W.M. Salemink

IBM Research Division, Zurich Research Laboratory,
CH-8803 Rüschlikon, Switzerland

In this paper the surface analysis with tunneling microscopy is summarized with emphasis on the following aspects: 1) the physics of vacuum barrier tunneling, 2) instrumental techniques, 3) metal and semimetal surfaces, 4) semiconductor surfaces and spectroscopy, and 5) selected applications.

1. Physics of Vacuum Barrier Tunneling

The wide application of scanning tunneling microscopy (STM) in surface science is based on tunneling phenomena, which are well known in solid-state physics. These are in particular: 1) the quantum-mechanical tunnel effect [1], 2) the semiconductor tunnel diode [2], and 3) electron tunneling into superconductive metal films via very thin (1 nm) oxide barrier junctions [3].

Common features of the tunnel structures are 1) a barrier width of comparable dimension as the relevant electron wavelength, 2) available electron density of states (occupied or unoccupied) on both sides of the barrier to provide initial and final states, and 3) an energy (voltage) difference over the barrier which drives the electron tunneling (Fig. 1). In the case of oxide coupled junctions, the energy barrier is formed by the barrier height and thickness of the oxide layer. For vacuum tunneling, the barrier consists of the electron workfunctions (ϕ_1, ϕ_2) of the two electrodes and their separation distance *(s)*. This variable electrode separation and the resulting electron current flow was demonstrated in early experiments, when "surface images" were obtained by operating one electrode in the form of a sharp tip in the field-emission regime (typical voltages larger than 5 eV over the barrier) [4]. The advantage of the sharp electrode is the high localization of the electron emission source, which results in a high spatial resolution over the surface structure. Clear vacuum tunneling experiments provided the exponential current increase as a function of electrode separation [5], as expected by theory. Successive experiments, tunneling into Au metal [6] and silicon single crystal surfaces [7] demonstrated that atomic resolution could be achieved. If a voltage difference *(V)* is applied over the barrier, the resulting tunnel current (*I*) is described as

$$I \propto V \exp(-2\phi^{1/2}s), \qquad (1)$$

where ϕ is the average workfunction of the two electrodes. Note that on the atomic scale, the concept of ϕ loses its classical meaning. The exponential relation of I on barrier width is shown in Fig. 2; these data provided the first evidence of vacuum tunneling and represent a major improvement over the earlier attempts, in which electron emission from larger distance in the field-emitting regime was investigated. The slope of the curve in Fig. 2 gives ϕ = 3.5 eV for the average workfunction, close to

Springer Series in Materials Science, Vol. 13
Semiconductor Silicon Editors: G. C. Harbeke and M. J. Schulz

Fig. 1

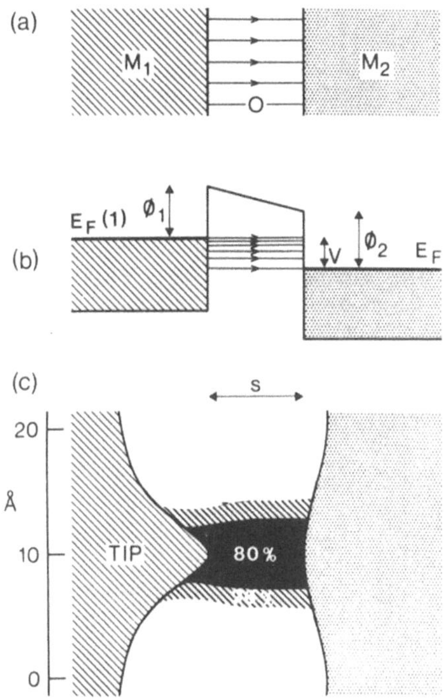

Fig. 2

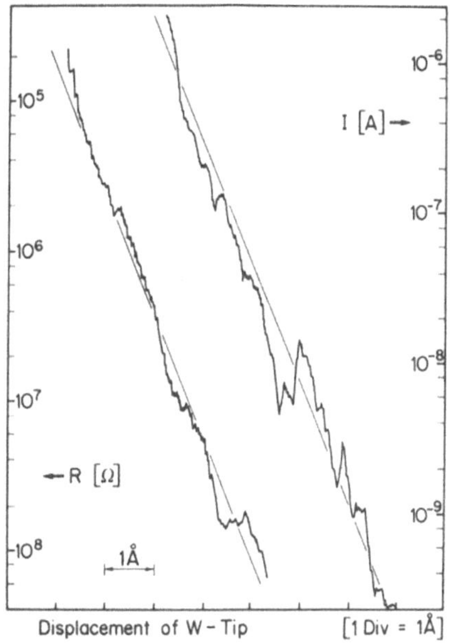

Figure 1. Scheme of electrodes and electron barrier in vacuum tunneling experiment

Figure 2. Exponential dependence of tunneling current on gap width, demonstrating the tunnel effect (from Ref. 5)

the expected value for a clean metal; the value is very dependent on cleanliness of the surface and can decrease to typically 0.1 eV for contaminated or adsorbed surfaces.

The local electron density of states near the Fermi level influences the tunnel current, which is determined by the density of initial and final states. For metal on both sides of the barrier (metal tip and surface), the electron density is nearly flat at the Fermi energy and the tunnel current vs. voltage curve (I-V) will show a near linear behavior. This is very different from the case with a semiconductor as surface; there the semiconductor energy gap will show up in the I-V curve, indicating the energy positions of both the conduction and valence bands with respect to the Fermi level [8]. Theoretically it can be shown that the electron density of states at the surface $N(E)$ is related to the local I-V tunneling curve by

$$N(E) \propto d(\ln I) / d(\ln V) = (dI/dV) / (I/V). \tag{2}$$

Thus the tunneling I-V curve can be used to perform local electron spectroscopy for energies near the Fermi level. The spectroscopy of semiconductor surfaces is complicated by the presence of surface states, which show up in the density of states near the Fermi level (see Sec. 4). Theoretical treatments on these subjects include the tunnel-current distribution in a free electron model [9], tunneling via a single atom [10], and charge density calculations for energies close to the Fermi energy [11]. Extensive reference to theory and examples of the topics discussed can be found in recent reviews on tunneling microscopy and spectroscopy [12-14].

2. Instrumental Techniques

All STM instruments rely on positioning a metallic electron emitter (tip) in a tunneling distance (typically 1 nm) from the surface under investigation. The probe material is mounted on a approaching mechanism. Typical values for gap voltage and current are 0.1 V and 1 nA, respectively, for (semi-)metallic surfaces. For semiconductors and small molecular complexes, these values might be 1 V and 0.1 nA. The sample carrier can be a micro-mechanical or a piezo-electric positioner with a step size of 0.1 to 1 μm. Scanning of the tip over the surface is done by actuation of three orthogonal *(X,Y,Z)* piezoelectric elements [6,15]. In more recent designs, these three elements are combined into a single tube with segmented, circular electrodes [16]. Typical piezoelectric sensitivities are in the range of 0.5 to 5 nm per Volt of applied voltage, and a noise level of 0.005 nm in *X,Y,Z* can be obtained. The tip-to-surface distance is kept constant by a feedback circuit which senses the actual tunnel current and maintains it to a preset value by adjusting the *Z*-piezoelectric. Owing to the exponential relation between I and s (Fig. 2), a very sensitive distance regulation is obtained (Fig. 3).

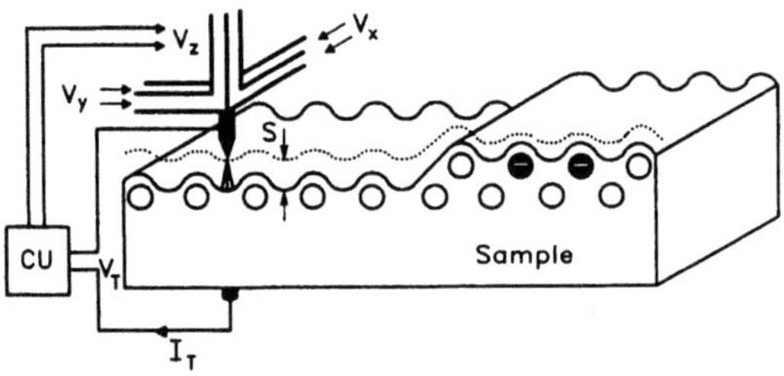

Figure 3. Schematic view of experimental configuration with probe surface on carrier mechanism, tip and piezoelectric tripod for *X,Y,Z* positioning of the tunneling gap (see also Fig. 5)

The "topographic" image of a surface is acquired by recording the *Z*-position voltage, necessary for maintaining a preset tunnel condition (tip voltage and tunnel current), whilst scanning *X* and *Y* in a given frame; thus the tip traces the surface at a specific distance *s*, as long as the fixed tunnel conditions are fulfilled [Eq. (1)]. A typical scan rate for a 5 × 5 nm window is 30 sec. Work-function measurements are obtained by modulating the gapwidth *s* for 0.01 nm and recording the resulting current change at the modulation frequency (dI/ds); this signal is proportional to the slope in Fig. 2.

The spectroscopic data are usually obtained by freezing the *Z*-tip feedback circuit for a short time (0.05-0.1 sec) and applying a voltage waveform during this period [17,18]: the resulting current response is recorded simultaneously with the voltage sweep. This results in the acquisition of a local I-V curve at a specific distance from the surface, set by the tunnel conditions during the usual regulating mode outside of the hold time (Fig. 4).

The above-mentioned operating modes of the STM are routinely applied on surfaces where a given tunnel condition can be maintained with sufficient stability. An alternate way of acquiring images at a higher rate (1 frame/sec) is to use a fixed gap width and to record the resulting variations in the tunnel current during *X-Y* scanning; of course this technique can be applied only on small frames or on very flat surfaces, where the risk of crashing the tip into the surface is small.

The data acquisition and processing equipment usually has tree parts:

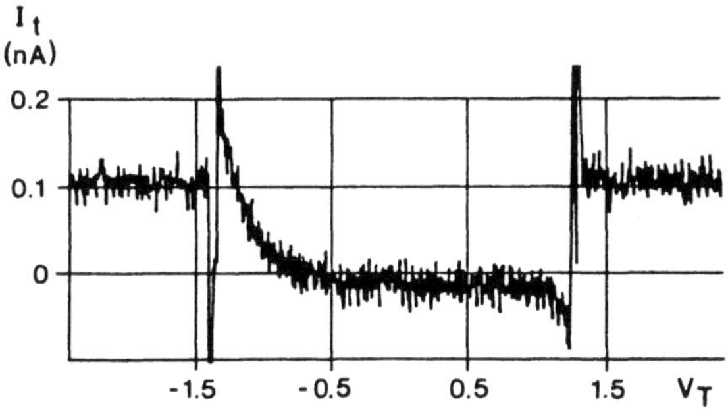

Figure 4. *I-V* characteristic of GaAs(110) surface, obtained during short interruption of usual gap regulation mode. DC tunneling conditions are -1.2 V and 0.1 nA (segment 1 and 5); gap regulation is interrupted in segments 2 - 4 and the tip voltage is swept as indicated. Each segment is 8 msec wide. GaAs gap is seen between -0.7 V and +0.8 V

a) Analog circuitry for maintaining gap conditions (*V*, *I*, *s*) and interrupting the feedback loop to impose voltage sweeps. Typical frequency bandwidth is 10 kHz in the current acquisition amplifier at a gain of 10^8 V/A and 3 kHz in the feedback circuit.
b) Digital circuitry – A/D converters and timers – to record *X,Y,Z* positions from the piezoelectric drivers; digital storage to record additional signals at specified positions (current in spectroscopy, modulation response for derivative signals d*I*/ds, d*I*/d*V*, photon response).
c) Post-processing like filtering and contouring to make "topviews" or coding the frame with the additional information mentioned in b).

The current response at fixed mean gapwidth at the higher frame rates usually is displayed directly on slow-scan tv monitors or recorded in a digital frame storage.

The present generation of STM instruments is rather compact and fit on ultra-high vacuum (UHV) flanges of 63 to 100 mm diameter (Fig. 5). As the gap stability depends largely on the relative displacement of tip and sample, instruments generally benefit from small size and a minimum amount of hardware for support: to this respect, the construction should be as sturdy as possible, but still comply with requirements for exchanging tip and samples. Several completely mechanical constructions are devised to handle the approach of tip to sample; many use leveraged micrometers for step-size reduction [19]. The small size of the instruments allow these to be incorporated into existing analysis equipment, like Auger or electron microscopes [15].

Finally, it should be mentioned that these instruments also are used in air or controlled atmospheres to study samples or surfaces in a more natural environment; this goes at the expense of the knowledge of the barrier conditions, as in all these cases gaseous or liquid adsorption is either wanted or unavoidable. Examples are the studies on thin Langmuir-Blodgett films, small molecules and molecular chains.

3. Metals and Semimetals

Metals satisfy the basic requirements for vacuum barrier tunneling well, and hence were used to prove the atomic spatial resolution once the basic concepts had been demonstrated. The reconstruction on the Au(110) (Fig. 6) provided the first evidence of the atomic resolution of STM [6]. The periodicity and orientation of the atomic rows

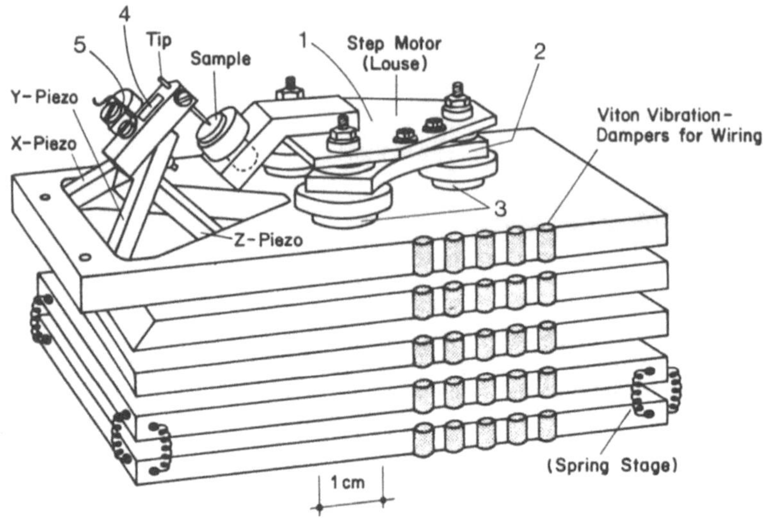

Figure 5. View of "pocket size" STM for UHV or air operation. The sample is mounted on "louse" carrier mechanism for coarse approach to within 1 μm. More recent versions of this instrument have current amplifiers on the louse as well as removable tunneling tips, in addition to exchange of samples

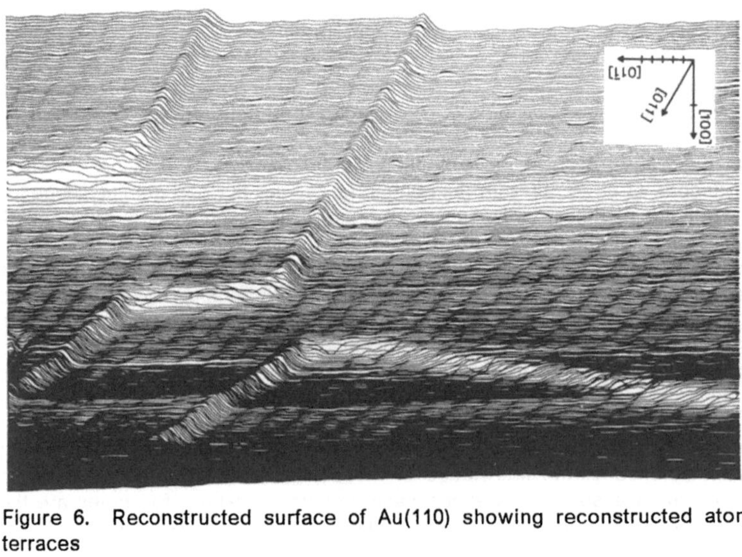

Figure 6. Reconstructed surface of Au(110) showing reconstructed atomic rows and terraces

were compared with the LEED image; in this case the STM delivered a confirmation based on atomic resolution, whereas other surface-science techniques up to then were averaging their information from areas determined by beam diameter or diffraction limit. Subsequently, chemisorption and physisorption experiments demonstrated the induced surface reconstructions by oxygen on Ni(110) [20] or CO on Pt [21].

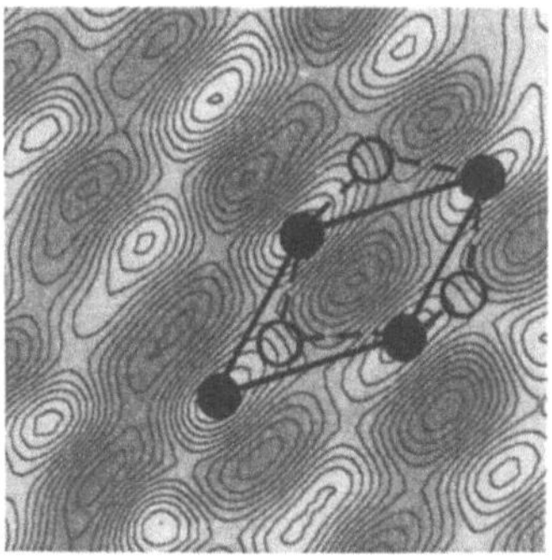

Figure 7. Topview of highly ordered pyrolytic graphite (001) surface

The transition from the tunneling regime towards a metal-metal point contact has been studied on an Ag surface, and provides a nice way of studying the barrier as a function of electrode separation [22].

The (001) surface of highly oriented pyrolytic graphite (HOPG), a prototype semimetal, is of interest for STM studies for the following reasons. The interatomic carbon bond length is 0.146 nm in the plane, with 2 atoms in an 0.254-nm-wide unit cell and 0.33 nm between the weakly coupled layers. The surface atoms are grouped in a hexagonal structure, similar to those in the bulk layers (Fig. 7). HOPG has a low gas-adsorption coefficient, which is convenient for experiments. The electronic surface structure has been calculated and compared with STM data [23,24]; a maximum corrugation in the charge density near the Fermi energy of 0.03-0.1 nm is expected, most experimental data find a (much) larger value for gap voltages smaller than 0.3 eV [25]. This is in contrast to the behavior on pure metals, where the corrugation is weakly dependent on tunnel conditions. Features of the electron-state density are reflected in the tunnel spectroscopy and confirmed by data from inverse photoemission experiments [26]. For low gap voltages, anomalously high apparent corrugations (over 0.1 nm) are reported [25,27,28], which are ascribed to atomic-force interaction between tip and sample. Also dislocations in the surface plane were seen, disturbing the crystal symmetry [28]. All data on graphite essentially display a threefold symmetry, which is explained by the inequivalency of the two atoms in the unit cell, although any charge transfer to the second layer is very small as a result of the relatively large interlayer distance. Another explanation for a lower symmetry might be a tip with nonrotational symmetry at the apex; however, such tip influences and their impact on symmetry have received little attention up to now.

In connection with layered structures, investigations have been made to directly observe charge density waves in layered compounds of TaS_2. The charge density wave is observed with a periodicity larger than the atomic periodicity of the host material [29].

4. Semiconductor Surfaces and Electronic Spectroscopy

Most surfaces of semiconductors have a tendency to reconstruct their atomic positions at the terminated plane in order to minimize the surface energy [Si(111), Si(100), Ge(100), GaAs(001)], and classically the resulting larger surface unit cells have been studied with electron diffraction (LEED). The STM provides a tool to study these unit cells in detail with local atomic resolution, as well as to perform spectroscopy at

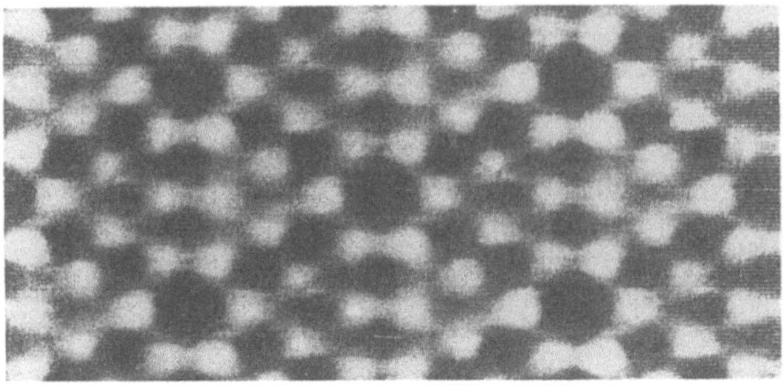

Figure 8. Topview of the reconstructed surface of Si(111) 7x7: white hills indicate the adatoms in the unit cell

positions in the cell. The topography of the reconstructed 7x7 unit cell on the clean Si(111) surface [7,18,30,31] finally decided the validity of the atomic arrangement of the DAS model for the surface, featuring dimers, adatoms and a stacking fault (Fig. 8) [32]. Subsequent spectroscopic data revealed the energy levels of the bonds within the cell [18]. In a similar way, results are reported for the reconstructed faces of Si(001) 2x1 [33], Si(111) 2x1 [34] and for thin epitaxial Ge overlayers on Si(111) [35]. Interesting is also the observation of several different $m \times n$ cells, which are simultaneously present on the surface of Si and Ge [36,37]; clearly such multiple reconstructions will mix into the results of spatially averaging techniques, such as electron beam diffraction. Similar observations of reconstructions have been reported for terraced areas of the arsenic-rich GaAs(001) surface, which shows many small "stepped" islands; these steps are thought to be important nucleation sites [38].

The GaAs(110) face is constructed with both Ga and As atoms present at the surface in a 1x1 unit cell: depending on the polarity of the tunnel voltage (into the empty or out of the filled states), either the Ga or the As atoms (Fig. 9) dominate the "topographic" image [17]. This means essentially that for these surfaces always spectroscopic features are imaged, a property which is useful for obtaining a chemical specificity. The local spectroscopic *I-V* curves reveal a 1.43 eV energy gap and both the conduction and valence band edges [17]. When the chemical composition of the surface is altered by adsorption of oxygen [39] or a semimetal (Sb) [40], the resulting

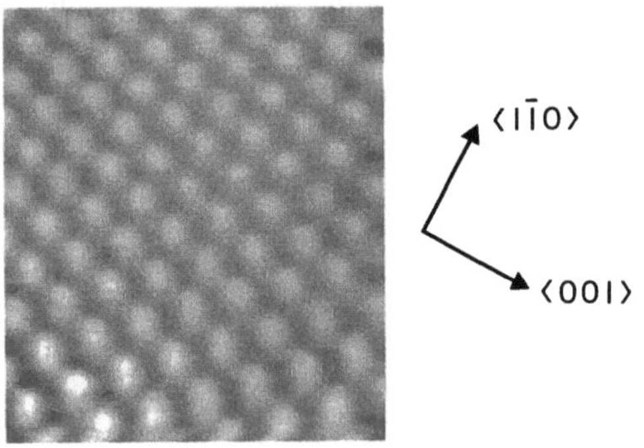

Figure 9. Undisturbed surface of GaAs(110) with individual Ga atoms in 1x1 arrangement

changes in the conduction and valence band onsets are clearly seen in the local spectroscopy. The shifts in the band onset result from the local change in electronic structure due to the chemical coverage: both the spatial influence (screening length) and the changing band structure are observable on uniform or partially covered surfaces (clusters).

5. Special Studies and Applications

A major drawback of STM is that a conductive sample electrode is necessary to detect the tunneling current. The Atomic Force Microscope (AFM) overcomes this limitation [41]. In its simplest concept it consists of a very thin conductive lever, which is inserted into the tunneling gap (Fig. 1). The lever touches the isolating sample surface (pushed by means of a piezoelement) and an STM tip is used to tunnel on the other side of the lever. With this method, a resolution of 3 nm was initially obtained on Al_2O_3 [41]; it was later improved to resolve unit cells on graphite and boron nitride.

Optical photon generation with STM at tip voltages of up to 20 eV shows a remarkably high efficiency [42] on Ag film surfaces; this is attributed to a resonance phenomenon in the metallic film with a characteristic spectral signature (plasmon): the

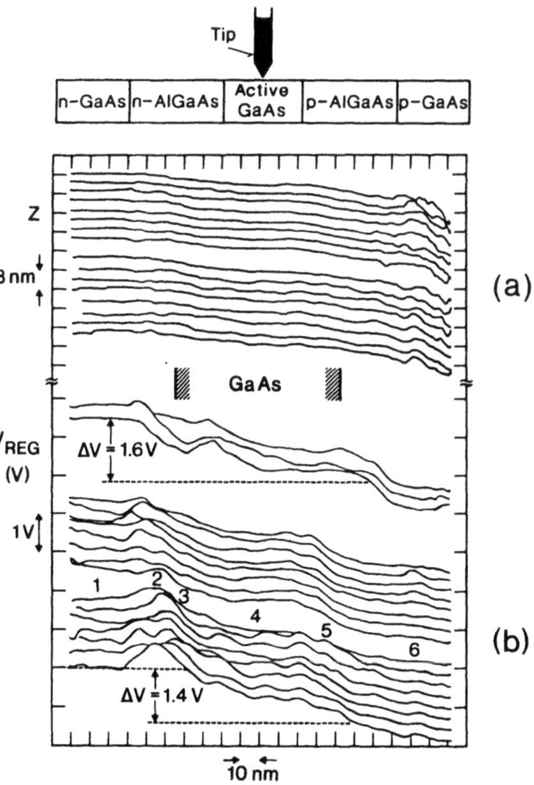

Figure 10. Topography and potentiometry on (110) face cleaved AlGaAs heterojunction. Upper traces (a): topography over junction

operation resembles an inverse photoemission experiment with high spatial localization and surface enhanced efficiency.

Tunneling spectroscopy might be used to study electronic properties in epitaxially grown heterostructures (Fig. 10), such as the steepness of *p-n* junctions in AlGaAs-GaAs interfaces [43]. The layers were grown with MBE along the GaAs <001> direction and cleaved in UHV to expose the (110) surface to the STM. In this way a cross-section through the multilayer stack can be investigated. To position the STM tip over the 0.1 μm thick layers, the STM is built inside an UHV SEM [15]. Such STM/SEM combinations will be very useful for spectroscopy on small artificial semiconductor structures.

6. Summary

A review is given of the utilization of scanning tunneling microscopy on different types of surfaces (metals, semimetals and semiconductors). Apart from the observation of topography, with atomic resolution, the spectroscopic information is uniquely obtained on a similar scale. This gives access to small molecules and artificial structures in the nanometer regime.

Acknowledgements are due to E. Stoll for image processing and to O. Albrektsen for contributing to the experiments.

7. References

1. *Tunneling Phenomena in Solids*, ed. by E. Burstein and S. Lundquist (Plenum Press, New York 1969)
2. L. Esaki: Phys. Rev. **109**, 603 (1957); I. Giaever: Phys. Rev. Lett. **5**, 147 (1960)
3. L. Solymar: *Superconductive Tunneling and Applications* (Chapman and Hall, London 1972)
4. R. Young, J. Ward, F. Scire: Rev. Sci. Instrum. **43**, 999 (1972), and Phys. Rev. Lett., **27**, 922 (1971)
5. G. Binnig, H. Rohrer, Ch. Gerber, E. Weibel: Physica **109**, & **110**B, 2075 (1982), and Appl. Phys. Lett. **40**, 178 (1982)
6. G. Binnig, H. Rohrer, Ch. Gerber, E. Weibel: Phys. Rev. Lett. **49**, 57 (1982)
7. G. Binnig, H. Rohrer, Ch. Gerber, E. Weibel: Phys. Rev. Lett. **50**, 120 (1983)
8. R. Feenstra and J. A. Stroscio: J. Vac. Sci. Technol. **B5**, 923 (1987)
9. E. Stoll, A. Baratoff, A. Sellomi, P. Carnevali: J. Phys. C **17**, 3073 (1984)
10. N. D. Lang: Phys. Rev. Lett. **55**, 230 (1985)
11. J. Tersoff, D. R. Hamann: Phys. Rev. B **31**, 805 (1985) and Phys. Rev. Lett. **50**, 1998 (1983)
12. IBM J. Res., Develop. **30**, No. 4 & 5 (1986): Proc. STM Workshop, Oberlech, Austria (1985)
13. Surface Science, **181**, Special Issue (1986), ed. by N. Garcia: Proc. STM'86, Santiago de Compostela, Spain (1986)
14. J. Vac. Sci. Technol., **A6**, No. 2, ed. by R. Feenstra: Proc. STM'87,
15. Ch. Gerber, G. Binnig, H. Fuchs, O. Marti, H. Rohrer: Rev. Sci. Instrum. **57**, 221 (1986)
16. D. Smith, G. Binnig, Ch. Gerber: Rev. Sci. Instrum. **57**, 1688 (1986)
17. R. Feenstra, J. A. Stroscio, J. Tersoff, A. P. Fein: Phys. Rev. Lett. **58**, 1192 (1987)
18. R. J. Hamers, R. M. Tromp, J. E. Demuth: Phys. Rev. Lett. **56**, 1972 (1986)
19. M. D. Pashley, K. W. Haberern, W. Friday: J. Vac. Sci. Technol. **A6**, 448 (1988)

20. A. Baró, G. Binnig, H. Rohrer, Ch. Gerber, E. Stoll, A. Baratoff, F. Salvan: Phys. Rev. Lett. **52**, 1304 (1984)
21. W. Hosler, R. J. Behm, E. Ritter: IBM J. Res. Develop. **30**, 403 (1986)
22. J. K. Gimzewski, R. Moeller: Phys. Rev. B **36**, 1284 (1987)
23. A. Selloni, A. Carnevalli, E. Tosatti, C. D. Chen: Phys. Rev. B **31**, 2602 (1985)
24. I. P. Batra, N. Garcia, H. Rohrer, H. Salemink, E. Stoll, S. Ciraci: Surface Sci. (1987)
25. G. Binnig, H. Fuchs, Ch. Gerber, H. Rohrer, E. Stoll, E. Tosatti: Europhys. Lett. **1**, 31 (1985)
26. B. Reihl, J. K. Gimzewski, J. M. Nicholls, E. Tosatti: Phys. Rev. B **33**, 5770 (1986)
27. J. M. Soler, A. M. Baro, N. Garcia, H. Rohrer: Phys. Rev. Lett. **57**, 444 (1986)
28. H. Salemink, I. P. Batra, H. Rohrer, E. Stoll, E. Weibel: Surface Sci. **181**, 139 (1987)
29. A. G. Slough, W. W. McNairy, R. V. Coleman, B. Drake, P. K. Hansma: Phys. Rev. B **34**, 994 (1986)
30. R. J. Hamers, R. M. Tromp, J. E. Demuth: Phys. Rev. Lett. **56**, 1972 (1986)
31. R. S. Becker, J. A. Golovchenko, D. R. Hamann, B. S. Schwarzentruber: Phys. Rev. Lett. **55**, 2032 (1985)
32. Y. Takanayagi, Y. Tamashiro, M. Takahishi, S. Takahishi: J. Vac. Sci. Technol. **A3**, 1502 (1985)
33. R. M. Tromp, R. J. Hamers, J. E. Demuth: Phys. Rev. Lett. **55**, 1303 (1985)
34. R. M. Feenstra, W. A. Thompson, A. P. Fein: Phys. Rev. Lett. **56**, 608 (1986).
35. R. S. Becker, J. A. Golovchenko, B. S. Swartzentruber: Phys. Rev. B **32**, 8455 (1985)
36. R. S. Becker, J. A. Golovchenko, B. S. Swartzentruber: Phys. Rev. Lett. **54**, 2678 (1985)
37. R. S. Becker, J. A. Golovchenko, G. S. Higashi, B. S. Swartzentruber: Phys. Rev. Lett. **57**, 1020 (1986)
38. M. Pashley, K. W. Haberern, J. M. Woodall, P. D. Kirchner: Phys. Rev. Lett. **60**, 2176 (1988); M. Pashley, K. W. Haberern, J. M. Woodall: J. Vac. Sci. Technol. B **6**, 1468 (1988)
39. J. A. Stroscio, R. M. Feenstra, A. P. Fein: Phys. Rev. B **36**, 7718 (1987)
40. R. M. Feenstra, P. Martensson: Phys. Rev. Lett. **61**, 447 (1988)
41. G. Binnig, C. Quate, Ch. Gerber: Phys. Rev. Lett. **56**, 930 (1986)
42. J. H. Coombs, J. K. Gimzewski, B. Reihl, J. K. Sass, R. R. Schlittler: Proc. STM'88, to be published in J. Microscopy.
43. P. Muralt, H. Meier, D. Pohl, H. Salemink: Appl. Phys. Lett. **50**, 1352 (1987)

Scanning Minority Carrier Transient Spectroscopy: A Method to Investigate the Lateral Distribution of Defects in Semiconductors

A. Mesli and T. Heiser

Centre de Recherches Nucléaires (IN2P3),
Laboratoire PHASE (UA du CNRS no 292), 23, rue du Loess,
F-67037 Strasbourg Cedex, France

ABSTRACT

A scanning minority carrier transient spectroscopy (SMCTS) method with optical injection and full transient analysis is developed to obtain the lateral distribution of electrically active defects. The technique is applied to study the gold diffusion along the grain boundary in a silicon bicrystal. We show how important it is to control the shape of the transient in order to have credible results.

1. INTRODUCTION

Since the development of the deep level transient spectroscopy (DLTS) technique [1], two major improvements have been brought out. First the so called DDLTS [2] which yields the defect profile perpendicular to the surface. Later on, PETROFF et al. [3] proposed in their scanning DLTS to use an external focused excitation source in order to reduce the active volume to smaller dimensions. By moving the source along the surface, the two dimensional defect profile can be obtained.

Unfortunately, both techniques use classical correlation methods for signal processing. The resulting main limitation comes from the assumption that the transient is a pure exponential which is not always the case. In our technique, we develop a new signal processing with full transient analysis which enables a check of the exponential character of the transient. This technique is applied to show an enhanced diffusion of gold at the grain boundary of a silicon bicrystal.

2. THEORETICAL MODEL AND SIGNAL PROCESSING

Figure 1 shows the basic model of our technique where the analyzed zone is reduced to the illuminated part. For a defect distribution N_T (r), the SMCTS signal S_m (s) is given by :

$$S_m(\mathbf{s}) \propto \int_D N_T(\mathbf{r})\, F(\mathbf{r} - \mathbf{s})\, dr, \tag{1}$$

where s is the position of the beam, D the depletion region and F(r) is the occupancy function of the defects. According to fundamental considerations [4], we must take into account both the emission (e_p) and capture (c_n) rates in the expression F(r). For an n-type Schottky barrier it follows :

Springer Series in Materials Science, Vol. 13
Semiconductor Silicon Editors: G. C. Harbeke and M. J. Schulz

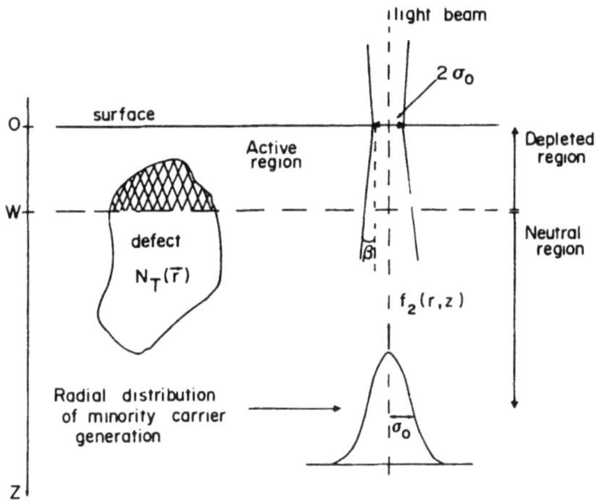

Fig. 1 : Schematic illustration of the model used to derive the SMCTS response

$$F(r) = \frac{c_p\, p(r)}{c_p\, p(r) + e_p} \{1 - \exp [- (e_p + c_p\, p(r))\, t_F]\}, \qquad (2)$$

where t_F is the optical filling pulse. p(r) is the distribution of the excess minority carriers reaching the depletion region. Unfortunately, at low temperature or at high injection level ($c_p\, p(r) \geq e_p$) the tail part of p(r) becomes important in the capture-emission process, extending the electrically active volume in SMCTS. Thus, in this technique, the spatial resolution is determined by F(r). By solving the diffusion equation with adequate boundary conditions for a Gaussian incident beam, we can obtain an expression for p(r). Then we estimate the spatial resolution, defined as the full width at half maximum (FWHM) of F(r), as well as its dependence on the beam diameter and on the minority carrier diffusion length (L_p). In a previous work [4], we have discussed in detail all these parameters. Here we restrict ourselve to the role of the saturation level. In Figure 2, the spatial resolution is shown to degrade when the ratio $R_S = c_p\, p(o)/e_p$ is increased. There is an optimum value R_S^{opt} below which the signal amplitude is considerably reduced without improving the resolution [4].

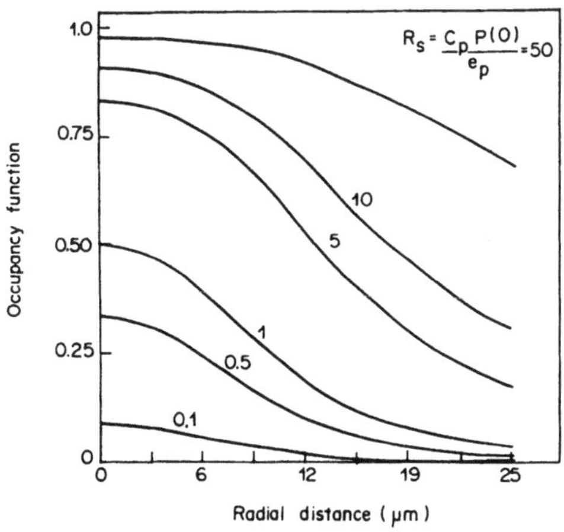

Fig. 2 : F(r) for different values of R_S

Generally, the signal processing in DLTS consists of a correlation of the transient with an analog weighting function [5]. However, the output of such correlator will yield only a single value for each transient. Thus, the important information which is the shape of the transient is lost. As a consequence, the interpretation of the data in SMCTS may be seriously altered if a deviation from the exponential behaviour occurs, for example, when two levels overlap. We will show in the experimental results that advantage can be taken of the digitalization of the full transient. The operations consist first in averaging the data over several pulses during acquisition. It results in an increase of the signal to noise ratio. Further, the signal is numerically filtered via a recursive technique. The details of this method are discussed in ref. 4. Figure 3 shows an example of a noisy transient before and after numerical filtering. In the case of multiexponential transients, fitting by a modulation functions method [6] is applied, the time constant being known from the global MCTS analysis. The goodness of the fit is characterized by $\kappa^2 = \Sigma_n (X_n - S_n)^2$ where S_n is the result of the fit and X_n the experimental transient.

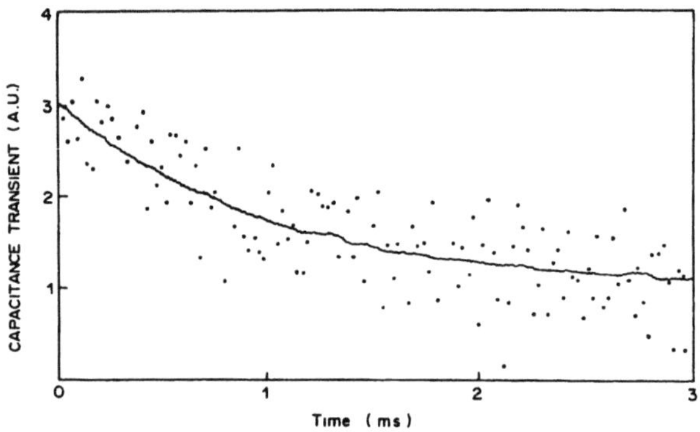

Fig. 3 : Noisy and filtered transient

3. EXPERIMENTAL CONDITIONS

The SMCTS system is described in ref. 4. In this work, we use this technique to illustrate the enhanced diffusion phenomena of gold along the grain boundary of a silicon bicrystal. This effect has been described phenomenologically by Fisher [7]. He assumes the grain boundary to be a region of limited thickness through which the impurity diffuses with a higher diffusion coefficient than in the bulk. It results in a non uniform lateral distribution of the gold across the grain boundary. To observe this phenomenon the diffusion conditions were fixed at 850°C for 2 hours. The SMTCS spectra are recorded from the backside of the n-type silicon doped at 5.10^{14} cm^{-3}. In order to avoid any contamination of the backside during the annealing, we previously imbedded the gold source into the wafer by recoil implantation, and placed it between two silicon samples during the thermal process.

4. EXPERIMENTAL RESULTS

Figure 4 shows the SMCTS response of the gold acceptor level. The spectrum corresponds to a single exponential treatment. The quality of the fit was not satisfying and we had to consider the response of a second deep level present in the MCTS spectra which could be approximated by a linear term, ($A \exp[-e{\cdot}t] \approx A [1-e{\cdot}t]$ for $e{\cdot}t \ll 1$). Figure 5a shows the final distribution. The decrease of the value of κ^2 reveals a net improvement of the quality of the fit. As expected from

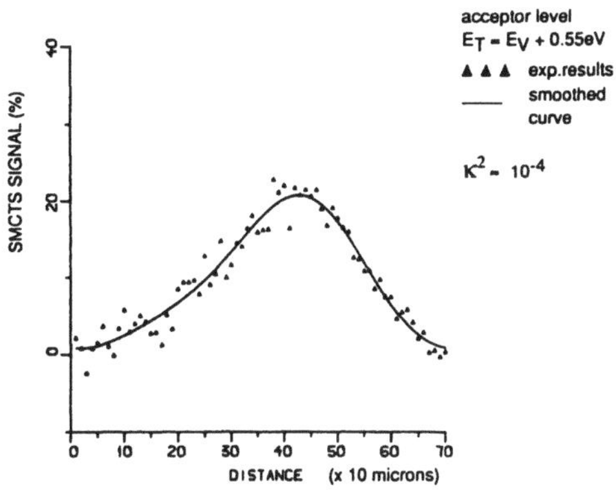

Fig. 4 : SMCTS response of the gold acceptor through the grain boundary

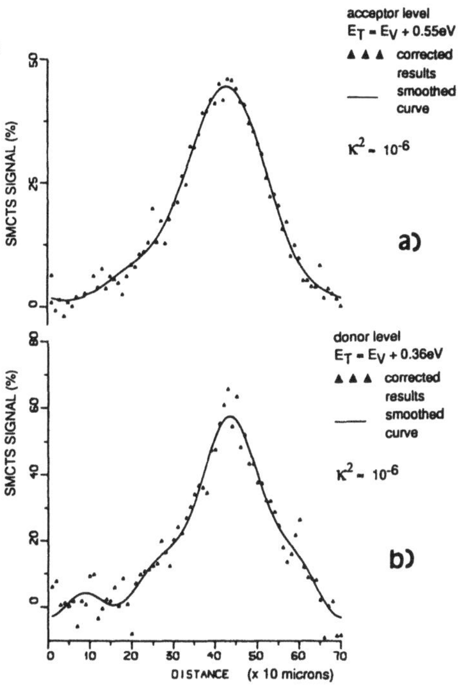

Fig. 5 : (a) distribution of the acceptor level
(b) distribution of the donor level

Fisher's model, a strong increase of the gold concentration near the grain boundary is observed. Light beam induced current (LBIC) measurements on the same line revealed no recombination activity of the grain boundary, even though the acceptor level is known as a recombination center. This result illustrates the difference between the sensitivities of LBIC measurements and the SMCTS technique. It also implies that, even if a grain boundary seems to affect only slightly the minority carrier lifetime, it can play an important role as a short circuit for diffusion of impurities. Figure 5b shows the distribution of the donor level along the same line as in figure 5a. Its increase at the grain boundary is stronger and more confined than that the acceptor level. The disparity between the spatial distribution of the gold acceptor and donor levels can be explained if we attribute different gold related defects to the two deep levels. Actually, even though gold is one of the most studied impurities in silicon, doubts remain about the relation between the acceptor and donor levels. In particular, Lang et al. [8] have proposed the acceptor level to be a complex including vacancies associated with oxygen and dopant atoms, in order to explain its behaviour in various kinds of silicon. If such a model is correct, the gold acceptor level must be influenced in a different manner than the gold donor level by a non uniform distribution of oxygen and dopant atoms around the grain boundary. This is coherent with our SMCTS observations. Precipitation of oxygen and dopant atoms at the grain boundary has been reported in the literature [9]. However, supplementary analysis should be done to confirm these observations.

5. CONCLUSION

To conclude this work, we can state that both the theoretical model and the new signal processing were necessary to make SMCTS analysis under optimum conditions and to relate the profiles obtained to the distribution of the analyzed deep levels. The study of the diffusion of gold in the silicon bicrystal has shown this new technique to be a sensitive tool for local analysis which has yielded new results on some fundamental aspects of defects in semiconductors.

REFERENCES

1. D.V. Lang : J. Appl. Phys. 45, 3014 (1974).
2. H. Lefèvre and M. Schulz : Appl. Phys. 12, 45 (1977).
3. P.M. Petroff and D.V. Lang : Appl. Phys. Lett. 31, 60 (1977).
4. T. Heiser, A. Mesli, E. Courcelle and P. Siffert : to be published in J. Appl. Phys., October (1988).
5. G.L.. Miller, J.V. Ramirez and D.A.H. Robinson : J. Appl. Phys. 46, 2638 (1975).
6. B. Valeur : Chem. Phys. 30, 85 (1978).
7. G. Martin and B. Peraillon : J. de Phys. C4, 34 (1975).
8. D.V. Lang, H.G. Grimmeiss, E. Heijer and M. Jaros : Phys. Rev. B22, 3917 (1980).
9. L.L. Kazmerski and P.E. Russel : J. de Phys. C1, 171 (1982).

Optical Characterization of Silicon Materials and Structures

G. Harbeke

Paul Scherrer Institut, c/o Laboratories RCA Ltd.,
Badenerstr. 569, CH-8048 Zürich, Switzerland

1. Motivation

With the increasing complexity of present-day silicon technology there is a growing demand for non-destructive probing techniques for quality assessment and quality control. There are other powerful techniques such as transmission electron microscopy (TEM), secondary ion mass spectroscopy (SIMS) and Rutherford back-scattering (RBS) (to name just a few) but they are of destructive nature and require large amounts of time for sample preparation and measurement.

Many optical techniques have proven useful as rapid, contactless, non-destructive and clean methods for characterization of semiconductor materials, surfaces, films and composite structures. For these reasons, they lend themselves to applicatons in a development or even production environment for quality assurance and testing.

This chapter is devoted to the description of such optical techniques based on transmission and reflection as well as on elastic and inelastic light scattering. After a discussion of the basics of the optical properties of solids in section 2 the following three sections deal with the results obtained in a number of applications of the characterization techniques to bulk silicon, polycrystalline silicon and silicon on insulators.

2. Optical Properties of Solids

The linear response of a nonmagnetic medium to transverse electromagnetic radiation is completely described by the dielectric function $\epsilon(\omega)$ with real and imaginary part $\epsilon_1(\omega)$ and $\epsilon_2(\omega)$, respectively. These two quantities are related to the complex refractive index $N = n + ik$ by

$$\epsilon_1 = n^2 - k^2 \qquad (1)$$

$$\epsilon_2 = 2\,nk \qquad (2)$$

with the refractive index n and the absorption index k. The absorption

Springer Series in Materials Science, Vol. 13
Semiconductor Silicon Editors: G. C. Harbeke and M. J. Schulz

coefficient $K = 4\pi k/\lambda$ describes the attenuation of a plane electromagnetic wave and its penetration depth $d = K^{-1}$. Causality imposes a relationship between the real and imaginary part of either $\epsilon(\omega)$ or $N(\omega)$ in the form of the so called Kramers-Kronig relations [1].

The most straightforward optical techniques to determine the optical constants n and k are based on the measurement of reflection and transmission. One path requires measuring the reflectivity over as large a spectral region as possible (ideally from $\omega = 0$ to $\omega = \infty$) and employing the Kramers-Kronig relations in the form connecting the reflection intensity R and phase angle Θ [1]. The knowledge of these two quantities is equivalent to that of n, k or ϵ_1, ϵ_2. In reality, there will, however, always be some spectral regions missing in the experiment which have to be substituted by extrapolation or other good guesswork.

It is therefore preferable to determine two optical entities at each wavelength. That can be done by measuring reflection from and transmission through a plane-parallel plate of thickness t and calculating n and k. This method is no longer applicable when k is so large that extremely thin samples have to be prepared for taking transmission data which may no more be representative of bulk properties. In this situation one is well advised to take resort to the technique of spectroscopic ellipsometry which has been developed over the last decade [2]. In this technique, incident linearly polarized light is elliptically polarized upon reflection. Phase and amplitude of the reflected light are measured by a rotating-analyzer system and converted to the complex dielectric function.

In addition to the linear response of a medium there are nonlinear effects to higher order in the electric field associated with electromagnetic radiation. In light scattering experiments one field is associated with the exciting light wave of frequency ω_o and a second one with the scattered light wave of frequency ω_s. In inelastic light scattering events ω_o and ω_s differ by the frequency ω of the wave excited ($\omega_s = \omega_o - \omega$) or annihilated ($\omega_s = \omega_o + \omega$) in the crystal. The part of scattered light of lower frequency than that of the incident light is called the Stokes component while the part of the scattered light with higher frequency is called the anti-Stokes component. Since momentum has to be conserved in the scattering process, the phonons scattered in a first-order process must have a momentum comparable to that of the photons, i.e. they are located very close to the centre of the Brillouin zone. In second-order processes, pairs of phonons of opposite and almost equal absolute-value wave vector adding up to the wave vector difference of incoming and outgoing photons fulfill the momentum condition. Phonons throughout the whole Brillouin zone can thus participate provided the symmetry requirements are properly met. This leads to fairly broad second-order bands which are dominated by contributions from critical points where the dispersion curves are horizontal and the density of phonon states is highest.

Traditionally the distinction is made between Raman scattering from optical phonons and Brillouin scattering from acoustic phonons. Only lattice vibrations having certain types of symmetry are Raman active. In centrosymmetric crystal structures such as the diamond structure of silicon Raman activity and infrared absorption are mutually exclusive for a given lattice vibration for symmetry reasons. We shall see that Raman scattering from optical phonons is very sensitive to deviations from crystallinity as well as to strain in the scattering volume.

Elastic light scattering, i.e. $\omega_s = \omega_0$, results from local and temporal variations of the dielectric function. Such variations can be of intrinsic nature such as temperature or entropy fluctuations or of extrinsic origin caused by imperfections on a microscopic scale such as strain, stacking faults, dislocations, grain boundaries as well as surface irregularities.

3. Bulk Silicon

The dominant intrinsic contribution to the dielectric function of semiconductors arises from electronic interband transitions between occupied valence band and empty conduction band states. Conservation of momentum imposes direct transitions without change of wave vector for a first-order process. Figure 1 shows the dielectric function of silicon at room temperature in the spectral region of interband transitions as obtained by spectroscopic ellipsometry [3]. The data have been obtained on silicon after an optimized wet-chemical surface processing in order to minimize the thickness of surface transition regions between the bulk silicon and vacuum (or nitrogen gas). The structure in $\epsilon(\omega)$ is caused by critical

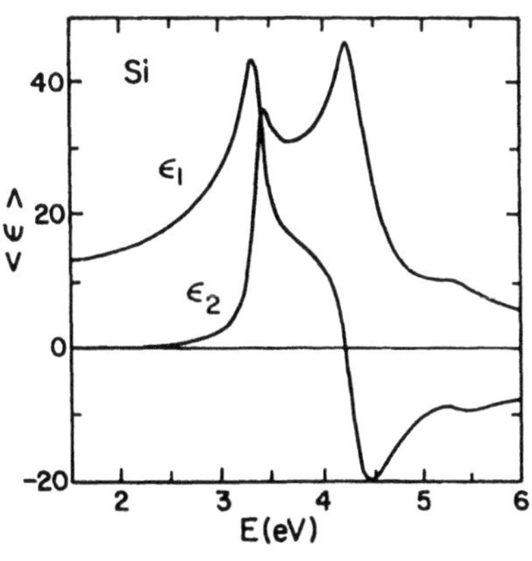

Fig. 1. Dielectric function of crystalline silicon versus energy (from Ref. 3)

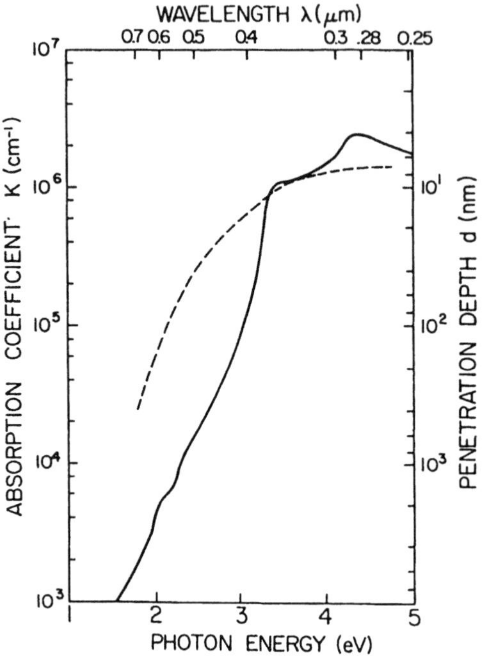

Fig. 2. Absorption coefficient of crystalline silicon (solid line) and LPCVD amorphous silicon (dashed line) versus photon energy

points (van Hove singularities) where high values of the joint (valence and conduction band) density of states are reached.

As described in section 2, the absorption coefficient $K(\omega)$ can be derived from $\epsilon(\omega)$. The solid line in Fig. 2 for single crystal silicon shows that $K(\omega)$ increases beyond 10^6 cm^{-1} corresponding to a penetration depth smaller than 10 nm. In this spectral region the optical properties of silicon are extremely sensitive to the state of the surface in terms of morphology, structure and overlayers as well as to changes induced by processing. Moreover, the spectral dependence enables the researcher to probe the silicon sample to increasingly greater depths by decreasing the photon energy of the incident radiation from 4.5 eV on. The dashed line in Fig. 2 gives for comparison the absorption coefficient of amorphous silicon deposited from silane in a glow discharge. The absence of lattice periodicity causes the pronounced structure to disappear and induces a strongly enhanced absorption at energies below 3 eV. Obviously, the absorption and, correspondingly, the reflection of silicon will be sensitive to the state of crystallinity.

Once the dielectric function of crystalline silicon had been obtained on surfaces free of overlayers [3], it became possible to judge the corrections which have to be made for the presence of surface layers on well polished wafers exposed to air, a realistic situation for the silicon technology. It could be shown that it is not meaningful to assume the overlayer known as "native oxide" to consist of stoichiometric SiO_2 but that its first 0.6 to 1.5 nm are rather deficient in oxygen [3].

One of the basic ingredients of the silicon technology is the doping process either by diffusion, in-situ doping in chemical vapour deposition or ion implantation. In doped surface layers the dopant concentration can either be uniform or in the form of a depth profile. The main effect of high doping concentrations on the dielectric function in the interband region is to broaden the structures due to critical points shown in Fig. 1 and to shift them to lower energies, noticeably at concentrations above 10^{19} cm^{-3}. Both, red shift and broadening parameters have been measured by spectroscopic ellipsometry [4] and found to reach more than 100 meV for the highest concentration of $3x10^{21}$ cm^{-3} achieved by ion implantation of doses of several 10^{16} cm^{-2} and subsequent laser annealing. The dependence of these two parameters on the doping concentration was independent of the polarity of the doping and has been interpreted by considering the carrier-impurity interaction in first- and second-order perturbation theory [4].

Optical analysis is also useful to monitor the disorder induced by medium- and high-dose implants which are increasingly used in device processing. The damage is a function of many parameters such as energy and mass of the implanted ion, dose rate and sample temperature. The characterization of the as-implanted wafer is relevant for the subsequent processing since the residual disorder will depend on the initial one. Figure 3 shows the reflectance spectra of silicon implanted with various doses of arsenic [5]. After implant, at a dose of 10^{14} cm^{-2} the E_1 and E_2

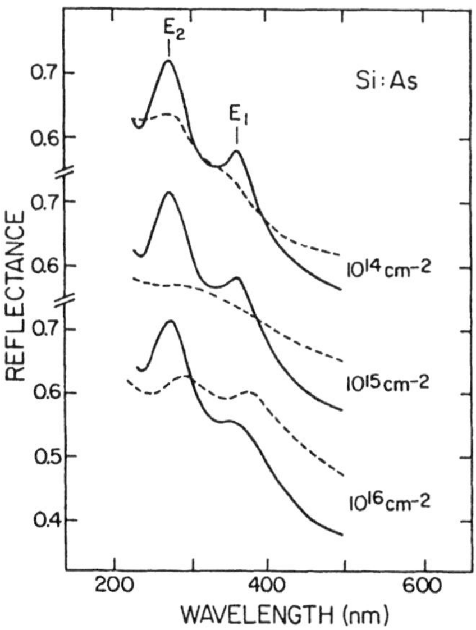

Fig. 3. Reflectance of arsenic implanted silicon before (dotted line) and after (solid line) laser annealing versus wavelength (from Ref. 5)

structures (corresponding to the two peaks in ϵ_2 in Fig. 1) are strongly attenuated, at 10^{15} cm^{-2} the reflectance is featureless but at a dose of 10^{16} cm^{-2} E_1 and E_2 reappear albeit red-shifted. The reappearance of the structure is caused by self-annealing since the wafer temperature increases with inreasing dose to about 500°C for the ion beam current corresponding to 10^{16} cm^{-2}. The solid lines in Fig. 3 give the reflection after laser annealing of the implanted samples where the E_1, E_2 structure is fully recovered except for the broadening of the E_1 peak for the highest concentration as discussed in the preceding paragraph. The optical properties after implant will also depend on the mass of the implanted ion. For a given implant energy, lighter atoms such as phosphorus or boron will damage the silicon in greater depths than arsenic such that only light of photon energy below that of the E_1 peak "sees" the damage [6].

The infrared part of the spectrum contains information on the effective mass of carriers introduced by doping - also for high-dose implantation [6] - provided their concentration is determined by other techniques [6]. IR measurements at low temperatures also reveal the nature and concentration of donors and acceptors by virtue of optical transitions from their shallow - i.e. close to either band edge - energy levels into the continuum [7].

The intrinsic absorption due to lattice vibrations in silicon is also in the infrared spectral region but it does not contain a strong Reststrahlen band since the optical mode does not carry a dipole moment and is therefore not infra-red active. The absorption is thus resulting from weaker two-phonon and even weaker multi-phonon sum - and difference processes reflecting the density of states of phonons in the dispersion curves [7]. Localised vibrations involve silicon and dopant or impurity atoms, each delivering its specific signature in the spectra. The quantitative determination of the carbon and oxygen concentration is routinely done on this basis after initial calibration with, e.g. neutron activation analysis.

The oxygen atoms occupy interstitial positions in the Si host lattice. Si-Si bonds are broken and Si-O-Si clusters with bond angles of 162° are formed without any significant distortion of the host lattice. The Si-O-Si "defect molecule" has three infrared active normal modes of vibration, a symmetric stretching mode at 515 cm^{-1}, a symmetric bending mode at 29 cm^{-1} and an asymmetric stretching mode at 1106 cm^{-1}. The absorption associated with the latter one is the strongest and has been used for many years to characterize the concentration of interstitial oxygen in silicon [8]. The absorption band has a width of about 30 cm^{-1} and splits into three features at low temperatures with strongly temperature dependent relative ratios of oscillator strength. It has been shown [8] that infrared spectroscopy can also monitor the amount of oxygen in silica precipitates produced in special high-temperature annealing cycles for the purpose of internal gettering.

Raman scattering has been used as a rapid non-destructive method of analysis for processed silicon. The work has focused on the position and width of the first-order Raman active optical phonon band in the vicinity of 520 cm^{-1} [9]. Brillouin scattering from opaque materials has become a useful technique by the development of the tandem multipass interferometer by Sandercock [10]. Figure 4 shows a Brillouin backscattering spectrum of silicon, both Stokes and anti-Stokes part. In the order of decreasing Brillouin shift the longitudinal (L) and transverse (T) bulk phonons, Rayleigh (R) surface phonons and a broad central peak can be seen, the latter one probably due to two-phonon difference processes [10]. The technique of Brillouin scattering has also been applied to study the damage and amorphization of silicon induced by irradiation with high-energy ions. One of its advantages over conventional ultrasonic methods is that the elastic properties of thin films or near-surface layers of the thickness of 1 μm can be studied [11].

The use of elastic light scattering for non-destructive assessment and quality control of silicon wafers and films is widespread. A number of commercially available instruments serve for rapid testing for dust particles and surface or sub-surface defects. More recently, it has been established that the technique can be used to determine "bulk" structural perfection [12]. The intensity signal across the wafer shows in addition to peaks originating from individual defects a background level that is

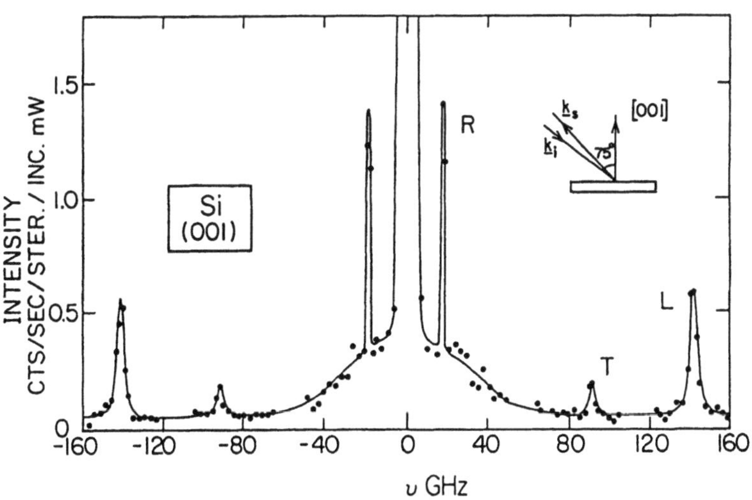

Fig. 4. Brillouin backscattering spectrum of crystalline silicon showing bulk (T, L) and Rayleigh surface phonons (R) together with a central peak (from Ref. 10)

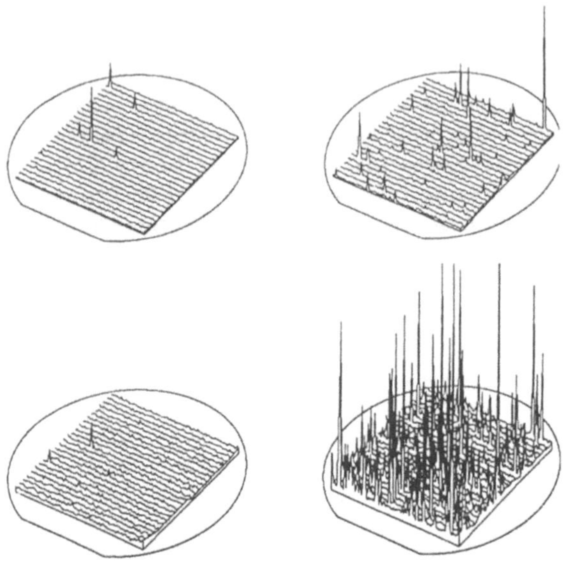

Fig. 5. 3D light scattering topography plot obtained on silicon wafers from two different manufacturers A (top) and B (bottom) before (left) and after (right) denuding (from Ref. 13)

directly related to the overall wafer quality. It is produced by light scattering due to local variations of the refractive index. These experiments have led to the development of the light scattering topography (LST) [13] technique in which a wafer is scanned with a spot size of 2 mm x 40 μm using 100 mW incident laser power at 514.5 nm wavelength. A photomultiplier with a low dark-count rate is used, the equipment is computer controlled and the data are stored on disk. Figure 5 shows LST plots of (100) oriented CZ wafers before and after a denuding cycle, A and B denote wafers from different manufacturers. Before denuding one can clearly see the difference in absolute level and the difference in local fluctuations. After denuding, the increase in background level and fluctuations is noticeable as well as an increase in the number of individual defects, though to a different degree in the two wafers.

4. Polycrystalline Silicon

The optical properties of polycrystalline silicon have been described to some extent in the Proceedings of a previous Erice Course [14]. In particular, the advantages in terms of surface roughness and structural perfection from the deposition in the amorphous phase and subsequent thermal recrystallization to polycrystalline silicon had been demonstrated. One of the main benefits of obtaining extremely smooth amorphous films has been exploited in improving the insulating properties of SiO_2 films thermally grown on such silicon films. The reduced texture of the silicon surface leads to a reduced roughness at the interface between polycrystalline silicon and its oxide and thus to less electric-field enhancement at the asperities than for strongly textured surfaces of growth-crystallized films. This growth-roughness relation gives also the knowledge base for those device situations where a well-defined surface texture has to be reliably manufactured. This is the case in certain floating-gate nonvolatile memory structures where the interlevel oxide is required to carry high tunneling currents and be able to withstand the passage of large charge densities before failure. It has been shown that these requirements can be met by controlled texturing of the extremely smooth surfaces obtained by the amorphous deposition. The texturing procedure entails growing a predetermined thickness of "texturing oxide" and then its removal prior to formation of the actual tunneling polyoxide of fixed thickness. The electrical properties of the sacrificial interlevel oxide have been correlated to the optically determined interface roughness using the surface plasmon technique [14] after successive removal of the upper poly layer and the oxide. Figure 6 shows the vertical surface roughness of both interfaces as a function of thickness of the polyoxide in the thickness range from 16.7 to 165 nm grown on untextured polysilicon deposited in the amorphous phase [15]. The roughness increases with oxide thickness from values well below 1 nm to 1.8 nm for the lower interface. The increase is slower for the upper interface in agreement with the experimental result that polyoxides have a higher destructive breakdown field and lower Fowler-Nordheim tunneling currents when the electron injection occurs across the upper interface. It means that the

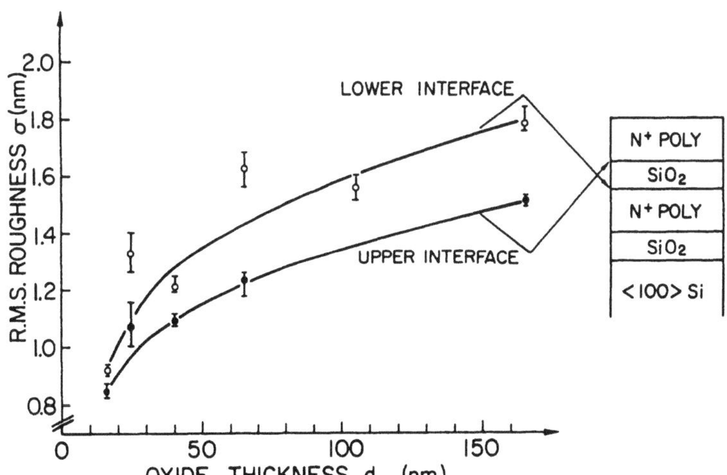

Fig. 6. Measured rms roughness of both oxide/polysilicon interfaces for thermally grown polyoxides on polysilicon deposited in the amorphous phase versus oxide thickness (from Ref. 15)

growth-induced roughening at the lower interface is not fully transmitted through the entire layer because the growth kinetics has again a smoothening effect.

Figure 7 shows the effective barrier height for Fowler-Nordheim tunneling of the set of oxide samples of Fig. 6 as a function of roughness of the injecting interface. It can be seen that the relation is essentially independent of the voltage polarity, i.e. the effective barrier height appears to be uniquely determined by the interface roughness. The device performance of certain nonvolatile memory structures can thus be influenced by control of the interface morphology. It should be noted that these results hold for phosphorus doped layers in the concentration range of 10^{20} to 10^{21} cm^{-3} due to the requirements of the device structure. For undoped polysilicon deposited in the amorphous phase the growth-induced roughening is about half as strong as for the highly phosphorus doped ones [16].

Optical characterization of SiO_x (0.1< x < 0.92), also known as semi-insulating polycrystalline silicon (SIPOS) and used as passivation layer in high-voltage power devices or as wide band-gap emitter material in bipolar devices, has been carried out along the lines as before for polycrystalline silicon [17]. The binary polycrystalline system allows a great variability of the materials parameter such as grain size and electrical conductivity. It was shown that the advantages of the amorphous deposition in terms of low surface roughness carry over from x = o to the entire range of x-values under investigation.

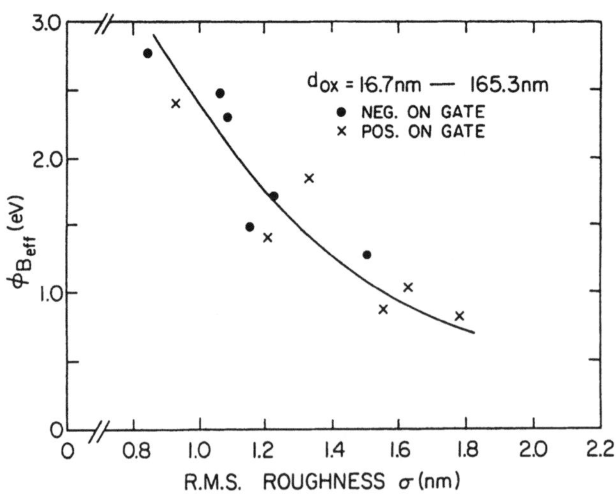

Fig. 7. Effective barrier height for Fowler-Nordheim tunneling under both positive and negative gate-bias polarities for polyoxides on polysilicon deposited in the amorphous phase versus rms roughness (from Ref. 15)

5. SIMOX

Among various SOI (silicon on insulator) approaches such as ELO (epitaxial lateral overgrowth), porous silicon, recrystallisation by laser or electron beam annealing, and the SIMOX (separation by implantation of oxygen) technology the latter one is currently attracting considerable scientific interest as a suitable substrate for future generations of small geometry integrated circuits. Recently, test circuits fabricated on SIMOX substrates have confirmed and in some cases exceeded the expectations of the technology [18]. As the process, however, moves from the research into the production stage it is becoming increasingly apparent that reliable methods of quality assessment need to be established to support the development of circuit processing for the production of commercial circuits.

The fabrication of SIMOX structures involves two steps, namely implantation of high doses of energetic oxygen ions (typically $\geq 1.4 \times 10^{18}$ cm^{-2} at 150-200 keV) at elevated temperatures (typically 450-700°C) and subsequent annealing at very high temperature, typically above 1200°C, sometimes even at 1405°C, i.e. a few degrees below the melting point.

The oxygen must be implanted at energies of at least 100-200 keV so that the SiO_2 layers will be buried under 100-300 nm of crystalline Si. Moreover the implantation must be carried out in the 450-700°C range, to preserve the crystallinity of the Si surface layer. Nuclei of SiO_2 are formed at the lowest doses but the implantation profile has the overall Gaussian shape. At higher doses, the nuclei coalesce and a layered structure forms at the SiO_2 stoichiometric level. There is a considerable chemical driving force for the formation of SiO_2 causing diffusion of oxygen within the SiO_2 layer to the wings of the distribution where more SiO_2 is formed. The critical dose, for example, for forming a continuous SiO_2 layer at 200 keV is $1.4 \times 10^{18}/cm^2$. The wafers must be annealed following the implantation to anneal damage in the overlying Si and clean up the Si/SiO_2 interfaces.

Since the quality of the final product depends sensitively on the implant conditions it is important to monitor the wafer already in the as-implanted state. Spectroscopic ellipsometry [19], ultraviolet reflectance, infrared transmission and Raman scattering have been used to investigate the structure and the strain in both the top silicon and the buried oxide layer. Figure 8 shows the reflectance of a SIMOX substrate from 200 nm to 500 nm together with the reference spectrum for bulk silicon. Above 400 nm, where the penetration depth of the light becomes comparable to the thickness of the silicon overlayer of 200 to 300 nm, interference with light reflected from the two interfaces of the buried oxide dominates the spectra. These interference patterns (over a larger wavelength scale) can be evaluated in terms of the thicknesses of both the top silicon layer and the buried oxide [20]. Meaningful near-surface quality information is contained in the spectral region below 400 nm. The observed deviations from the bulk silicon reflectance are caused by

(i) surface contamination induced in the implantation process, e.g. SiO_x or silicon carbide. It is reasonable to assume this contribution (denoted C in Figure 8) to be wavelength independent.

(ii) Rayleigh scattering caused by surface roughness due to sputtering increases as $B\omega^4$ and therefore reduces the reflection mostly at short wavelengths.

(iii) amorphization reduces the intensity by Y and X at the reflection maxima at 367 nm and 270 nm, respectively, corresponding to the peaks in ϵ_2 in Figure 1.

The amorphization or damage (in arbitrary units, related to reflectance) can thus be determined after subtraction of the other two contributions. It was shown for a set of samples to increase with increasing dose and decreasing temperature and implant voltage, the latter dependence being caused by the shorter distance between the damage region and the surface. Also the minimum channeling yield, as determined by Rutherford back-scattering correlates well with the amorphization as measured by UV reflectance [21]. It should be noted that optimized implant conditions minimize the contributions due to surface contamination and Rayleigh scattering such that the degree of damage can be directly deduced from the reflectance curves.

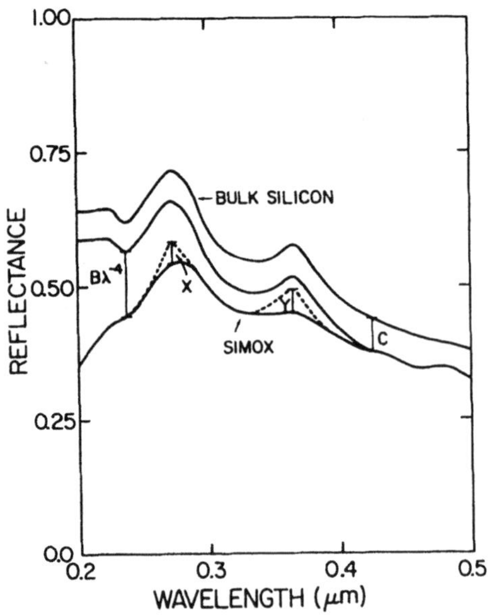

Fig. 8. Reflectance spectra of bulk silicon and a SIMOX sample in the UV range (see text). (from Ref. 21)

The antisymmetric stretching mode of the Si-O-Si unit, which causes the 1106 cm^{-1} absorption band for interstitial oxygen in crystalline silicon, gives also rise to the strongest band in amorphous SiO_2. In Figure 9 (solid line) the spectrum is shown for SiO_2 grown by thermal oxidation of bulk silicon at temperatures well above the temperature region of viscous flow, i.e. about 1000°C. The position of the strong band is only slightly shifted to 1088 cm^{-1} but it is broadened to a half width of 75 cm^{-1} due to the distribution of bond angles and distances in the amorphous network.

It has been shown that a range of lower peak frequencies can be due to compressive stress as it occurs in low-temperature thermal oxides or to the formation of sub-oxides, i.e. SiO_x with $x < 2$. In the latter case, as the oxide moves off stoichiometry, the silicon atoms in an Si-O-Si configuration have a higher probability of having one or more silicon atom neighbours and this change in chemical environment shifts the stretching frequency. The lowering of the frequency is then accompanied by an increase in linewidth due to the statistical distribution of different bonding arrangements at each silicon site whereas stoichiometric low-temperature SiO_2 under compressive stress can suffer frequency shifts of up to 15 cm^{-1} essentially without line broadening [22].

It is believed that both of these effects influence the peak position, lineshape and half width of this band in the buried oxide in as-implanted SIMOX structures (see Figure 9). The buried oxide is under compressive

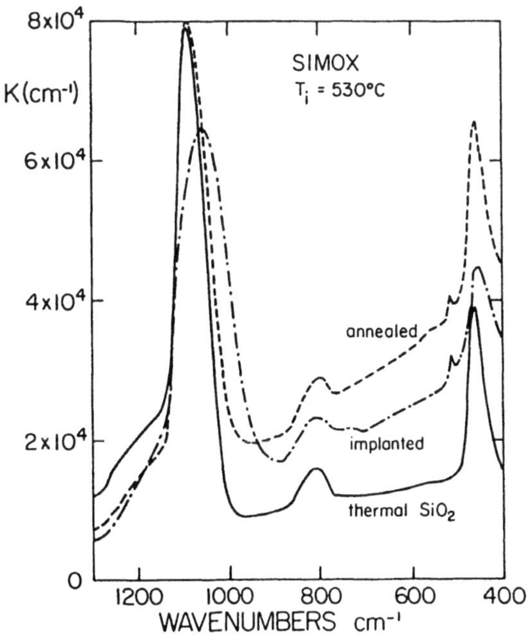

Fig. 9. Absorption coefficient in the IR range of thermally grown silicon oxide (solid line) and a SIMOX sample implanted to a dose of 1.4×10^{18} cm^{-2} before (dashed-dotted) and after (dashed) annealing. (from Ref. 23)

stress since the volume occupied by the SiO_2 is about 2.2 times greater than that occupied by the silicon. Some of this "accommodation volume" is provided by vacancies and by emission of silicon self-interstitials, the latter ones migrating towards the surface, their nearest available sink. In most samples the compressive stress is not completely relieved as it is evidenced by compensating tensile stress in the silicon overlayer seen by Raman scattering (see below). Also the second effect influencing the lineshape, i.e. sub-stoichiometry is certainly operative in the wings of the distribution, even if for supercritical doses the core region consists of SiO_2. As a result, peak position and lineshape depend - for a given dose - on implant temperature [23] as well as - for a given temperature - on implant dose as shown in Figure 10 [24].

The peak frequency is found to increase almost linearly with increasing dose. The half width is initially found to decrease with increasing dose, however for doses above 1.6×10^{18} cm^{-2} it appears to saturate at a value of 125 cm^{-1}. For the lower doses $< 1.4 \times 10^{18}$ O^+ cm^{-2} the oxide layer is known to be non-continuous being interwoven with fine lamellae of silicon, the thicknesses of which decrease with dose. Consequently as the dose increases from 0.25×10^{18} O^+ cm^{-2} the degree of non stoichiometry within

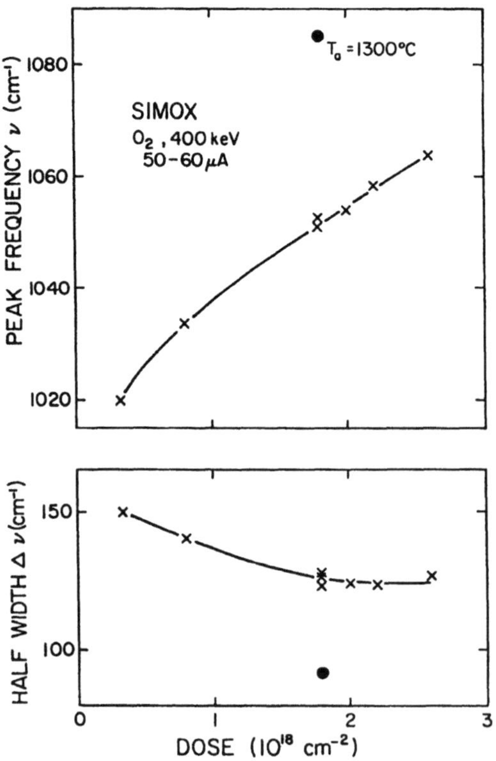

Fig. 10. Peak frequency (ν) and half width (Δν) of the antisymmetric Si-O-Si vibration versus implanted dose. x = 'as implanted', o = annealed 1300°C 5 hours (from Ref. 24)

the oxide layer decreases, as the volume concentration at the peak of the implanted distribution approaches that in the stoichiometric compound. For implanted doses > 1.5 x 10^{18} O^+ cm^{-2}, when a continuous oxide layer is produced upon implantation, the increase in the peak frequency and the saturation in the value of the half width can be attributed, in part, to a constant volume of substoichiometric oxide located in the wings of the distribution. One of the specimens is also shown, after annealing at 1300°C for 5 hrs. For this specimen the peak frequency ν = 1085 cm^{-1} is fairly close to that for thermally grown SiO_2 (1087 cm^{-1}) however, the half width $\Delta\nu$ = 91 cm^{-1}, although strongly reduced form that for the "as implanted" specimen, is still considerably higher than that for a thermal oxide ($\Delta\nu$ = 75 cm^{-1}). It means that the stress in the buried layer is relaxed by viscous flow but that there are still small regions with substoichiometric oxide.

As already mentioned, Raman scattering from the silicon overlayer yields the phonon frequency and linewidth. Both these quantities give information whether the SIMOX material properties are close to bulk silicon or deviate from it, i.e. how much lattice strain and distortion is introduced into the silicon overlayer by implantation and to what extent this has been annealed out in the annealing process. [23]. It is obviously advantageous to use a laser frequency with a penetration depth smaller than the thickness of the silicon overlayer. This condition is essentially fulfilled by the near ultraviolet 413.1 nm line of a Krypton ion laser which was used to obtain the results shown in Figure 11 [25].

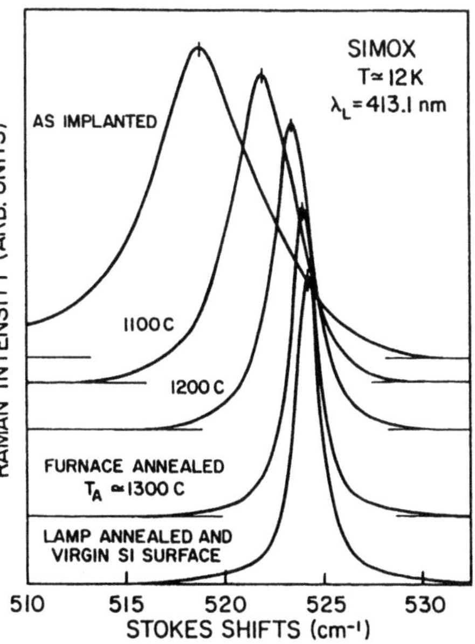

Fig. 11. First-order Raman spectra of the top Si layers of SIMOX structures for different annealing treatments. The temperature in the lamp-annealing case was 1405°C. For comparison, data are included for an as-implanted (not annealed) and virgin Si surface (not implanted, not annealed) (from Ref. 25)

The implanted and not-annealed sample shows a Raman band with a pronounced shift to lower frequencies (red shift) and a clear asymmetrical broadening compared to the reference Si surface. The low-frequency wing rises above the background level faster than the high-frequency one. The spectra of the furnace-annealed samples are also red shifted and broadened with respect to the virgin case. However, the red shifts and broadenings decrease with increasing annealing temperature T_A. These broadenings are symmetrical, as opposed to the as-implanted case. The symmetrical Raman line shapes confirm the crystalline nature of the top Si layer. It is argued [25] that the presence of microcrystalline grains in the network of the as-implanted layer causes the asymmetric broadening due to non-conservation of wave vector. The position and linewidth of the Raman peak corresponding to the lamp-annealed sample are the same as those of the virgin Si sample. The frequency shift is attributed to the presence of tensile stress in the crystalline layer induced by oxide precipitates [25, 26].

The technique of light scattering topography (LST) [13, 24] is an excellent tool to map the implant damage of as-implanted wafers and so to distinguish non-uniformities as well as to monitor the removal of implant damage in various annealing treatments. Figure 12 is a three dimensional LST plot taken with 325 nm radiation of a He-Cd laser which shows the implant damage and/or strain, in the region of a wafer implanted with an oxygen dose of 1.8 x $10^{18} O^+ cm^{-2}$ at 200 keV, prior to annealing. Due to the beam induced thermal gradient across the wafer, the centre of the implanted region, where the temperature was highest, shows a lower signal than at the edge. This is a consequence of increased *in situ* annealing in the centre of the wafer.

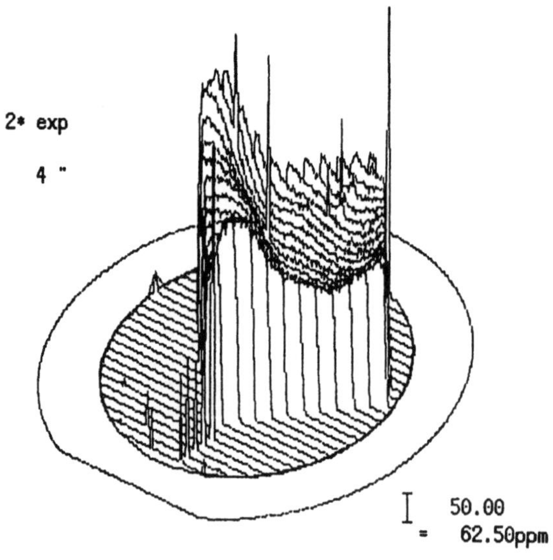

Fig. 12. 3D light scattering topography for a wafer implanted with an oxygen dose of 1.8×10^{18} $O^+ cm^{-2}$ at 200 keV, prior to annealing. (from Ref. 24)

From this type of measurement it is clear that non uniformities across a wafer, whether they be due to temperature, channelling or dose variations, can be monitored by the use of LST. As this technique is quick, clean, non destructive and non contacting it clearly has applications in a production environment for the quality assurance and testing of SIMOX substrates prior to device fabrication.

References

1. See, e.g., D.L. Greenaway and G. Harbeke: In Optical Properties and Band Structure of Semiconductors, Pergamon Press, Oxford, 1986
2. D.E. Aspnes and A.A. Studna: Appl. Opt., 14, 220 (1975); Rev. Sci. Instrum. 49, 291 (1978)
3. D.E. Aspnes and A.A. Studna: Phys. Rev. B 27, 985 (1983)
4. L. Vina and M. Cardona, Phys. Rev. B 29, 6739 (1984)
5. A. Borghesi, Chen Chen-Jia, G. Guizetti, L. Nosenzo and A. Stella, J. Appl. Phys. 58, 2773 (1985)
6. S.U. Campisano, E. Rimini, A. Borghesi, G. Guizetti L. Nosenzo and A. Stella, Physica Scripta T 19, 544 (1987)
7. In: Landolt-Börnstein: Vol. III/17a, Semiconductors, ed. by O. Madelung, Springer-Verlag, Berlin 1982
8. B. Pajot, H.J. Stein, B. Cales, and C. Naud: J. Electrochem. Soc. 132, 3034 (1985)
9. see, e.g. J.B. Hopkins and L.A. Farrow: J. Appl. Phys. 59, 1103 (1986)
10. J.R. Sandercock: In Light Scattering in Solids III, ed. by M. Cardona and G. Güntherodt, Springer, Berlin 1982
11. M. Grimsditch: Physica Scripta, in print
12. E.F. Steigmeier and H. Auderset: J. Electrochem. Soc. 131, 1693 (1984)
13. E.F. Steigmeier and H. Auderset: RCA Review 46, 3 (1985)
14. G. Harbeke: In Polycrystalline Semiconductors - Physical Properties and Applications, ed. by G. Harbeke (Springer, Berlin 1985) p. 156
15. L. Faraone and G. Harbeke: J. Electrochem. Soc 133, 1410 (1986)
16. G. Harbeke, A.E. Widmer and L. Faraone: In Proc. 18th Intern. Conf. "The Physics of Semiconductors", ed. by O. Engström (World Scientific, Singapore, 1987) p. 259
17. A.E. Widmer, G. Harbeke, L. Krausbauer and E.F. Steigmeier: J. Electrochem. Soc. 133, 1880 (1986)
18. J.R. Davis, F. Hopper, P.L.F. Hemment and K.J. Reeson, IEEE Tran. Electron Dev. 34, 1713 (1987)
19. F. Ferrieu, D.P. Vu, C. D'Anterroches, J.C. Oberlin, S. Maillet and J.J. Grob: J. Appl. Phys. 62, 3458 (1987)
20. S.N. Bunker, P. Sioshansi, M.M. Sanfacon and S.P. Tobin: Appl. Phys. Lett. 50, 1900 (1987)
21. G. Harbeke and L. Jastrzebski: submitted to J. Electrochem. Soc.
22. G. Lucovsky, M.J. Manitini, J.K. Srivastava and E.A. Irene: J. Vac. Sci. Technol. B 4, 530 (1987)

23. G. Harbeke, E.F. Steigmeier, P. Hemment, K.J. Reeson and L. Jastrzebski: Semicond. Sci. Technol. 2, 687 (1987)
24. E.F. Steigmeier, G. Harbeke, K.J. Reeson, P.L.F. Hemment and A.K. Robinson In: Proc. of Intern. Conf. on Ion Implant Technology, Kyoto, 1988, in print
25. D.J. Olego, H. Baumgart and G.K. Celler, Appl. Phys. Lett. 52, 483 (1988)
26. J. Takahashi and T. Makino: J. Appl. Phys. 63, 87 (1988)

Oxygen-Free Silicon; How Does IR See It?

B. Pivac

R. Boskovic Institute, P.O. Box 1016, YU-41000 Zagreb, Yugoslavia

1. Introduction

Polycrystalline silicon with large grains and columnar structure is still one of the prospective materials for the production of cheap solar cells. A wide range of various techniques is currently employed for solar grade polycrystalline silicon wafer production, among which the ribbon pulling and casting processes seem to be promising [1,2,3].

Due to the different conditions in the material preparation of the mentioned techniques, various concentrations of different impurities will be introduced into the samples in each specific case. One would therefore expect that the photoelectric behaviour of each material is affected in a different way.

The concentration of light elements, like oxygen and carbon, is typical for each process and differs significantly from one to another.

The importance of oxygen and carbon and the influence of thermal treatments on their behaviour, and various complex formations were studied in monocrystalline silicon for a longer period, but recently the knowledge of these processes is somewhat more systematised [4,5].

In polycrystalline silicon materials, the behaviour of these impurities is even more complicated due to a large number of structural defects and especially to grain boundaries. Therefore, to understand and/or predict the behaviour of oxygen upon thermal treatment, it is necessary to determine its concentration present in untreated material.

For the oxygen determination in silicon, infrared (IR) spectroscopy is established as a quick, reliable and nondestructive method unlike some nuclear ones. The only disadvantage of this method is that it can quantitatively detect oxygen placed only in the interstitial position, forming an Si_2O molecule. For this position the vibrational spectra are well known and the calibration constant is determined in comparison with some nuclear methods. Oxygen present in other positions is either IR invisible or gives rise to bands that are not undoubtfully attributed to specific and well defined forms of precipitates.

In this work we want to show that knowing the thermal history of wafers is important for the oxygen concentration determination.

Springer Series in Materials Science, Vol. 13
Semiconductor Silicon Editors: G. C. Harbeke and M. J. Schulz

2. Experimental

Our experiment has been performed with SILSO-type polycrystalline silicon wafers and with EFG silicon ribbon grown in the nonagon form. All samples were p-type with 5-10 Ω cm resistivity. Before the thermal treatments, the samples were degreased and SILSO-type samples were mirror polished on both sides with Al_2O_3 abrasive.

The samples were thermally treated at 1250 C for 6 h in dry N_2 atmosphere, and the samples were subsequently dipped in an HF solution to remove the oxide layer at the surface.

All the IR spectra were taken at room temperature with a FTIR Perkin-Elmer 1710 spectrometer, with 4 cm^{-1} resolution. A differential technique was used with a float-zone "oxygen-free" sample as a reference.

3. Results and Discussion

Figure 1 represents the spectra of as-received materials, where no interstitial oxygen could be detected, hence leading us to the assumption that samples are "oxygen-free". A sharp peak at 605 cm^{-1} indicates the presence of substitutional carbon in the concentration $c[C_s]$= $3.3x10^{17}$ atoms cm^{-3} for SILSO and $c[C_s]$= $9.1x10^{17}$ atoms cm^{-3} for EFG (the conversion factor is $1.1x10^{17}$ atoms cm^{-3} ANSI/ASTM F 123-74).

After the thermal treatment at 1250 C, as shown in the Fig. 2 a strong oxygen peak at 1107 cm^{-1} was observed The oxygen concentration

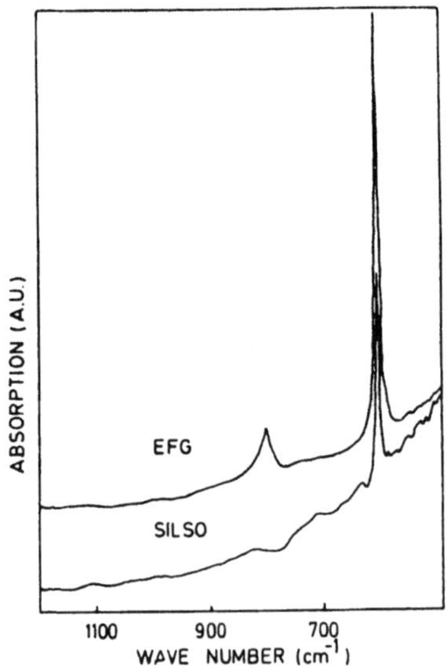

Fig. 1 IR spectra of as-received EFG and SILSO poly-Si samples

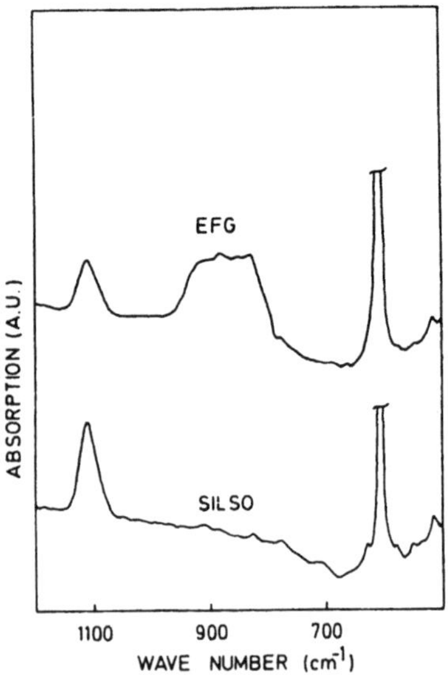

Fig. 2 IR spectra of EFG and SILSO samples after heating for 6 h in dry N_2 atmosphere

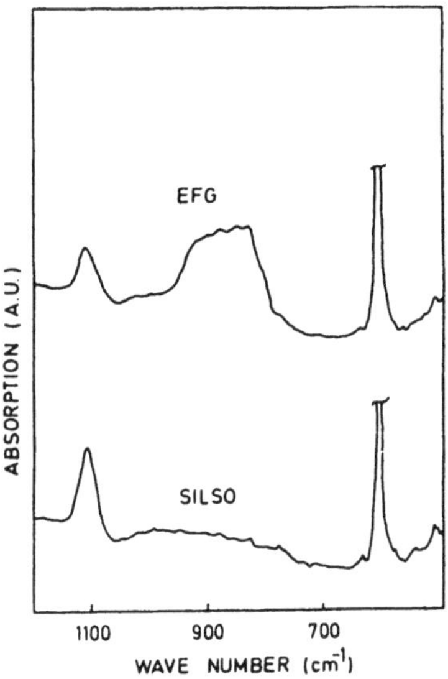

Fig. 3 IR spectra of EFG and SILSO poly-Si samples after heat treatment and etching about 20 μm of the surface layer

calculated from the peak was $c[O_i]= 3x10^{17}$ atoms cm^{-3} and $c[O_i]= 2.3x10^{17}$ atoms cm^{-3} for SILSO and EFG samples respectively (the conversion factor is $2.45x10^{17}$ atoms cm^{-3} ANSI/ASTM F 121-80).

Figure 3 shows spectra taken after we had etched away about 20 μm of the surface layer, after the thermal treatment. The oxygen concentration measured after this step was $c[O_i]= 2.5x10^{17}$ atoms cm^{-3} and $c[O_i]= 2x10^{17}$ atoms cm^{-3} for SILSO and EFG wafers, respectively.

It could be seen that the information we obtained for the as-received samples by using IR spectroscopy did not provide a realistic picture of the concentration of oxygen dissolved in the bulk of material. The high temperature treatment (1250 C and higher) is known from the literature to be a dispersion treatment [6-8], and is currently used in mono-Si technology. At this temperature oxygen precipitates are dissolved and oxygen atoms take interstitial positions homogenously distributed troughout the bulk of the material. In our case, due to the high concentration of structural defects (grain boundaries and dislocations) in the poly-Si material, a large fraction of oxygen is expected to segregate close to these defects, as it was found earlier [9].

As shown in the Fig. 2 a large amount of oxygen was present in the as-received material, but in the IR-invisible form. After the dispersion treatment, oxygen took the interstitial position and was easily detected by IR spectroscopy.

Furthermore, to obtain a more realistic picture of the concentration of oxygen present in the bulk of the material, we etched away about 20 μm of the surface layer. Etching was done in order to avoid any confusion and misinterpretation that may arise from the oxygen segregation close to the surface [10]. Any further etching did not change the measured bulk oxygen concentration by more than was estimated to be within the accuracy of the measurements.

4. Conclusion

In conclusion, our experiment has shown that by performing IR measurements at room temperature on as-received poly-Si samples, one cannot obtain a realistic information about the quantity of oxygen dissolved. However, this information can be obtained using the appropriate thermal treatment, e.g. oxygen dispersion and homogenisation.

Literature

1. D. Helmreich; Silicon Processing for Photovoltaics II, ed. by C.P. Khattak and K.V. Ravi, (North-Holland, Amsterdam 1987) p. 97
2. C.P. Khattak and F. Schmid; Silicon Processing for Photovoltaics II, ed. by C.P. Khattak and K.V. Ravi, (North-Holland, Amsterdam 1987) p. 153
3. J.P. Kalejs; Silicon Processing for Photovoltaics II, ed. by C.P. Khattak and K.V. Ravi, (North-Holland, Amsterdam 1987) p. 185
4. A. Bourret; Proc. 13th ICDS, Coronado, CA, (The Metallurgical Soc. of AIME, New York 1985) p. 129
5. R.C. Newman; Oxygen, Carbon, Hydrogen and Nitrogen in Crystalline Silicon, ed. by J.C. Mikkelsen Jr, S.J. Pearton, J.W. Corbett and S.J. Pennycook, (MRS, Pittsburgh 1986) p. 403
6. H.M. Liaw; Semiconductor Characterisation Techniques, ed. by P.A. Barnes and G.A. Razgonyi, (The Electrochem. Soc., Pennington 1978) p. 190
7. K. Sumino; Semiconductor Silicon 1981, ed. by H.R. Huff, R.J. Kriegler and Y. Takeishi, (The Electrochem. Soc., Pennington 1981) p. 208
8. M. Stavola and L.C. Snyder; Defects in Silicon, ed. by W. Murray Bulis and L.C. Kimerling, (The Electrochem. Soc., Pennington 1983) p. 61
9. S. Pizzini, P. Cagnoni and A. Sandrinelli; Appl. Phys. Lett. 51, 1987, 676
10. B. Pivac and N. Urli; J. Mater. Sci. Lett., 7, 1988, 9

Part V

Insulating Films

Characterization of the SiO_2-Si Interface

M. Schulz

Institut für Angewandte Physik, Universität Erlangen-Nürnberg,
Glückstr. 9, D-8520 Erlangen, Fed. Rep. of Germany

1. INTRODUCTION

The Si-SiO_2 interface is of central importance to the operation of a (metal oxide semiconductor) MOS device. Various uses of the insulator SiO_2 and the occurrence of its interface to silicon are illustrated on the schematic of the MOS memory cell in Fig. 1. A recent device design based on a trench capacitor is presented by S.Röhl in a separate book chapter. Silicon dioxide is utilized for device isolation and surface passivation, as a capacitor dielectric, and most importantly, as the gate dielectric for the Si MOSFET. Progress in the Si MOS technology therefore corresponds with an increasing understanding of the Si-SiO_2 interface. This interface is probably the most intensively studied material system; however, the physical and chemical properties of the interface are still not well understood. If one considers that the insulator SiO_2 on silicon is simply fabricated by oxidation of the Si substrate at temperatures around 1000 °C, it seems a miracle that this interface works so well in devices. In fact, there is no peer among other semiconductors where an insulator

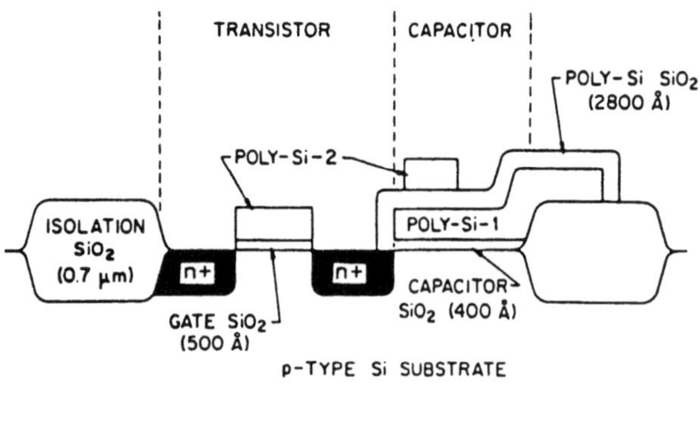

Fig. 1: A standard memory cell illustrating the occurrence of SiO_2 and its interfaces to Si

Springer Series in Materials Science, Vol. 13
Semiconductor Silicon Editors: G. C. Harbeke and M. J. Schulz

interface like that of SiO_2 on Si can be prepared. The predominance of Si for the device fabrication is based on the excellent interface properties to silicon. The most important features are briefly reviewed in this paper.

2. INTERFACE ROUGHNESS

The Fig.2 shows a recent high-resolution electron micrograph of an SiO_2-Si interface /1/. The resolution can be envisaged by the lattice planes in the crystalline Si substrate. The irregular pattern in the overlayer represents the amorphous network of the oxide. The interface appears smooth within one or two monolayers.

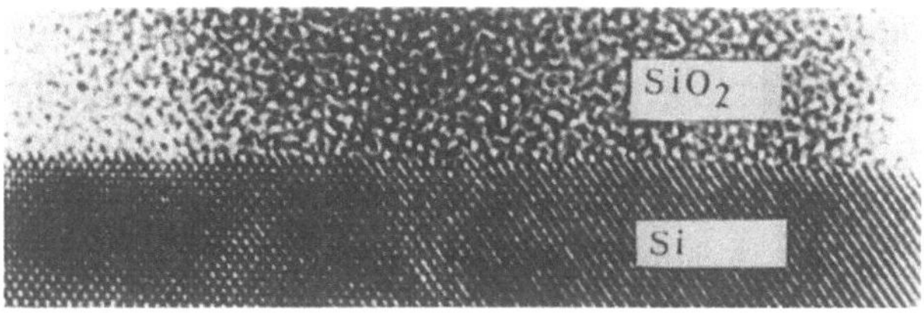

Fig. 2: High-resolution electron micrograph of the SiO_2-Si interface /1/

This result is confirmed by other measurement techniques/3/. During oxide growth there seems to be a mechanism which smoothes out any interface roughness /2/. This is the first miracle which only occurs for silicon and which makes this system SiO_2-Si so useful in devices. Other oxides e.g. on germanium or GaAs, are rough and leaky insulators.

3. ELECTRONIC STRUCTURE OF THE INTERFACE

At the SiO_2-Si interface of metal-oxide semiconductor (MOS) device structures, the change in the lattice and the electronic structure of the material cause trap levels. Similar to bulk traps in semiconductors, these interface traps rapidly interact with free charge carriers in the semiconductor substrate, as indicated in Fig. 3. The traps are

filled with majority carriers when the interface is biased into accumulation. When the interface is biased into depletion or inversion, the charge in the interface states is changed. In contrast to these so-called "fast" interface states, the fixed and mobile charge centers that are located in the bulk of the oxide do not interact with the silicon substrate except after charge carrier injection across the high interface barrier, e.g. by photoexcitation or by strong heating of charge carriers in silicon in a high electric field.
Only interface states that have an energy position within the bandgap of the silicon substrate affect devices -and are therefore considered here- because charge trapping is possible at interface states in the same manner as it is at bulk traps in the forbidden gap of silicon. As shown schematically in Fig.4, the density of the interface state peaks at the SiO_2-Si interface. This trap density, which is continuously distributed in the energy gap of silicon, is denoted by D_{it} in units of $cm^{-2}eV^{-1}$. It is not clear today whether the interface states are located in a monolayer at the interface or whether they are distributed within a tunneling depth of a few angstroms into the SiO_2.

Traps in the bulk silicon substrate may also originate from the SiO_2-Si interface. Such traps are treated as bulk traps at

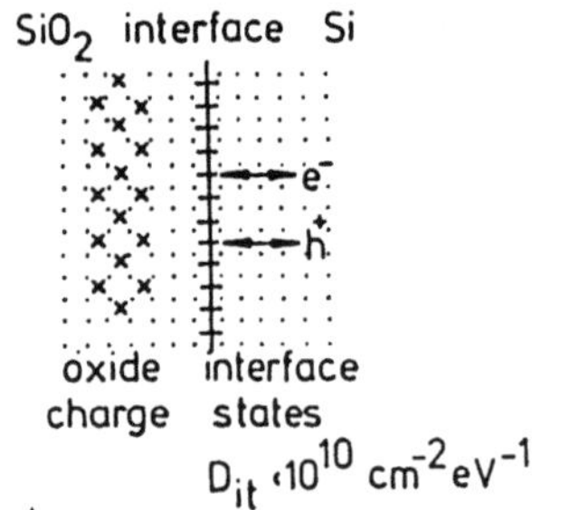

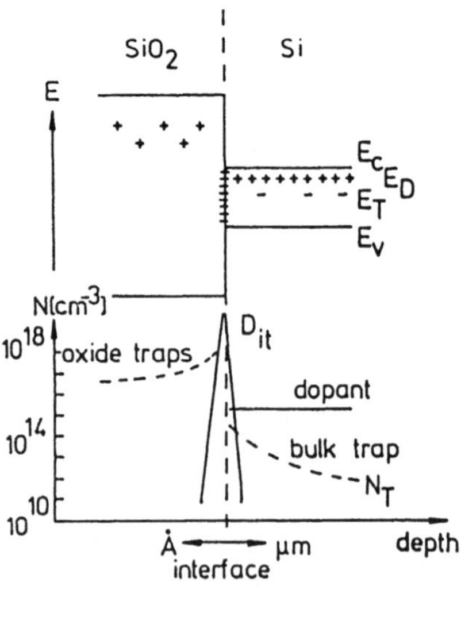

▲ Fig. 3: Definition of interface states at the SiO_2-Si interface

► Fig.4: Energy distribution and spatial distribution of interfacial traps concentration $N_T(x)$. The concentration of bulk traps is generally less than the dopant concentration. They are distributed throughout the space-charge-layer width (of the order of micrometers). Their total concentration in the space charge layer may be comparable to the interface state density.

Interface states at the SiO_2-Si interface have received strong scientific attention since the advent of the first MOS transistors in the early 1960s. At that time, interface states still limited device performance. Since then, the measured interface-state density has continuously dropped as shown in Fig.5. The observed improvement from $D_{it} = 10^{12}$ cm^{-2} eV^{-1} to $D_{it} = 10^{8}$ cm^{-2} eV^{-1} (four orders of magnitude) has been achieved entirely by empirical methods. Today interface states are barely detectable in uniform MOS structures. They are, however, still of importance in relation to submicron structures for large-scale integration, radiation effects, and impurity contamination. Many review papers deal with various aspects of interface states /4-7/. Studies on interface states continue to be the topic of conferences /8,9/.

The main reason why the interface state density is so low, is the strong affinity of oxygen to silicon. Most of the bonds are saturated and bonding and antibonding states are removed from the bandgap region in silicon to energy positions deep in the valence band or well above the conduction edge. Theoretical calculations are still limited in accuracy and

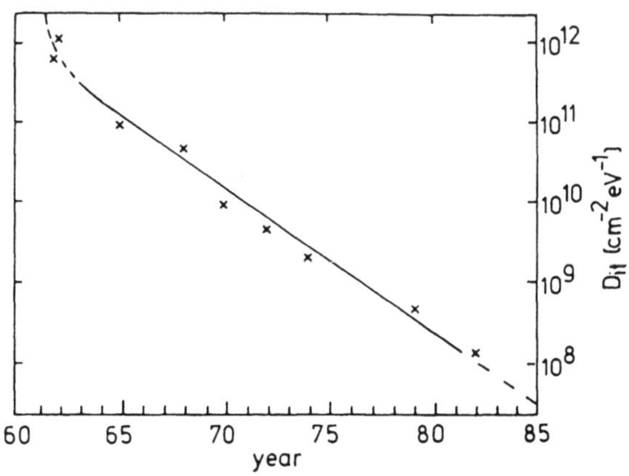

Fig. 5: Drop of interface state density versus time. The points represent the lowest interface state density as reported in each year

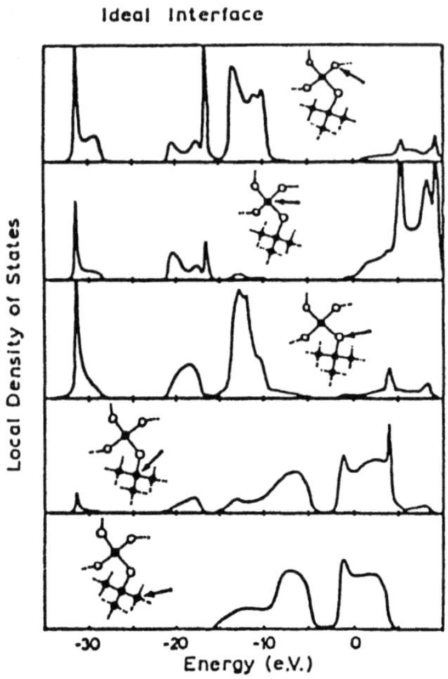

Fig. 6: Local densities of states versus energy for an ideal interface; starting at the top with SiO_2 and proceeding through the interface layer by layer into Si

reliability, because the model assumptions are fairly crude; however, removal of interface states from the bandgap region is generally accepted /10-12/. A typical result is shown in Fig.6 where the density of states are calculated per layer /10/. A bulk silicon layer density is shown in the bottom figure. The semiconductor bandgap between the valence band is visible. In the intermediate layers from the bulk silicon through to the SiO_2 (top figure), states never appear within this bandgap. Bonding or antibonding states only occur at low or high energies well outside this bandgap region. This is the second miracle which favors silicon and its interface to SiO_2. Other semiconductors, e.g. GaAs are not free of interface trap levels.

4. INTERFACE DEFECTS

Interface traps, between SiO_2 and Si are only caused by deviations from the perfect saturation of interface dangling bonds. Various possibilities of such defects are shown in the schematic Fig.7. Due to the mismatch of the bond density in the interface of SiO_2 and Si (roughly a factor 2), dangling bonds must occur at Si with 3 bonds back to the Si substrate. These defects are identified by ESR /3,4/ in freshly oxydized silicon at densities of up to $D_{it} \approx 10^{13}$ cm^{-2}. The dangling bond is amphoteric. It may have 3 charge states, positive, neutral or negative by capturing one or two electrons into the empty defect. The occupation may be controlled by the gate bias. The unoccupied bond has no spin and therefore is not visible in ESR. The spin of one electron is active in ESR and 2 electrons again are invisible. The ESR signal therefore passes through a maximum when the dangling bond is occupied on the various charge states by increasing the gate voltage (Fig.8). The energy levels associated are 0.36 eV above the valence band or 0.35 eV below the conduction band as measured by deep level transient spectroscopy. The dangling bond and the level distribution is similar to that of amorphous bulk silicon except that the defects are distributed in a plane rather than in the three-dimensional space. This case is shown in Fig.9b. The level density in device-grade MOS structures is much less (Fig.10) than for a freshly oxidized interface (fig.9a). The electronic level density is strongly reduced by saturating the dangling bonds with hydrogen atoms as already indicated in

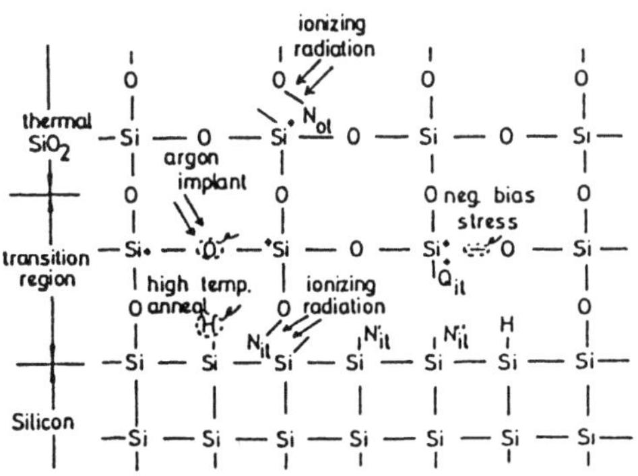

Fig. 7:Schematics of the microscopic and chemical SiO_2-Si interface structure showing possible defects and the mechanism introducing them /20/

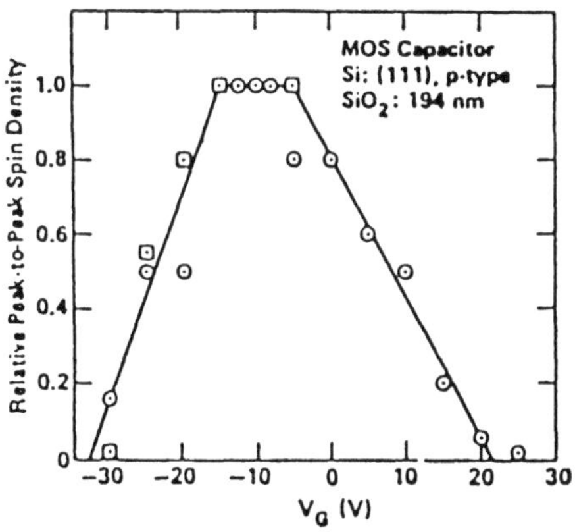

Fig.8: ESR measurement of the spin density in dangling bonds of a MOS capacitor vs. gate voltage /14/

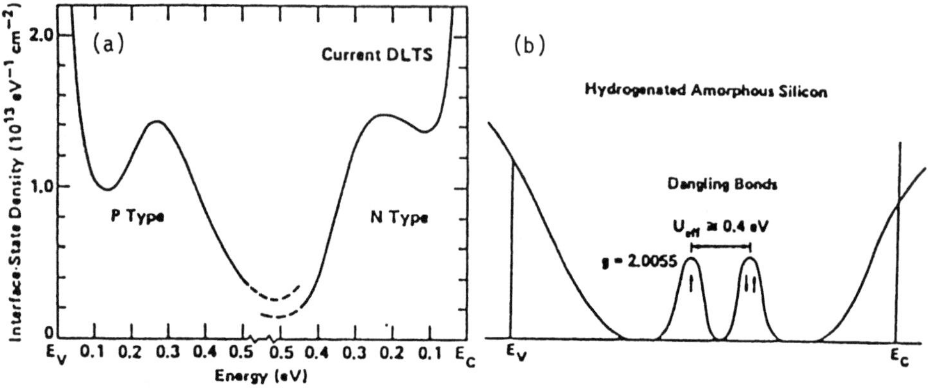

Fig, 9: Density distribution of defect states in a) a MOS capacitor and b) in hydrogenated amorphous Si. The peaks are explained by dangling bonds /14/

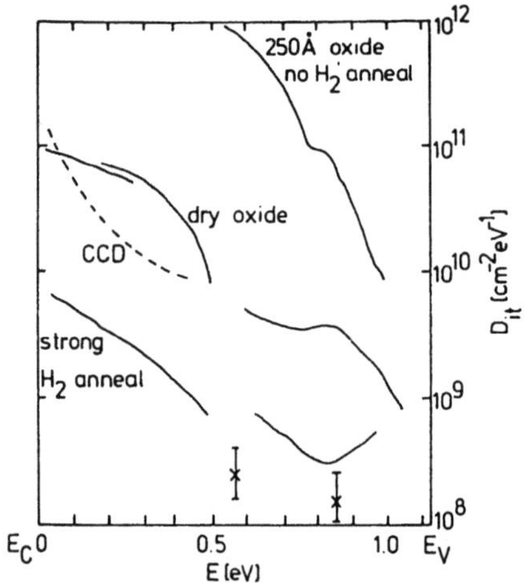

Fig. 10: Distribution of interface state density versus energy. A continuous drop from the conduction band edge is usually observed. The density is strongly reduced after hydrogen annealing

Fig.7. This chemical bonding also removes the level from the bandgap region in silicon and thus inactivates the defect electrically. The hydrogen bond, however, is much weaker than the oxide bond. Dissociation occurs at temperatures above 500 °C. The bond saturation is achieved by annealing at temperatures around 450 °C. This is the final processing step in device fabrication. Typical level density distributions are shown in Fig. 10. The lowest density is observed after strong hydrogen annealing. Level densities as low as 10^8 cm^{-2} eV^{-1} may be obtained. The mean distance between electronic levels then is approx. 1 μm.

5. INDIVIDUAL TRAP LEVELS

The low interface trap density can be directly observed in modern micron-size devices, i.e. in a MOSFET. For this experiment, the MOSFET is biased into the threshold region with a low density of approx. 10^{10}e/cm², i.e. 100 electrons in the channel area 1x1 μm². The source drain voltage (10mVolts) is small enough for a uniform field in the channel. When one of the few electrons is trapped, the number of carriers in the channel is reduced in the order of 1%. This change in density is visible as a sudden statistical change in the source drain current /15/ (Fig.11). Reemission of the electron increases the current again. The signal observed in the current is a random telegraph signal with a statistical distribution (Poisson distribution) of the switching events. The distribution of the pulse length is exponential (Fig.12). The ratio of the average trapped and free duration times of the electron can be directly measured. It yields the Fermi occupation factor of the trap.

The traps observed are slow traps (τ = ms ...100s) somewhere located in the oxide film close to the interface. These traps show a strongly temperature dependent capture rate which is related to relaxation processes in the oxide /16,17/. The energy position of these traps is close (<250meV) to the conduction band edge. Measurements on individual single traps are still a topic of recent research /21/.

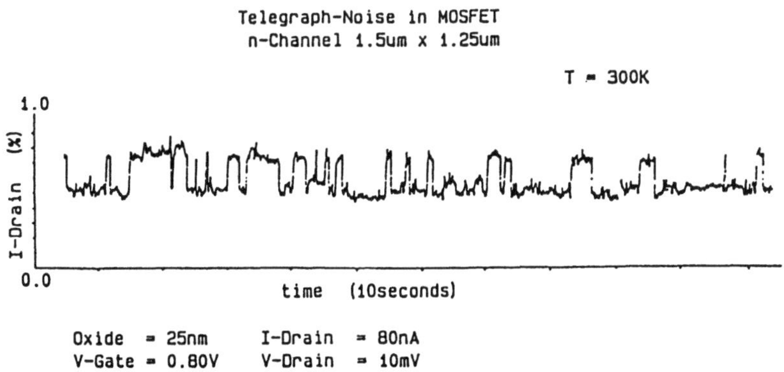

Fig. 11: Random telegraph signal observed for single electron trapping events in a micron-size MOSFET

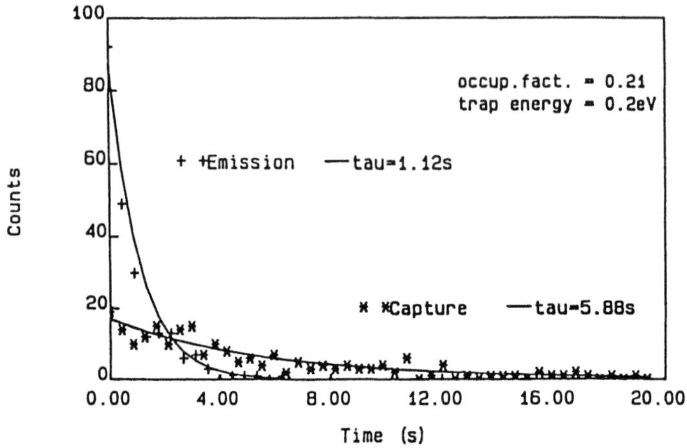

Fig. 12: Duration distribution of the random telegraph signal in Fig. 11. Times measured for captured and emitted electrons are shown

6. CONCLUSIONS

The almost ideal properties of the SiO_2-Si interface are the basis for the dominance of the silicon device technology. The interface is abrupt within a few angstroms and the interface state density is so low that today single trapping events may be observed in micron size device structures.

The limit of SiO_2 for the use as a gate dielectric in devices is the high temperature of around 1000 °C which has to be applied for an extended time period. Dopant diffusion during this processing limits the further reduction of the device size into the submicron range. Modifications of the processing (high-pressure oxidation, plasma or other stimulated oxidations) are presently under investigation. The amorphous soft structure of the SiO_2 make it prone to radiation and bias stress defects which create space charge in the insulator and deteriorate the device characteristics. However, the SiO_2 presently still is the best suited insulator for device fabrication. Alternative insulators e.g. CaF_2, BaF_2 /18, 19/ or epitaxial films grown by MBE /18/, or Langmuir films /19/ have not yet reached a grade comparable to that of SiO_2.

ACKNOWLEDGEMENTS

The author is grateful to A.Karwath for giving permission to present her random telegraph noise results of Fig.11 and 12.

REFERENCES

1. Goodvik,S.M., O.L.Krivanek: J.Vac.Sci.Tech. B-1(1983)803
2. Carim,A.H., R.Sinclair: "The semiconductor silicon 1986" Proc. of the Int.Symp. Boston 1986 The Electrochem. Soc. 86-4 (1986) 458
3. Helms,C.R.: in "Insulating Films on Semiconductors" Springer Series in Electrophysics 7, M.Schulz, G.Pensl eds. Springer Verlag, Heidelberg 1981, page 19
4. Goetzberger A., E.Klausmann, M.Schulz: CRC Critical Reviews in Sol.St.Science 6(1976) 1-43
5. Schulz,M.; Surface Science 132(1983)422
6. Nicollian E.H., J.Brews: "MOS: Physics and Technology" Wiley, New York 1982
7. Schwarz, S.A., M.Schulz: in VLSI Electronics, Vol.10, N.G.Einspruch ed., Academic Press (1985) 30-70
8. Simmonne, J.J., J.Buxo editors: "Insulating films on Semidonductors (INFOS 85) Proceedings of the Int. Conference Toulouse 1985, North-Holland 1986
9. DeKeersmaecker, R. ed: "Insulating films on Semiconductors (INFOS 87) Proc. of Int.Conf., Leuven 1987, North-Holland 1987
10. Laughlin, R.B., R.B.Joannopoulis, J.Chadi: in Proc.Top.Conf. on the Physics of SiO_2, Yorktown Heights, N.Y. 1978, Pergamon Press, page 321
11. Yndurian,F., J.Rubio: Phys.Rev.Lett. 26(1971)1389
12. Sugano, T.: Insulating Films on Semiconductors (INFOS 83) J.Verwey ed. North-Holland 1983, page 1
13. Caplan, P.J., E.H. Poindexter, B.E. Deal, R.R. Razouk,: J.appl.Phys. 50(1979)5847
14. Poindexter, E.H., P.J.Caplan, N.M.Johnson; D.K.Biegelsen, M.D.Moyer, S.T.Chang: in "Insulating Films on Semiconductors (INFOS 83) J.Verwey ed. North Holland 1983, page 24

15. Karwath,A., M.Schulz: Appl.Phys.Lett. 52 (1988)634
16. Uren, M.J., D.J.Day, M.J.Kirton: Appl.Phys.Lett. 47 (1985) 1195
17. Kirton, M.J., M.J.Uren: Appl.Phys.Lett. 48(1986) 1270
18. Poate, J.M.: in "Insulating Films on Semiconductors" (INFOS 85) J.J.Simmonne ed. North-Holland 1986, page 41
19. G.G.Roberts: in "Insulating Films on Semiconductors" (INFOS 81) M.Schulz, G.Pensl eds. Springer 1981, page 56
20. Deal B.E.: J.Electrochem. Soc. 121 (1974) 198c
21. Karwath A., M.Schulz: in Proc. of the Electrochem. Soc.Meeting, Atlanta 1988

Advanced Silicon on Insulator Materials: Processing, Characterization and Devices

S. Cristoloveanu

Laboratoire de Physique des Composants à Semiconducteurs (UA CNRS), Institut National Polytechnique de Grenoble, ENSERG, 23 rue des Martyrs, F-38031 Grenoble Cedex, France

Abstract. The major attributes of silicon on insulator structures for VLSI are discussed by referring to the conventional bulk Si technology. The preparation and properties of various SOI materials (SIMOX, SOS, SOZ, ZMR, FIPOS, wafer bonding) are presented and their advantages or drawbacks outlined. Relevant methods used for electrical evaluation of very thin films are analysed with emphasis on SIMOX behaviour. Distinction is made between the properties of as-fabricated layers (carrier transport, impurity activation) and those deduced from the operation of fully processed devices (interface defects, lifetime, inhomogeneities). The specific aspects related to the operation of SOI transistors are underlined. It is demonstrated that the domain of SOI extends from almost conventional devices with improved performances to totally new structures and components.

1. INTRODUCTION

Silicon On Insulator (SOI) structures had been proposed very early on as advanced substrates for chip integration, essentially because the conventional Si devices were intolerant to radiation effects. About 25 years later, the development of the bulk Si technology exceeded the most optimistic predictions whilst SOI was still just a very good candidate. This period was characterized by an euphoric increase in the density and complexity of monolithic circuits based on the shrinkage of the dimensions. The consequence is that Si will soon be confronted with other fundamental limitations occuring in submicron components.

These facts reinvigorate the interest in SOI alternatives which have been demonstrated to support better the constraints of scaling-down rules. We are probably going to assist in a dramatic challenge : will silicon still be able to go into the deep submicron range (channel lengths below 0.3-0.4 μm) by further sophisticating the processes and keeping prices reasonably low or will other technologies, including SOI, be in a better position to take the lead ?

This fundamental dilemma explains the recent growth of the research in SOI. At present, new SOI materials have been proposed and successful demonstration chips have been fabricated. SOI has overcome the stage of being a subject for intellectual controversy and is now the material on which the most rapid Si devices are processed.

The questions "SOI-why ?" and "SOI-how ?" will be addressed in the next two sections. The electrical properties represent the key criteria for deciding on the real chance of a new material entering into the micro-

Springer Series in Materials Science, Vol. 13
Semiconductor Silicon Editors: G. C. Harbeke and M. J. Schulz

electronics family. The difficulty in characterizing very thin films and the basic properties are discussed in section 4. Section 5 is related to the special operation of MOS/SOI components. Another fundamental aspect developed here is the critical analysis and comparison of various techniques (conventional or very recent) for interface evaluation in SOI. Finally, three dimensional integrated circuits and other new devices will be described. Although the methods and principles are rather general for any SOI structure, the experimental data is used to provide a clear image of SIMOX material which now appears to be the most promising variant.

2. CAPABILITIES OF SOI STRUCTURES

In a Si wafer containing MOS devices only the surface region, about 0.2 μm thick, is really useful for electron transport. As the residual material (99.9 % !) causes parasitic effects to occur, an obvious idea is to isolate the active overlay. Therefore, SOI consists of a thin Si film either deposited on a thick insulator substrate or separated from the Si substrate by a buried dielectric layer. The major advantages of SOI are illustrated by the schematic comparison (Fig.1) of CMOS transistors fabricated on bulk Si and SOI substrates.

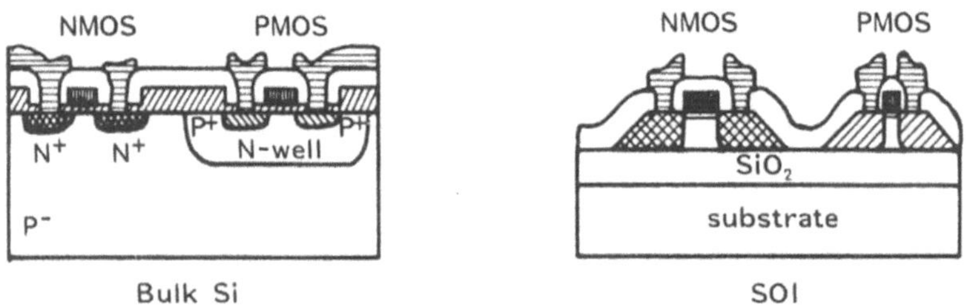

Fig.1. Schematic representation of CMOS transistors in bulk Si and SOI

2.1 Vertical Junctions

In thin SOI films, the heavily doped source and drain regions extend until the insulator. Only the active lateral sides of these regions are used to form "vertical" junctions. The total junction surface is, therefore, greatly reduced in comparison with the case of bulk Si (Fig.1). This results in a significant lowering of parasitic capacitances due to diffusion regions and interconnections. A shorter RC time constant means higher speed of operation and lower dynamic power consumption. This is essential for VLSI since, for a constant level of power dissipation, SOI devices are much faster and a higher packing density of integrated circuits can be achieved in SOI chips.

2.2 Full Dielectric Isolation

Another argument for using SOI in VLSI circuits is the minimum separation distance between devices (about 1 μm in SOI and 3-4 μm in Si) [1]. The limitation in bulk Si is due to the proximity of diffused regions belonging to neighbouring components which leads to unintentional turn-on of parasitic devices (latch-up). For instance, in Fig.1, a NPN bipolar transistor can be activated between the N^+ drain of the N-channel MOSFET, the P-type substrate and the N-well of the P-channel MOSFET, while a PNP transistor is formed by the substrate (P), well (N) and source (P^+) of the P-channel MOSFET.

Superposition of these NPN and PNP transistors with common diffused regions represents a parasitic thyristor which has high current drive capability and causes circuit failure.

In contrast, each SOI device is made in an individual island, fully isolated from both the substrate and adjacent islands. This avoids any proximity effect and confers a **natural** immunity to latch-up.

2.3 Fabrication Simplicity and Flexibility

The CMOS process in SOI does not require any well formation. This allows a simpler design with optimum placement of individual devices and minimum interconnections. Moreover, the number of masking and technological stages is significantly lower in SOI. Figure 2 shows the main steps of a CMOS process already used to fabricate half-micron circuits. It has recently been demonstrated that channel implants become unnecessary since very good performance is obtained with totally depleted transistors [2].

The lateral separation between SOI devices is made by LOCOS (Fig.2) or mesa isolation. Existing solutions for achieving an "artificial" lateral isolation in bulk Si are based mostly on trenching techniques. This implies a rapidly increasing degree of sophistication and complexity as dimensions are scaled down. By anticipating the commercial fabrication of half-micron circuits, it is speculated that the cost of the bulk Si technology will exceed that of SOI, in spite of a much higher initial price of SOI wafers. Additional flexibility for SOI processes is provided by the possibility of fabricating high voltage circuits and combining CMOS and bipolar components for analog applications.

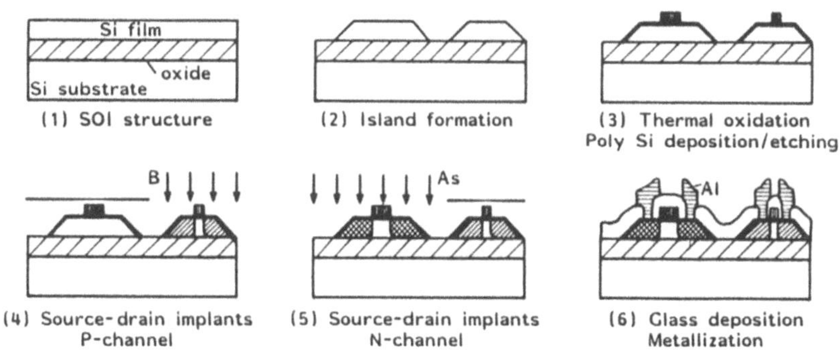

Fig.2. Possible sequence for processing CMOS integrated circuits in SOI

2.4 Radiation Tolerance

This is the advantage which historically initiated the research in SOI. Photocurrents are generated by various sources of radiation (protons, neutrons, α particles, X and γ-rays, etc) in proportion with the active volume of the device. For instance, alpha particles penetrate over about 10-100 μm in Si substrates and create many electron-hole pairs. This results in significant leakage currents and charge collection in MOS devices, leading to inacceptable soft errors.

In SOI memory cells, logic upsets are greatly reduced since the available volume for carrier generation is 2 or 3 orders of magnitude lower. On the other hand, latch-up paths cannot be formed by well-junction activation, as in bulk Si, because the word latch-up simply does not exist in the SOI terminology.

Besides these **transient** effects affecting mostly memories, **permanent** damage is produced by pair generation in the gate oxide and side-walls. After an initial recombination and electron collection, the holes migrate towards an interface thus building up a positive charge. This fixed charge and its partial reconversion in interface states lead to threshold voltage shift, transconductance degradation and enhanced leakage. This type of damage depends on the cumulative dose of ionising radiations and the advantage of SOI is not as obvious as for transient effects.

2.5 Short Channel Effects

In small geometry MOS transistors, the depletion regions induced by source and drain junctions become non-negligible so that the gate is no longer able to control the whole space charge region. This "charge sharing" between gate and junctions is the main reason for a number of short channel effects in bulk Si : threshold voltage reduction, degradation of the slope in weak inversion, drain-induced barrier lowering, punchthrough, activation of the parasitic bipolar transistor, etc. These unsuitable effects are greatly alleviated in SOI where the extension of junction-dominated depletion regions is restricted ; junctions are indeed smaller and vertical, and the surface potential is controlled at both the front and back interface [3].

The most prohibitive short channel consequence is the performance degradation induced by hot carrier injection into the gate oxide. Again, the use of SOI substrates has been demonstrated to be a successful solution which avoids the use of LDD structures and the lowering of the bias supply.

2.6 Novel Structures

The particular configuration of SOI material (very thin Si overlay, multi-interface topology, variable oxide thickness, adjustable back interface quality, etc) offers the opportunity to envisage new components including power devices, transducers and 3-dimensional circuits. Several options for 3-D integration as well as a new concept for transistor operation will be discussed in section 5.3.

The above enthusiastic enumeration of SOI advantages may appear to be in contradiction with the relatively poor use of SOI in commercial applications. Until now, only Silicon On Sapphire integrated circuits have been fabricated in a few countries, mostly for space and defense electronics. This ambiguous situation is due, firstly, to the fact that a number of capabilities are closely connected with VLSI and have been fully understood at a later time. Secondly, the quality of SOI materials still suffers from one or several drawbacks : crystalline defects, lattice stress, lateral and vertical inhomogeneities, defective back interface, various contamination sources.

The complexity of SOI preparation is also prohibitive in terms of cost, yield and substrate availability. Furthermore, the specific double or triple interface structure requires new simulation tools and evaluation techniques to be conceived. These problems have generated competitive research in several SOI technologies, which will be briefly described now.

3. ALTERNATIVES FOR SOI

3.1 Silicon On Sapphire (SOS)

Silicon On Sapphire (SOS) is the senior member of the SOI family. It can also be speculated that SOS was launched as a distress signal of the conventional Si technology ! The choice of sapphire (Al_2O_3) as the insulating substrate was a hopeful compromise regarding the matching of its thermal and crystallographic properties with those of the epitaxial Si film. SOS is fabricated by CVD using silane. The hetero-epitaxial process starts with the formation of small isolated Si islands and the progressive surface coverage by coalescence [4]. This transition region is about 20 nm thick and contains crystallographic defects (due to residual lattice mismatch) and Al atoms incorporated from the substrate. The film quality increases rapidly during the subsequent growth. An annealing is performed with the aim of neutralizing Al atoms and improving the mobility.

The Si film, 0.5 μm thick, is submitted to a large compressive stress (10^{10} dyn/cm^2) caused by the difference in thermal expansion coefficients which induces differential contraction of Si and Al_2O_3. Crystalline defects (dislocations, stacking faults and microtwins) and lateral stress are responsible for poor values of carrier lifetime and electron mobility, essentially in the interfacial region [5].

Although this standard process is used for commercial applications, several methods have been proposed more recently for improving SOS. The film hydrogenation has a beneficial effect by suppressing deep levels and microscopic inhomogeneities [6]. The deposition of a thin amorphous Si buffer layer prior to the epitaxy results in a less defective interface region [7].

In the liquid-phase epitaxial regrowth (LPER), laser irradiation is used to melt the surface layer whilst the back interface zone serves as a seed [8]. The carrier mobilities are improved by LPER and the material is strain-free as a consequence of the thermal gradients existing in the Al_2O_3 substrate.

Another very attractive approach (Fig.3) consists of the elimination of structural defects by a complete disordering of the Si lattice, followed by solid-phase epitaxial regrowth (SPER). Silicon ions are implanted into the Si film in order to erase the memory of the lattice by amorphization, except for a thin surface layer which is used as a seed during furnace annealing. The compressive strain and leakage current at the back interface are greatly reduced [9]. In order to improve the front surface crystallinity as well, a shallower implant can then be performed. In this double SPER process (DSPER), the second epitaxial growth is directed from the already recrystallized deeper region towards the surface [10]. Good quality films 0.2-0.3 μm thick have been obtained by SPER (i.e. without any second Si implant) simply by etching [11] or oxidizing [12] the non-recrystallized surface layer.

The key aspect in all these new variants is to be able to avoid substrate damage and thus the release of electrically active Al atoms. However, CMOS

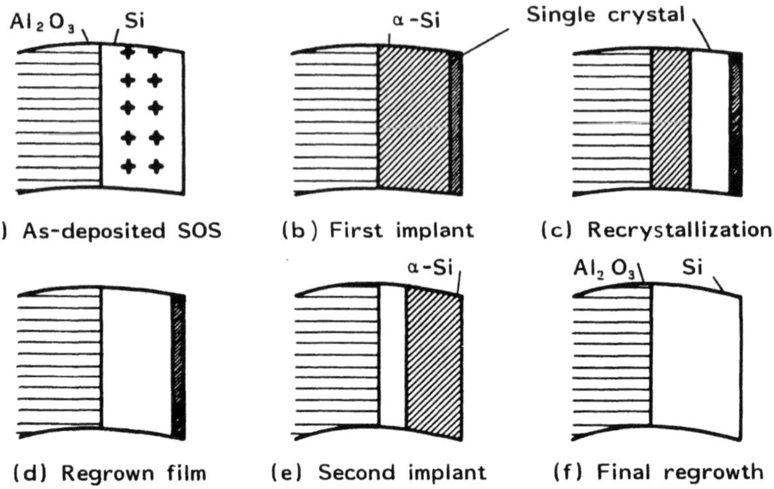

Fig.3. Double recrystallization of SOS by silicon implantation and solid-phase regrowth

integrated circuits, MESFET's and lateral bipolar transistors with good performance have been fabricated [13]. In conclusion, SOS is a good silicon on insulator material but certainly not the best. History is repeating itself : bulk Si deficiencies generated SOS a long time ago and, now, doubts accumulated over years on SOS have initiated intensive research in rival technologies.

3.2 Silicon on Zirconia (SOZ)

Yttria-stabilized Zirconia (YSZ) is a better substrate than sapphire in regard to the lattice mismatch and difference in thermal expansion coefficients. The heteroepitaxial process is similar to that of SOS. A seductive capability is that YSZ becomes an ionic conductor above 1200°C and allows oxygen atoms to penetrate through the substrate into the epitaxial film. In other words, the interfacial transition layer may be electrically screened by internal oxidation. Physico-chemical and microscopic analyses (RBS, TEM, SIMS, X-ray diffraction) as well as integrated test devices have demonstrated a reasonable crystalline quality of the film [14].

Nevertheless, the development of SOZ is inhibited by the difficulty of fabricating large diameter YSZ substrates. An interesting alternative is given by the SOZOS structure (Silicon on Zirconia on Silicon) : a thin zirconia layer is epitaxied on the Si wafer and then serves as support for the growth of the Si film. Further research is, however, necessary to explore all the possibilities of SOZ and SOZOS materials.

3.3 Zone Melting Recrystallization (ZMR)

The ZMR structure is fabricated by deposition of a polycrystalline Si film on an oxidized Si wafer. Subsequent liquid-phase recrystallization of the film is made using various heating sources : laser, electron beam, halogen lamp, graphite strip heater. An example of a **seeded**-ZMR process is shown in

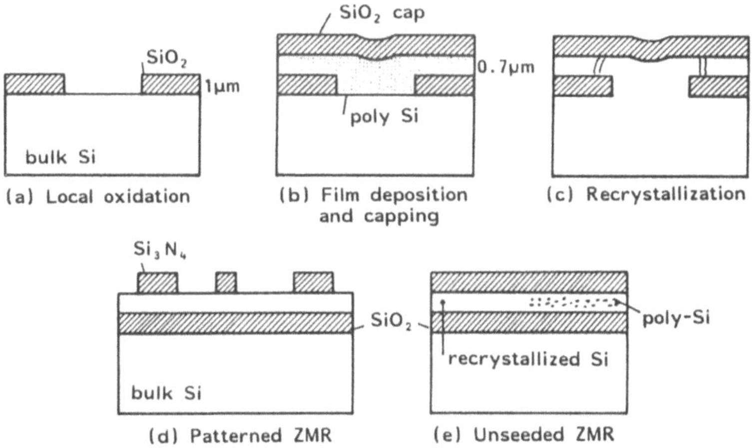

Fig.4. Schematic representation of seeded ZMR (a, b, c), patterned-ZMR (d) and unseeded ZMR (c) processes

Fig.4 : the Si wafer is oxidized over about 1 μm depth and patterned with either continuous seeding lines or discrete seeding windows (1x1 μm²). A poly-Si film, 0.7 μm thick, is deposited by LPCVD at 580°C and then a SiO_2 capping layer is formed by CVD at 450°C. Rapid Si melting is achieved at relatively low temperature (1470°C) by scanning a laser spot (20x100 μm²) at 45° from the seed direction [15].

Seeding is responsible for thermal and topological discontinuities which induce mass transport. The surface non-uniformity (Fig.4) is eliminated by planarization and thinning the Si film to about 0.2-0.3 μm. The rapid cooling of the melted zone causes defects to be generated near the seed. It is clear that SOI devices should be fabricated only in Si islands which are defect-free and situated above the oxide.

As the beam power increases there are several successive stages of heating : (1) silicon melting over oxide, (2) Si melting at the top of the seed, (3) substrate melting below the seed, (4) overheating of the SOI or capping layer and (5) substrate overheating below the oxide. It is obvious that stages (1) and (2) do not allow a good crystalline quality to be obtained, while situation (5) is prohibited for 3-D integration in order to avoid the degradation of the devices realized in the substrate. The optimum range (4-5) can be achieved with a 20 W argon laser scanned at 0.03-40 cm/s [16] or with a dual electron beam system (background heating at 700-1000°C with the first beam and melting with a 20 keV second beam) [17].

A drawback of this local recrystallization associated with the scanning of either the beam or the wafer : mass and impurity transport is assisted by lateral thermal gradients. Optimization parameters are the thicknesses of the oxide, Si and capping layers, scanning speed and direction. Seed interdistance exceeding 100 μm causes the formation of grain and subgrain boundaries, while below 40 μm it gives too small Si islands.

Regrowth of seeded material in liquid-phase can also be achieved by pulsing the energy over the whole wafer instead of scanning a continuous focused beam. In this case the solidification progresses laterally, from each seed, due to thermal gradients (LEGO : Lateral Epitaxial Growth over Oxide).

LEGO substrates suffer from grain boundaries generated by the intersection of solidification wings. Other approaches, more or less abandoned now, are the solid-phase epitaxial regrowth by furnace annealing and the epitaxial lateral overgrowth (ELO). In ELO, the epitaxy of a monocrystalline Si film proceeds from the seed and extends laterally over the oxide so that no further recrystallization is necessary. Unfortunately, the ratio between the lateral and vertical propagation of the epitaxial growth is too small (10:1) thus limiting the size of the monocrystalline Si islands.

Unseeded ZMR material normally results in large polycrystalline grains. The location of grain boundaries can be controlled by depositing a patterned antireflecting Si_3N_4 layer over the Si film (Fig.4d). As the discontinuity in the thermal gradient confines grain boundaries below the Si_3N_4 patterns, integrated circuits can be placed in good quality crystalline bands.

Very recently, wafer-scale monocrystalline ZMR substrates were fabricated without any seeding or pattern process. The buried oxide, Si film and capping layer were continuous (Fig.4e) and the recrystallization was performed by a tight control of thermal gradients [18]. This material appears to be a very serious SOI candidate since it avoids the basic limitation of ZMR, i.e. locating active components in the recrystallized region while reserving the defective zones for pads and interconnections.

3.4 Full Isolation by Porous Oxidized Silicon (FIPOS)

In a typical FIPOS sequence (Fig.5a), phosphorous islands are implanted in a P-type Si wafer. Anodic reaction in HF is used to transform the implanted superficial P-type region in porous silicon. As porous Si has a very large effective surface (several m^2/cm^3), it is selectively oxidized. The final structure is obtained after annealing (1100°C) for oxide densification.

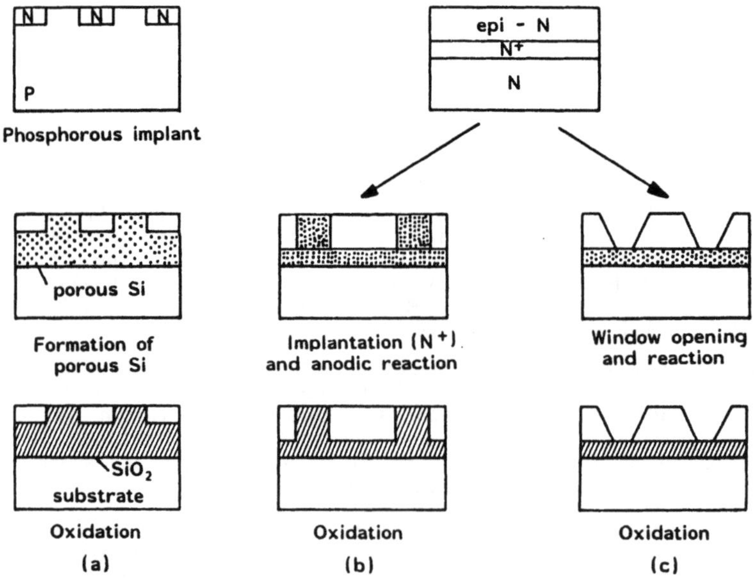

Fig.5. Conventional sequence of FIPOS (a) and recent alternatives for isolation by a lateral oxide (b) or mesa etching (c)

The key aspect in FIPOS is the formation of porous Si. The anodic reaction current is 10-100 mA and the pore diameter varies from 1 to 30 nm. The optimum porosity is 56% in order to accomodate the change in volume from Si to SiO_2 and thus to reduce the strain [19].

Selective formation of porous Si may be obtained by changing not only the type but also the concentration of the dopant. Abrupt NN^+N structures (Fig.5b) are prepared by homogeneous phosphorous implantation (10^{19} cm^{-3}) and epitaxial growth (10^{15} cm^{-3}). Contacts to the heavily implanted N^+ layer are made either by local implantation or etching. The N^+ regions are transformed in porous Si and, then, oxidized. According to the options of Figs.5b and 5c, the Si islands are separated by a lateral oxide or "mesa" isolated [19].

Although FIPOS stands as an inexpensive material which offers variable thickness of the SiO_2 insulator and good crystalline quality, it is still under optimization. 64 kbit RAM's have already been fabricated by NTT. Inconveniencies are mainly associated with the relatively large thickness of the film (0.5 μm), residual stress and contamination induced by the chemical process. A recent and interesting alternative is to form a thin porous layer **at the surface** of the Si wafer and to use it as a substrate for epitaxial growth of a thin and low-doped Si film. Windows are opened to allow the oxidation of the porous buried layer.

3.5 Wafer Bonding

The bonding of Si wafers is an operation well-known for the dielectric isolation of power components and fabrication of pressure and acceleration transducers. In the SOI process (Fig.6) an **active** wafer is prepared from a heavily doped P^+ substrate : a low-doped (P^-) film is epitaxially grown and partially oxidized. The **passive** wafer is simply oxidized. Direct oxide to oxide contact of the two wafers results in a strong Van der Waals bonding (50-100 erg/cm^2). After adhesion, the P^+ substrate of the active wafer is removed by mechanical thinning to about 25 μm followed by selective electrochemical etching which stops at the P^+P^- junction. The final SOI structure is, therefore, composed of the low-doped (P^-) Si film, separated from the substrate of the passive wafer by the bonded oxides.

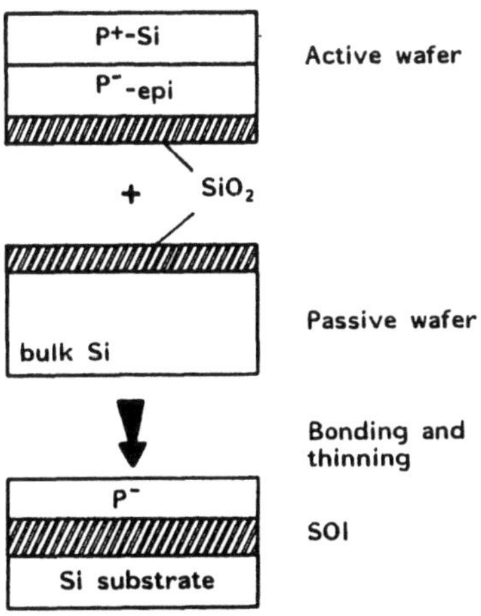

Fig.6. Wafer bonding process

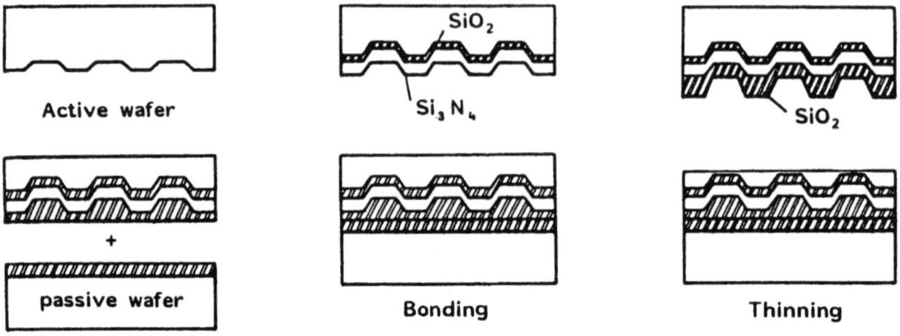

Fig.7. Formation of fully-isolated SOI islands by wafer bonding

The critical stages of this process are the bonding and etching. Wafers have to be dust-free and hydrophil. High temperature annealing improves the adherence by activating chemical reactions : for example, the quartz cohesion (2200 erg/cm²) was achieved after annealing above 1250°C [20].

Totally-isolated islands can also be fabricated by wafer bonding (Fig.7). Grooves are etched with normal lithography on the active wafer which is then thermally oxidized. A polish-stop layer is deposited (Si_3N_4 or densified SiO_2) and, finally, a thicker SiO_2 is superposed by CVD and planarized. This wafer is bonded to an oxidized passive wafer. Mechanical and electrochemical thinning is performed until the polish-stop layer is met [21]. This results in Si islands with selected size floating in an ocean of SiO_2.

The attribute of the wafer bonding is to provide inexpensive wafer-scale SOI monocrystals with variable thicknesses of the underlying oxide. This process needs further basic research regarding the uniformity and the unintentional doping of the Si film induced during electrochemical thinning. Although very optimistic results have been reported, it is not clear at present whether very thin films can be fabricated. We await the electronic properties and performance of integrated circuits with great interest.

3.6 Separation by Implantation of Oxygen (SIMOX)

The Curriculum Vitae of SIMOX is composed of successive stages of development (Fig.8) :

- shallow implantation of oxygen in Si wafers leading to oxygen precipitation (Fig.8a),
- very high dose implantation ($\geq 1.4 \times 10^{18}$ O^+/cm^2) forming, by internal oxidation, a stoichiometric SiO_2 layer at the surface (Fig.8b),
- high energy implantation (150 keV) allowing the synthesis of a buried oxide 0.3-0.4 µm thick and leaving at the surface a thin Si film (Fig.8c) which can be completed by epitaxy (Fig.8d),
- higher energy implantation (200 keV) leaving a Si overlay with adequate thickness (0.2-0.3 µm, Figs.8e,f),
- annealing above 1150°C in order to perfect the structure by crystal rearranging (Figs.8e,f).

The annealing represents a major aspect of SIMOX technology. After **low temperature annealing** (LTA : 1150-1200°C) many SiO_2 precipitates and defects subsist in a greatly disordered interfacial region the composition of which

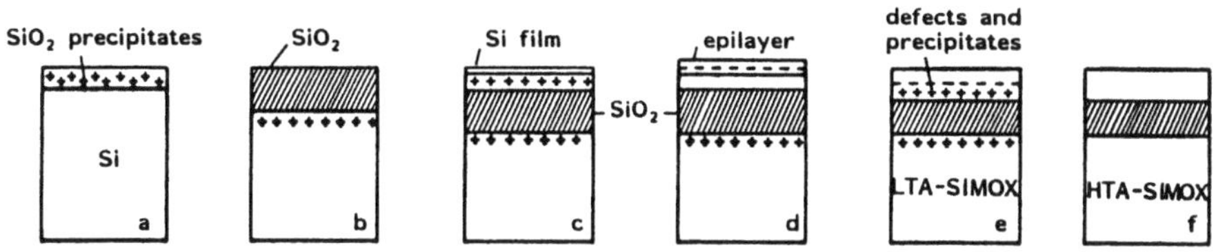

Fig.8. Successive stages in the development of SIMOX

(mono and polycrystalline) depends upon the implantation temperature and dose [22]. However, the top layer, 0.1 μm thick, is almost free of defects and supports successful fabrication of integrated circuits (Fig.8e). After **high temperature annealing** (HTA : 1300-1405°C) the vertical configuration of the SOI structure is simplified (Fig.8f), since interfaces become sharp and the high quality region extends over the whole film [23]. Smaller SiO_2 precipitates are dissolved and larger precipitates are formed closer to the interface. This cascade-like process results finally in the pumping of oxygen from the film (where the residual concentration drops below 10^{18} O/cm^3) towards the buried oxide which acts as an infinite radius precipitate [24]. HTA annealing is performed up to 1350°C in polysilicon furnaces [23] or at 1405°C with focusing lamp irradiation on the back of the wafer to achieve partial melting of the rear surface (1412°C) [25].

The Si overlay is free of stress [26] but contains a large density of dislocations which hopefully do not significantly affect the basic electrical properties. The density of dislocations can, however, be lowered by replacing the single-step implantation with a multi-step process composed of sequential implantations of lower doses ($\leq 0.3 \times 10^{18}$ O/cm^2) and intermittent annealings [27].

A number of alternatives for SIMOX fabrication have been explored. The implantation of nitrogen instead of oxygen results in sharper interfaces after LTA, but leads to donor-like residual doping and a "leaky" insulator. Mixed implantations of oxygen and nitrogen or oxygen and carbon were performed with the aim of improving the back interface and lowering the activation of oxygen donors. The implantation of higher doses of oxygen ($\geq 2.2 \times 10^{18}$ O/cm^2) paradoxically reduces the residual oxygen concentration in the film and improves the interface [28]. A better quality back interface was also obtained by increasing the implant temperature from 400°C to 600°C. In other words, the opportunity for "interface engineering" is offered by the adjustment of the dose, implantation temperature and annealing conditions [29]. By implanting "substoichiometric" doses ($\leq 1.4 \times 10^{18}$ O/cm^2) a thinner but still continuous oxide is synthesized after HTA [30].

A double SIMOX structure (Fig.9a) has been fabricated by two oxygen implantations at different energies [31]. The advantage comes from a better isolation and the possibility of using the sandwiched Si layer for buried electrical and optical interconnections. Totally isolated Si islands (Fig.9b) were processed by masked implantation through a patterned capping layer [32] : below the cap the oxide is synthesized at the surface while in the window region it is situated at normal depth.

It is pleasing to be able to conclude that the seemingly destructive process of SIMOX results in the Si overlayer being a wafer scale monocrystal with high structure quality and excellent electrical properties [29]. What was considered five years ago to be just an exotic and crude method of synthesis has now become a very exciting technology offering the best SOI

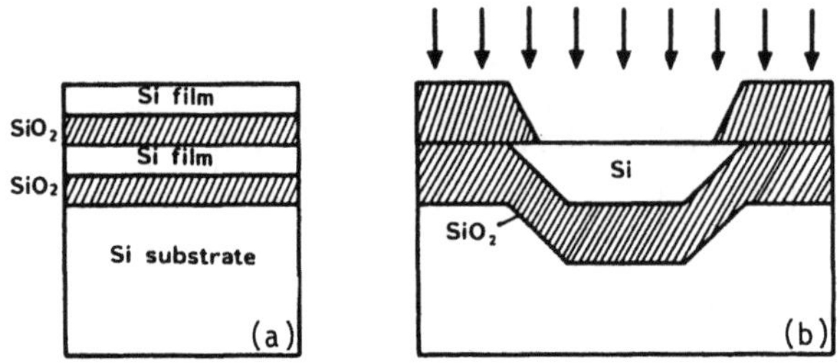

Fig.9. Double SIMOX structure (a) and totally-isolated SIMOX islands

material for microelectronics. The goal is so important that even a new generation of high current implanters (150 mA) has been specially designed to drastically reduce the implant duration (from about 24 hours/wafer with 10 μA beam to only 2 hours/batch). Most of the leading companies have research and development programs on SIMOX. The characterization work and the processing of integrated circuits is more advanced in SIMOX than in other SOI materials. These are very good arguments to focus the discussion of the elctronic properties of SOI materials and devices on the specific case of SIMOX.

4. EVALUATION OF THE BASIC ELECTRICAL PROPERTIES

Conventional techniques used for the characterization of semiconductor materials should be carefully adapted or reconsidered in the case of very thin SOI films. The sheet resistance can be extremely large and high impedance detection systems are necessary. Additional difficulties are related to the preparation of ohmic contacts, probe penetration into the film and extension of depletion regions (induced by the contacts or interfacial charges) which prevent current injection.

4.1 Resistivity, Photoconductivity and Hall Effect

Four-point probing is used on the as-formed SOI material to control the degree of contamination, homogeneity across the wafer and activation of dopant species [33]. The activation of As dopant in LTA-SIMOX has been investigated by annealing a number of samples between 300°C and 800°C for 30 minutes [34]. The sheet resistance presents a steep decrease between 350°C and 500°C, towards a saturation level which correlates well with the dose. This behaviour is similar with that known for bulk Si and suggests an easy control of this important stage of the integration process.

The **Hall effect** and Van der Pauw measurements provide a more detailed inspection of carrier mobility and concentration. In Fig.10 the variations of Hall mobility at low temperature for four typical SIMOX samples are compared. Sample 1 illustrates the excellent crystal quality after HTA (1300°C, 6 hours) in inhert Ar ambient [23,35]. The mobility is high, presents a maximum for about 100 K and is dominated above this by acoustic phonon scattering. Samples 2 and 3, annealed at 1300°C and 1405°C in nitrogen ambient, present a N-type contamination (2×10^{16} cm^{-3}) which results in a more intensive ionized impurity scattering and lower mobility.

Sample 4, annealed at only 1150°C, shows a dramatic mobility degradation and a shift of the mobility maximum above room temperature. This is due to

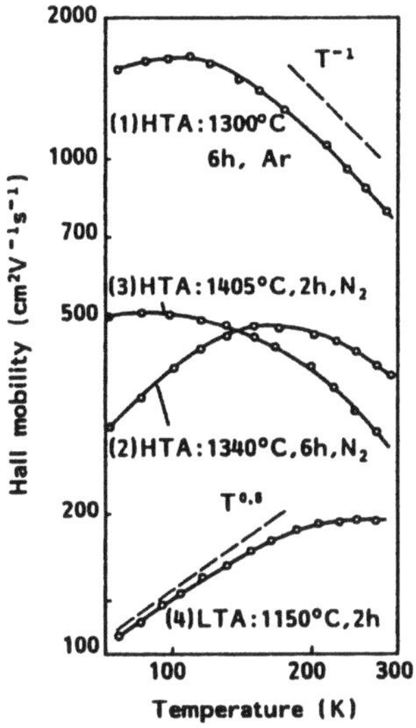

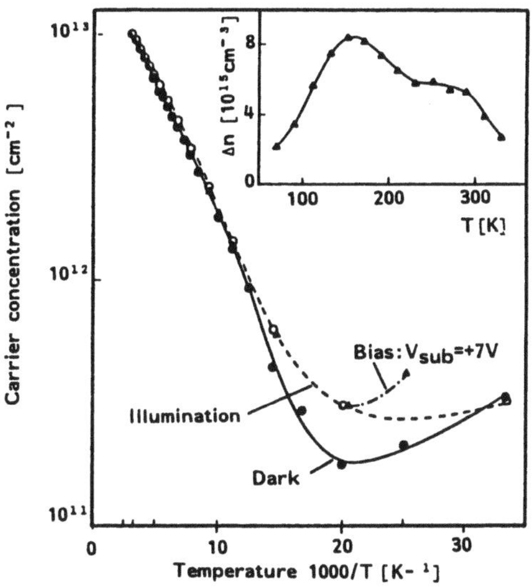

Fig.10. Hall mobility against temperature for SIMOX films ($1.8x10^{18}$ cm^{-2}, 200 keV) annealed under various conditions (after Cristoloveanu et al [35])

Fig.11. Sheet carrier concentration versus reciprocal temperature in HTA-SIMOX. The inset shows the excess concentration of light-induced electrons against temperature (after Papaioannou et al [36])

electrically active defects in the interface region which hopefully are eliminated after HTA. The control of the annealing temperature, ambient atmosphere and purity of the capping layer appear, therefore, to be crucial aspects in fabrication.

Photoconductivity and photo-Hall measurements, performed in the range 25-350 K on HTA-SIMOX [36], show that the carrier concentration normally decreases with decreasing temperature during the progressive carrier freeze-out, but increases below 40 K (Fig.11). This is explained by a two-layer model and by the transition to a hopping conduction in the interface region. The mechanism of electron hopping between impurities is revealed also by the rapid mobility decrease observed below 77 K. The illumination increases the portion of band carriers with higher mobility and, consequently, the transition to hopping is attenuated and the average mobility improves at very low temperature. In contrast, hopping is enhanced by applying a substrate bias which causes carriers to accumulate in the more defective interface region. The excess concentration of photogenerated carriers and the photoconductivity decay suggest that the carrier lifetime is high in HTA-SIMOX.

The **Photo-Induced Current Transient Spectroscopy** (PICTS) was used to characterize slow bulk traps in as-grown films. The method consists of filling interface states and bulk traps via pulsed photoexcitation. Under illumination only the minority carriers are trapped. As the carrier recombination and emission from interface states are very fast processes compared with the emptying of bulk traps, when the light is switched off the instantaneous concentrations of excess majority carriers and trapped minority carriers become equal. In other words, the emission rate of trapped carriers is the same as the the decay rate of the excess majority carrier concentration which is proportional to the current. This transient current is monitored while increasing the temperature and the signal is processed with a DLTS-like system. It is deduced that in SIMOX the average density of bulk traps is about 10^{12} cm^{-2} and has a non-uniform energy profile [37].

4.2 In-depth Profiles

Although the intrinsic limitations of **spreading resistance** are accentuated in thin films, this technique is helpful to verify the vertical configuration of SOI structures. Figure 12 shows that in a double SIMOX structure two distinct buried oxides, 0.2 μm thick, are indeed visible (see also Fig.9a). The sandwiched Si layer has poorer quality and higher resistivity than the Si overlay. On the other hand, the analysis of the redistribution of As implanted with a dose of 10^{15} cm^{-2} reveals an enhanced diffusion, probably assisted by dislocations and interface charges [38]. By increasing the dose, however, normal profiles were found (similar to those in bulk Si and simulated by SUPREM-3).

Accurate profiles of the transport parameters along the vertical axis were determined using depletion-mode **MOS-Hall** devices [39]. The Hall effect was measured while gradually varying the gate voltage to reduce the thickness of the conductive region of the SIMOX film (Fig.13). A differentiation technique provided the **local** values of carrier concentration and mobility in each point. The experiment shows that in the upper 100 nm of the LTA-SIMOX film the mobility is high, almost constant and governed by normal scattering mechanisms. Closer to the synthesized oxide the mobility falls rapidly. An almost flat profile is expected for HTA-SIMOX and ZMR. This anticipation is supported by the very large value of average mobilities and the small difference between the carrier mobilities in the inversion layers situated at the front and back interfaces [35]. At low temperature a discrepancy was observed between the profile of ionized impurities (deduced from capacitance curves) and that of the mobile carriers provided by the Hall effect. The explanation was based on the existence of potential fluctuations induced by crystal imperfections. The depth of potential walls increases towards the buried oxide but still remains much lower (≤10 meV) than in SOS [5].

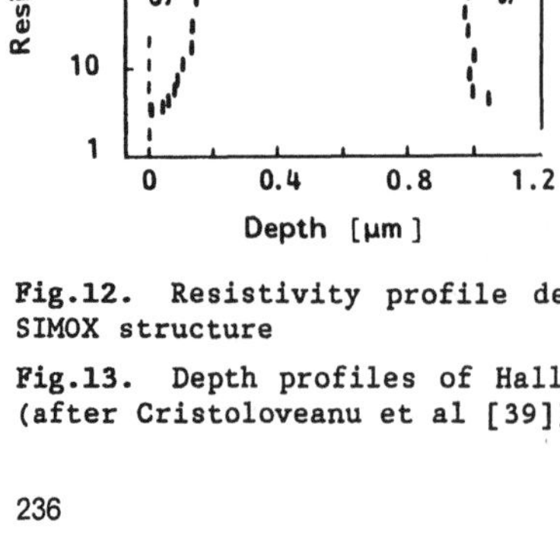

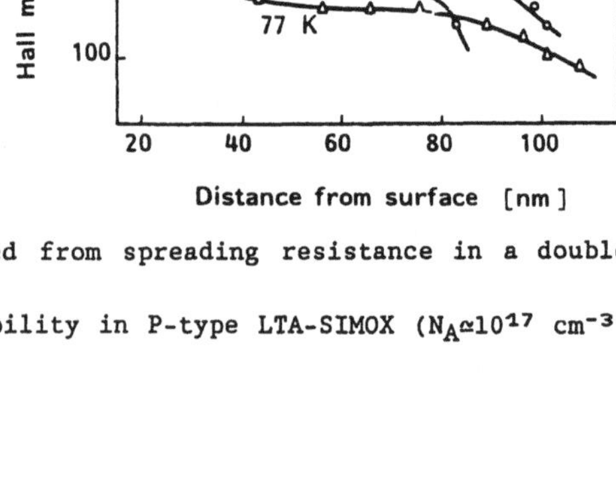

Fig.12. Resistivity profile deduced from spreading resistance in a double SIMOX structure

Fig.13. Depth profiles of Hall mobility in P-type LTA-SIMOX ($N_A \simeq 10^{17}$ cm^{-3}) (after Cristoloveanu et al [39])

The homogeneity of the **buried oxide** was evaluated by differential ellipsometry and **in situ** comparison with a thermal oxide [40]. The method consisted of thermal oxidation of the Si overlayer in order to obtain perfect superposition of the two oxides. The reflection angles were then measured while the oxide was progressively etched. The experiment did not reveal any difference between the optical properties of the thermal and synthesized oxides.

Recent variable angle ellipsometry [41] and wavelength dependent reflectance [42] confirmed that the buried oxide index is identical to that of the thermal oxide. A transition layer, 10-25 nm thick, was observed at the interface between the buried oxide and bulk Si substrate [40]. This layer has a higher index (2.5-3) than for SiO_2 and invalidates the use of standard ellipsometry tables. From the SIMS point of view the interface region is broader than for thermal oxides and has a complex configuration consisting of interpenetrating fingers of Si and SiO_2, also observed by TEM [43].

The dielectric integrity of the buried oxide was inspected by measuring the current flowing between film and substrate. The leakage depends on oxide thickness and is extremely low for normal voltages. Above 60-100 V however, a rapid and irreversible oxide deterioration occurs. Due to interface carrier trapping and depletion layers existing on both sides of the oxide, the leakage current is time-dependent and does not behave symmetrically when changing bias polarity.

4.3 Oxygen Donors in SIMOX

Oxygen is electrically activated in Si by annealing at 450-750°C and forms a number of donor species. As these free carriers remain mobile even after cooling down at room temperature, the intentional doping level is modified and the device performance degraded. This problem is of great concern in SIMOX due to the high dose of oxygen received. It has been demonstrated that in LTA-SIMOX, short additional anneals (30 minutes) are able to lower the wafer resistivity by orders of magnitude. Two types of donors are distinguished [33] : "thermal donors", which are double donors resulting from the formation of metastable oxygen-silicon aggregates between 450 and 550°C, and "new donors" generated at 750°C. At 650°C and above 900°C these

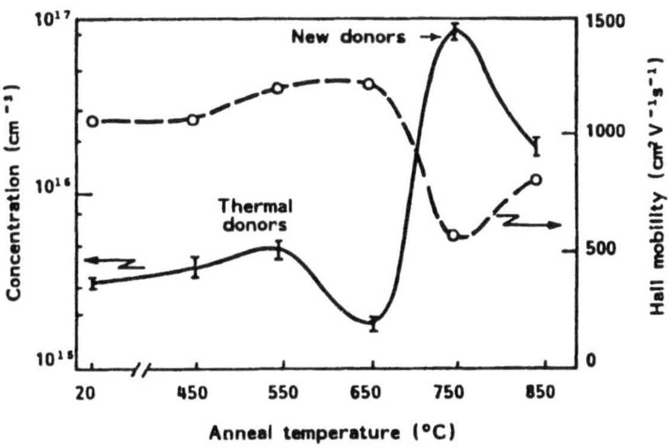

Fig.14. Carrier concentration (——) and Hall mobility (– – –) versus temperature of additional anneals in HTA-SIMOX samples containing oxygen donors (after Cristoloveanu et al [35])

donors are destroyed and the sample recovers its initial resistivity. This "normal" behaviour is modified according to the degree of contamination with carbon which greatly accentuates the formation of new donors.

Recent Hall effect analysis (Fig.14) shows that annealing at 1350°C in Ar ambient hopefully limits the thermal donor generation at a reasonable level of a few 10^{15} cm^{-3}. The kinetics of oxygen donors in SIMOX is more complex than in bulk Si. Isothermal anneals for various durations (15-200 minutes) showed that thermal donors with shallower energy levels and compensation centers are progressively formed, while new donors are essentially generated in the very early stage of the annealing [44].

After neutralization of thermal donors at 650°C, the background contamination is very low (10^{15} cm^{-3}) and the mobility extremely good (1250 cm^2/Vs). The high number of new donors (10^{17} cm^{-3}) generated at 750°C (Fig.14) is presumed to result from positively charged states existing at the interface between small SiO_2 precipitates ($\leq$5 nm) and the Si matrix. As a consequence small accumulation layers are formed and electron transport occurs by percolation. The main portion of these new donors are probably situated in a narrow region close to the buried oxide [35]. This model is supported by the variation of the Hall mobility at low temperature which shows degenerate behaviour below 50 K and interface-like scattering processes, dominated by phonons above 100 K, instead of the normal impurity scattering expected for such a high carrier concentration.

5. SILICON ON INSULATOR INTEGRATED DEVICES

5.1 Operation and Performance of SOI Devices

The schematic configuration of a MOS transistor fabricated in SOI substrates is given in Fig.15a. There are 3 interfaces separated by only about 0.6 μm, 2 oxides and 3 conductive regions (gate, film, substrate). This results in the operation of SOI-MOSFET's to be rather different from that in bulk Si. In normal conditions, the current flows at the front interface I_1 and is essentially governed by the gate. But in SOI the roles of the gate and substrate can be exchanged so that the substrate becomes then a secondary gate which controls the "back" inversion (or accumulation) channel formed at the buried interface I_2. SOI MOSFET's have normally 4 terminals (source,

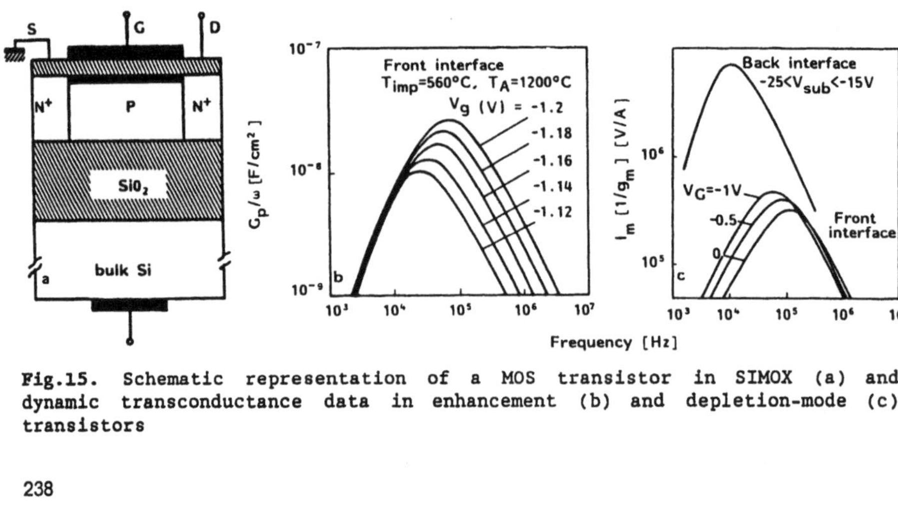

Fig.15. Schematic representation of a MOS transistor in SIMOX (a) and dynamic transconductance data in enhancement (b) and depletion-mode (c) transistors

drain, gate and substrate) without any contact to the Si island which is left "floating". This makes possible the coupling between front and back channels as well as the mutual influence of the gate and substrate on both channels. More or less accurate simulation can be achieved with PISCES, MINIMOS and other programs which solve the Poisson and continuity equations in the multilayer SOI structure.

The very good performance of MOSFET's fabricated on eather LTA or HTA-SIMOX has been emphasized early. Drain current characteristics from weak to strong inversion show that the carrier mobility in the inversion layer and the concentration of interfacial defects are almost the same as in control transistors fabricated on bulk Si wafers. Thus, the memory of the implantation-induced damage has been erased, at the front interface, during the anneal.

The existence of a floating island leads to a **kink** in the drain current (Fig.20b) due to impact ionization : majority carriers accumulate near the source thus increasing the potential until the diode is switched on and activates the parasitic bipolar transistor. The onset of the kink effect depends on the substrate voltage which is used to maintain the buried interface I_2 in the accumulation, depletion or inversion regions. Substrate biasing also controls leakage currents flowing mainly at the Si island sidewalls. Indeed in N and P-channel edgeless transistors with close geometry the leakage is very small (<10 pA/μm for L_{eff}=0.6 μm and V_D=5 V) [35]. On the other hand, the subthreshold swing improves from accumulation to depletion and becomes a minimum at total depletion (70-100 mV/decade).

The static characteristics of the **back channel** transistor greatly depend on the annealing temperature. After LTA the density of interfacial defects exceeds 10^{13} cm^{-2}, while the mobility of inversion carriers is very poor and decreases further (1 cm^2/Vs) with lowering implant temperature [45] and dose [28]. Scattering mechanisms are dominated by interfacial defects, not by surface roughness, and cause the mobility to have an anomalous temperature behaviour. The improvement of the back interface induced by HTA is emphasized by the lower density of fast interface states (<10^{12} cm^{-2}eV^{-1}). Consequently, the weak inversion swing is lowered to 0.6-1 V/decade and the threshold voltage is reduced from 50 V to 5-10 V (as in ZMR material). The mobility now becomes comparable with the front channel mobility, in magnitude (10-30% lower) and temperature dependence [35]. The back channel was found to be slightly longer than the front channel thus demonstrating the absence of any enhanced lateral diffusion at the buried interface [46].

Totally depleted MOS transistors and integrated circuits were recently fabricated in extremely thin SIMOX films (70-100 nm) and present excellent properties : no significant deterioration of mobility, improvement of the subthreshold slope and speed [47]. The suppression of the floating island effect results in kink-free characteristics, improved punchthrough and breakdown behaviour, reduced short channel effects, in particular the aging and threshold voltage shift. The aging is due to the injection of hot carriers into the gate oxide and formation of localized damage (interface states and trapped charges) near the drain junction. This degradation affects the transconductance, threshold voltage, weak inversion slope and leakage current. In totally depleted MOSFET's the lateral electric field is more homogeneously distributed, with a lower peak, so that the probabilities for impact ionization and carrier injection are reduced. Their immunity to aging is indeed better than in partially depleted MOSFET's and much better than in bulk Si transistors [48]. This advantage is certainly a strong argument for a more intensive development of SOI technologies.

SOI components are very fast and, in particular, the performances of SIMOX integrated circuits represent records for the Si family [2] : 27 ps delay/gate in ring oscillators, 2.5 GHz logic dividers, 30 MHz microprocessors, 16 and 64 kbit SRAM's with very good yield.

5.2 Device-Based Characterization Techniques

Unlike the case of bulk Si, the analysis of **MOS capacitance and conductance** in SOI is not straightforward. The measurement can be performed between gate and substrate, gate and film or substrate and film by keeping the third terminal grounded, open or independently biased. The problem is that each type of measurement is affected by high series resistances and parasitic capacitive links due to simultaneous contributions of the various interfaces, oxides and depletion layers. An appropriate equivalent circuit and a theoretical treatment have been proposed to explain the unusual aspect of high frequency C-V curves in SOI and to extract a number of interface properties [49]. In the case of totally depleted films, numerical simulations are required to calculate the coupling between front and back interfaces $I_{1,2}$.

As these various problems also limit the use of transient capacitance, MOS conductance and DLTS experiments, several modifications of the initial SOI structure have been explored. Etching the Si overlay and evaporating an Al gate directly on the buried oxide allowed the interface I_3 (buried oxide - substrate) to be characterized. Standard capacitance-based DLTS experiments are feasible if the thickness of the Si film is increased by epitaxy to a few microns, but unfortunately the data applies to the epitaxial layer and not to the original SIMOX film [50]. A more successful approach was the recent use of current-based DLTS in MOS transistors [51]. Similar information on slow traps is also provided by PICTS (see 4.1), but prior to any device processing.

Very relevant interface properties are obtained using the **dynamic transconductance** technique in SOI MOSFET's [52,53]. A small frequency-dependent signal is used to modulate the gate voltage in weak inversion. The imaginary part of the dynamic transconductance is proportional to the parallel conductance of interface states (also obtained by conventional conductance measurements $G(\omega)/\omega$ on MOS capacitors). There are two obvious advantages in performing this experiment. Firstly, the proportionality coefficient is very high and makes the measurement possible even in submicron transistors, instead of using large area capacitors. Secondly, as it is the current that is being measured, not the capacitance, any effect related to parasitic capacitors or series resistances is avoided and, in addition, there is no need for an electrical contact on the film.

The method has been applied to enhancement [52] and depletion-mode [53] SIMOX-MOSFET's and typical bell-shaped curves are plotted in Fig.15. The magnitude and position of the peak are related to the density and time constant of interface states, respectively. The gate voltage dependence of the peak (Fig.15b) is used to determine the energy distribution of fast states in the gap. Figure 15c indicates that the average density of fast states is 10^{11} $cm^{-2}eV^{-1}$ for the front interface of HTA-SIMOX. When the experiment is conducted from the rear face, the peak is one order of magnitude larger and suggests a more defective buried interface I_2. This peak does not change either in position or in magnitude and is presumed to be due to a dominance of bulk traps near the buried oxide [52]. In the case of LTA-SIMOX implanted at temperatures below 500°C, bulk traps can extend to the surface region.

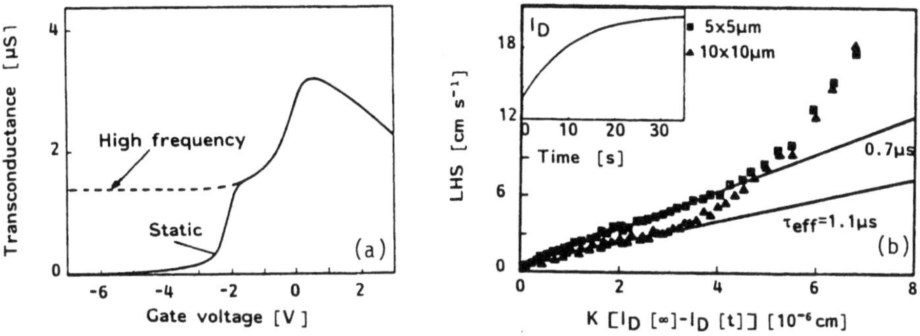

Fig.16. Measurements in depletion-mode SIMOX transistors. (a) Transconductance versus front gate voltage for static and high frequency operation [53]. (b) Transient current variation in deep depletion and Zerbst-type analysis [52]

Figure 16a shows that the **static** transconductance in depletion-mode transistors has a complicated variation with gate voltage which is explained by the successive predominance of the capacitances of the inversion layer, depletion region, interface states and oxide [53,54]. At very high frequencies, however, the inversion charge and interface states become unable to follow the gate signal and the real part of the transconductance gives the series combination of the capacitances of the depletion layer and oxide (Fig.16a). In other words, a conventional C-V curve is obtained, with a significant amplification, by simply measuring the drain current [53]. The oxide thickness, doping level and flat-band voltage can, therefore, be rapidly evaluated.

Charge pumping stands as a rapid and inexpensive evaluation method, but normally requires the existence of a film contact (5-terminal MOSFET's). Measurements were carried out in ZMR [55] and SIMOX [56] by monitoring the film current while continuously pulsing the gate from strong inversion to accumulation. In strong inversion, minority carriers are captured by interface traps. When the surface is pulsed into accumulation, the channel disappears almost instantaneously and trapped carriers recombine with majority carriers. This gives rise to a charge pumping current which is proportional to the density of fast states and can be used to determine their energy profile.

The need for a film contact can be elegantly avoided by using gated P^+PN^+ diodes (Fig.17). In this case, minority carriers are injected by the N^+ contact while the P^+ contact supplies majority carriers thus playing the role of the island contact [57]. The results obtained by pulsing both the gate and the substrate confirm the difference in quality between the front and back interfaces [56,57]. The hump in current, found in low frequency when the experiment is conducted from the substrate (Fig.17b), is due to the presence of bulk traps near the buried interface. This result is in very good agreement with dynamic transconductance data.

The **low frequency noise** in MOSFET's gives evidence for the carrier trapping in slow states situated within the oxide at about 2-3 nm from the interface. The fluctuations of the drain voltage were measured in the range 10^{-1}-10^{5} Hz using a digital spectrum analyser and then converted in a "normalized" drain current spectral density S_I/I^2. Front and back channels were investigated independently by creating an accumulation layer at the

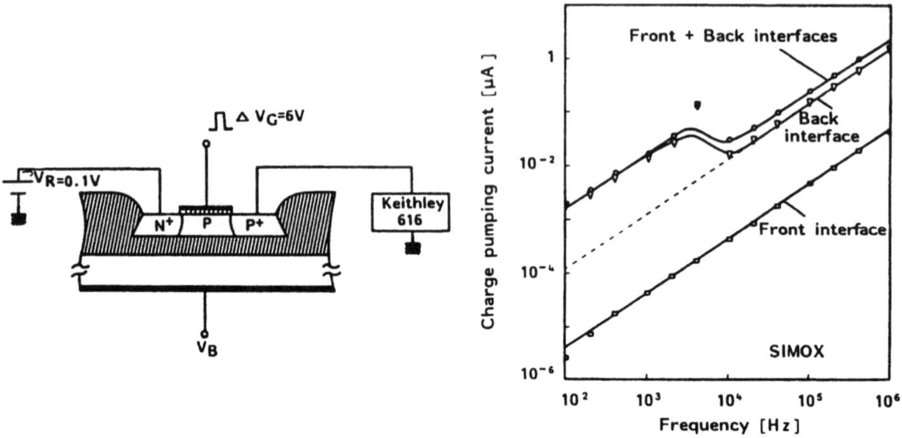

Fig.17. Charge pumping current versus gate voltage in gated P^+PN^+ diodes fabricated in HTA-SIMOX

opposite interface [58]. In LTA-SIMOX, the noise exhibits a "normal" 1/f spectrum and its magnitude shows that the concentration of slow states at the back interface exceeds that of the front interface. A smooth increase of the implantation temperature (from 490°C to 505°C) reduces the noise by one order of magnitude. This beneficial influence is also visible by TEM [59] and correlates well with the improvement in the back channel mobility [45]. In HTA-SIMOX, a $1/f^2$ spectrum was found, in relation with a strongly emerging burst noise probably induced by dislocations [58].

The minority carrier **lifetime** in SOI films is measured using depletion-mode transistors. The gate being pulsed in deep depletion an additional space charge region is formed and then disappears progressively due to the minority carrier generation. The transient drain current is proportional to the thickness of the undepleted region and increases, therefore, to a saturation value (insert of Fig.16b) [52,60]. The carrier generation properties are determined by using a numerical analysis of the transient current variation [52]. The depth profile of the carrier lifetime can also be extrapolated by varying the pulse magnitude. This technique is basically the same as that of Zerbst [61] but has the major advantage of avoiding the drawbacks of transient capacitance measurements in SOI (parasitic resistors and capacitors, need for large areas and film contact).

Carrier lifetime in LTA material is short, 10-100 ns, as in silicon on sapphire films [5,52]. The superior quality of HTA-SIMOX [52] and ZMR [60] is demonstrated by a very high value (1 μs), which is also supported by photoconductivity [36], transient capacitance measurements and magnetodiode effect [62]. The recombination velocity at the front interface is very small (0.2 cm/s) as it is in the system thermal SiO_2-bulk Si. In contrast, the generation velocity at the island edge, found by comparing normal and edgeless transistors (Fig.16b), is 3 orders of magnitude higher (1 m/s).

5.3 Novel Devices in SOI

Not only are conventional components expected to exhibit enhanced performance in SOI but also totally new devices can be envisaged. For instance, ZMR-SOI is the specific material having the capability to support **three**

dimensional (3-D) integration. The basic idea is to stack integrated circuits in SOI films which are successively deposited on previously completed levels. 3-D integration requires low temperature processing of SOI layers and circuits in order to avoid the thermal degradation of underlying circuits that occurs above 850°C. In this respect, lasers and electron beams are the best heating sources able to provide both local recrystallization and shallow in-depth penetration of thermal gradients. Other major difficulties are associated with the necessity of fabricating high quality SOI on the non-uniform topology of underlying layers and conceiving "vertical" interconnections resistant to a 800-900°C process.

The "mezzanine" structure combines the advantages of bulk silicon for power devices and SOI for CMOS and represents an intermediate solution for 3-D integration of smart power circuits. It consists of a power DMOS transistor fabricated in bulk Si connected with CMOS devices processed in a laser-recrystallized ZMR layer which is closely situated but not above the DMOS [15]. We think that the use of localized SIMOX material (Fig.19a) or SIMOX formed by masked oxygen implantation (Fig.9b) is probably more adequate for mezzanine structures. Patterned implanted SIMOX has also been prepared for the fabrication of special MOS transistors : the gate was used to limit the oxygen implanted area under only source and drain junctions [63]. This configuration reduces the junction capacitances, while avoiding any floating island effects.

Full 3-D circuits were processed according to the following sequence :

(1) use of bulk Si to fabricate uni-channel MOSFET's, since they do not suffer from latch-up problems,
(2) formation of the gate oxide, poly-Si gate and interconnections,
(3) deposition and planarization of an LPCVD oxide, 700 nm thick,
(4) deposition of a seeded poly-Si film, 500 nm thick,
(5) laser recrystallization,
(6) island isolation and normal implants of CMOS-SOI transistors (latch-up free),
(7) formation of gate oxide, poly-Si gate and CMOS interconnections,
(8) capping.

This **double-level** structure (SOI/bulk Si) was optimized by checking the invariability in the performance of bulk Si transistors after processing of SOI. The extension to a **triple-level** structure (SOI/SOI/bulk Si) was stra-

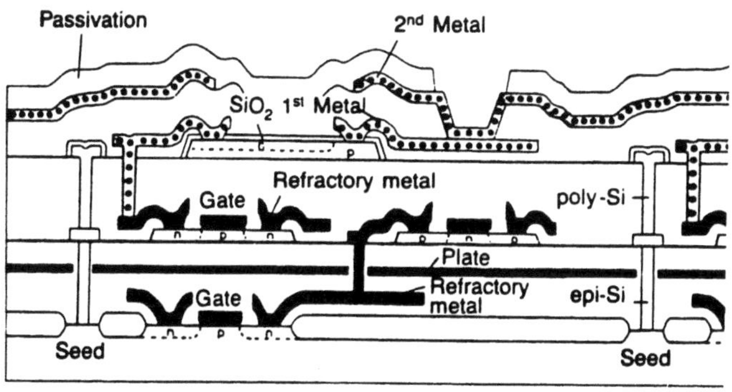

Fig.18. Schematic cross-section of a three dimensional circuit (after Nishimura et al [64])

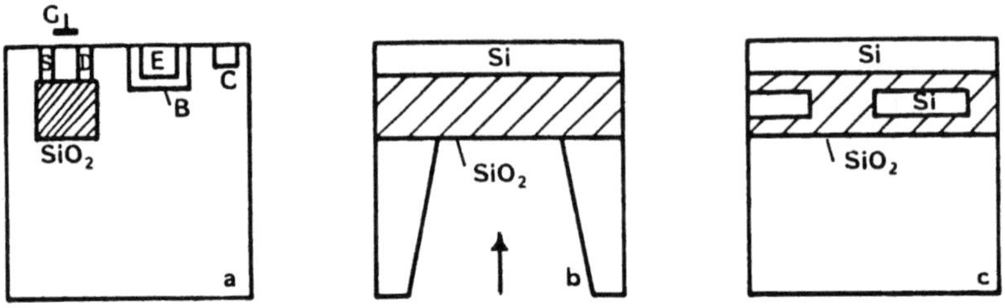

Fig.19. Possible SIMOX structures for new devices : (a) localized SIMOX for combining CMOS and bipolar circuits, (b) pressure sensor and (c) buried Si islands as gates for volume-inversion MOSFET's

ightforward by repeating one more time the stages (3)-(8) above. The electron mobility showed very reasonable values : 750 cm^2/Vs in bulk Si, 530 cm^2/Vs in the first level of SOI and 550 cm^2/Vs in the second SOI level [16]. A similar process was used recently to fabricate an exciting 3-D circuit for image sensing (Fig.18) : the photodiode was situated in the second SOI level, the CMOS processor in the buried SOI level and the NMOS circuit in the bulk Si substrate [64].

SOI offers a polyvalent basis for **transducer** applications. A pressure sensor is proposed in Fig.19b. A cavity is formed in the Si substrate by using the buried oxide as an ideal etch-stopping layer. Piezo-resistors or piezo-transistors are integrated into the membrane whose thickness can be adjusted to fit various ranges of pressure. A similar cavity can be etched in the substrate of a double SIMOX structure (Fig.9a). In this case, the sandwiched Si layer acts as an **optical fiber** whose properties depend on the pressure-induced bending of the diaphragm.

Notice that multiple buried SiO_2 layers can be produced by repeating the double SIMOX process once the thickness of the Si overlay is increased by epitaxy.

Magnetic transducers called **magnetodiodes** have been fabricated on SOS [65] and SIMOX [66]. This device is simply a gated (or ungated) P^+PN^+ diode (Fig.17a) forward biased to provide double-injection conditions. The forward current is dominated by the excess carrier densities and depends, therefore, on the electron-hole recombination in the film and at interfaces. Applying a transverse magnetic field shifts carriers towards an interface : the recombination rate is increased at this interface and reduced at the opposite interface. The current is modified according to the direction and intensity of the magnetic field [65]. High magnetosensitivities are obtained if the film is thin enough and the two surface recombination velocities are very different from each other. By chance these conditions are fulfilled naturally in LTA-SIMOX and SOS [66]. Moreover, **interface engineering** can be exploited in SIMOX to adjust the quality of the buried interface [29]. Indeed, although the HTA material is probably ideal for high speed CMOS circuits, it does not fit all applications. In particular, **high voltage** devices (>180 V) were conceived using the "electric field shielding layer", located at the back interface in LTA-SIMOX [67].

An outstanding SOI device is the **volume-inversion MOS transistor** (VI-MOSFET). The VI-MOSFET has the normal transistor configuration but is operated with simultaneous bias on the gate V_{G1} and substrate : $V_{G2} = KV_{G1}$, where K accounts for the difference in thickness between gate oxide and buried oxide. In thick films, the two inversion channels grow indepen-

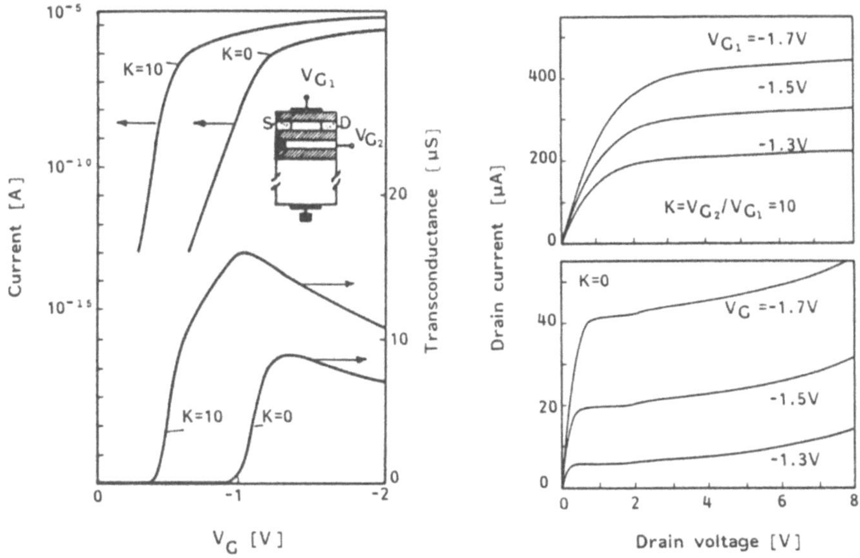

Fig.20. Experimental I-V characteristics and transconductance in the linear region (a) and drain characteristics in the saturation region (b) for a P-type transistor operated normally (K=0) or according to the concept of volume-inversion (VI-MOSFET : K=10) (after Balestra et al [68])

dently, whilst the middle of the film remains undepleted and governed by majority carriers. The current is obtained simply from the parallel combination of two transistors. The behaviour is different in low-doped SOI films, 0.1-0.2 μm thick, where the two interfaces interact. The inversion channels make by continuity the whole film **volume** to go in strong inversion as well [68].

The fact that with VI-MOSFET's minority carriers are no longer confined to an interface, is not only a new concept but leads to enhanced performance. In Fig.20 the experimental characteristics of a submicron VI-MOSFET (K=10) are compared with those of the normally operated transistor (K=0, i.e. inversion channel at the front interface only). The subthreshold swing is very good in the normal MOSFET (70 mV/decade) but is excellent in the VI-MOSFET (29 mV/decade). This is due to the reduced influence of interfacial defects on volume carriers and suggests a spectacular switching capability. As the carrier mobility is much higher in the volume than at the interface, the transconductance of the VI-MOSFET is enhanced by 80 % (Fig.20a).

In the saturation region (Fig.20b), the greatly increased number of carriers allows the current in VI-MOSFET's to be one order of magnitude higher than in normal MOSFET's [68]. Furthermore, the kink effect is eliminated and the avalanche voltage improved from 10 V (for K=0) to 15 V (for K=10). An enlargement of the linear region occurs in VI-MOSFET's since volume pinch-off requires a higher voltage than channel pinch-off at the surface. Finally, short-channel effects and hot carrier induced degradation are attenuated in VI-MOSFET's.

Similar improvements have been obtained for the case of depletion-mode transistors with volume accumulation. The SIMOX structure shown in Fig.19c could be used to provide independent secondary gates for **each** transistor. The buried Si islands are formed by localized double oxygen implantation or by additional mid-range oxygen implantation in the double SIMOX (Fig.9a).

6. CONCLUSION

After a long period of incubation and several years of latent research, SOI is now the object of renewed interest and intensive development. New materials have recently been promoted (SIMOX, FIPOS, wafer bonding) and optimization work is in rapid progress. Integrated circuits were fabricated at an early stage of development in order to compare the capability of various SOI alternatives with that of bulk silicon, and decide about the interest for a deeper investigation. Extremely successful and fast SOI circuits were realized. However, several SOI materials will certainly not be able to survive after this fratricidal competition. SIMOX now appears to be in the best position for CMOS applications. ZMR is unrivalled for three dimensional integration, while FIPOS and wafer bonding seem to be more adequate for power devices.

Although the role of SOI in the future of VLSI circuits can be forseen, further conditions have to be fulfilled before this actually becomes a reality. The wafer price should drop to a reasonable level by standardization and by wider distribution of the material. SOI integrated circuits should be supported by modelling, design and processing rules which are fully specific and not simply transposed from bulk Si. A number of appropriate characterization techniques have been addressed in this paper with the aim of illustrating the main points of interest, difficulties and particularities of SOI.

Finally, SOI has to overcome the historical stage of being an expensive material for very limited, sophisticated or special products. The question is whether SOI is definitively able to progressively occupy a significant portion of the commercial market, now covered by bulk Si chips (SRAM's, logic CMOS, power circuits), or to assist the development of a totally new field of applications with large potential in the future (3-D, transducers). Another challenge of SOI is to catalyse the reconversion of our colleagues working on III-V compounds back to silicon.

Acknowledgements. This work is based on the experience accumulated on SOI at the LPCS Laboratory. The author is pleased to thank his colleagues and in particular F. Balestra, M. Benachir, B. Boukriss, J. Brini, A. Chovet, T. Elewa, M. Gri, H. Haddara, G. Kamarinos, F. Mulholland. Gratitude for scientific interest, sample preparation and joint research is expressed to colleagues around the world : P. Hemment and K. Reeson from the University of Surrey (Guildford), C. Jaussaud, M. Bruel, J. Margail and A. Auberton from LETI (Grenoble), J. Davis and P. Arrowsmith from British Telecom, J. Kilner and R. Chater from the Imperial College (London), G. Celler from Bell Laboratories, I. Stoemenos from the University of Thessaloniki, G. Papaioannou from the University of Athens and D. Chen from Texas Instruments.

References

1. H.E. Oldham, S.L. Partridge: IEEE Trans. Electron Devices ED-29, 1593 (1982)
2. J.P. Colinge: European SOI Workshop, Meylan, France (1988)
3. J. Sturm: Mat. Res. Soc. MRS'87, Fall Meeting, Boston, USA (1987)

4. G.W. Cullen: in Heteroepitaxial Semiconductors for Electronic Devices (Springer, Berlin, 1978)
5. S. Cristoloveanu: Rep. Prog. Phys. 50, 327 (1987)
6. L. Jastrzebski, M.T. Duffy, J.F. Corboy, G.W. Cullen, J. Lagowski: Appl. Phys. Lett. 40, 713 (1982)
7. M. Ishida, Y. Yasuda, H. Wakamatsu, H. Abe, T. Nishinaga, T. Nakamura: Japan J. Appl. Phys. 22, L438 (1983)
8. I. Golecki, H.L. Glass, G. Kinoshita, T.J. Magee: Appl. Surf. Sci. 9, 299 (1981)
9. R.L. Maddox, A.L. Lin, I. Golecki, H.L. Glass: Microelectron. J. 15, 48 (1984)
10. T. Yoshii, S. Taguchi, T. Inoue, H. Tango H.: Japan J. Appl. Phys. 21, Suppl. 21-1, 175 (1982)
11. J. Amano, K.W. Carey: J. Crys. Groth 56, 296 (1982)
12. I. Golecki, R.L. Maddox, K.M. Stika: J. Electron. Mat. 13, 299 (1984)
13. A.C. Ipri: in Silicon Integrated Circuits, ed. by D. Kahng (Academic, New York, 1981)
14. D. Pribat, L.M. Mercandalli, J. Siejka, J. Perriere: J. Appl. Phys. 58, 3313 (1985)
15. J.L. Mermet, H. Achard, H. Bono, J.P. Joly, O. Rossetto: European SOI Workshop, Meylan, France (1988)
16. K. Sugahara, T. Nishimura, S. Kusunoki, Y. Akasaka, H. Nakata: IEEE Electron Device Lett. EDL-7, 193 (1986)
17. G.F. Hopper, J.R. Davis, R.A. McMahon, H. Ahmed: Electronics Lett. 20, 500 (1984)
18. D.P. Vu, L.T.P. Allen, W.R. Henderson, P.M. Zavracky, J.C. Fan: European SOI Workshop, Meylan, France (1988)
19. G. Bomchil, A. Halimaoui, R. Herino: European SOI Workshop, Meylan, France (1988)
20. W.P. Maszara, G. Goeta, T. Caviglia, A. Czerhati, J. Johnson, J.B. McKitterick: Mat. Res. Soc. MRS'87, Fall Meeting, Boston, USA (1987)
21. J. Haisma, E.M.L. Alexander, T.M. Michielsen: European SOI Workshop, Meylan, France (1988)
22. C.G. Tuppen, M.R. Taylor, P.L.F. Hemment, R.P. Arrowsmith: Thin Solid Films 131, 233 (1985)
23. J. Margail, I. Stoemenos, C. Jaussaud, M. Dupuy, P. Martin, B. Blanchard, M. Bruel: in Energy Beam Solid Interactions and Transient Thermal Processing, ed. by V.T. Nguyen and A.G. Cullis (Les Editions de Physique, Les Ulis, 1986) p. 519
24. C. Jaussaud: Mat. Res. Soc. MRS'87, Fall Meeting, Boston, USA (1987)
25. G.K. Celler, P.L.F. Hemment, K.W. West, J.M. Gibson: Appl. Phys. Lett. 48, 532 (1986)
26. G. Harbeke, E.F. Steigmeier, P. Hemment, K.J. Reeson, L. Jastrzebski: Semiconductor Science and Technology, in press (1988)
27. D. Hill, P. Fraundorf, G. Fraundorf: IEEE SOS/SOI Technology Workshop, Durango, USA (1987)
28. B.Y. Mao, P.H. Chang, C.E. Chen, H.W. Lam: J. Appl. Phys. 62, 2308 (1987)
29. S. Cristoloveanu: Mat. Res. Soc. MRS'87, Fall Meeting, Boston (1987)
30. P.L.F. Hemment, K.J. Reeson, J.A. Kilner, R.J. Chater, C. March, G.R. Booker, G.K. Celler, J. Stoemenos: Vacuum 36, 877 (1986)
31. P.L.F. Hemment, K.J. Reeson, J.A. Kilner, R.J. Chater, C. March, G.R. Booker, J.R. Davis, G.K. Celler, J. Stoemenos: Nucl. Instr. and Methods in Phys. Res. B21, 129 (1987)
32. J.R. Davis, A. Robinson, K.J. Reeson, P.L.F. Hemment: IEEE SOS/SOI Technology Workshop, Durango, USA (1987)
33. S. Cristoloveanu, J. Pumfrey, E. Scheid, P. Hemment, R. Arrowsmith: Electron. Lett. 21, 18 (1985)

34. S. Cristoloveanu: in The Physics and Technology of Amorphous SiO_2, Les Arcs, France, in press (1988)
35. S. Cristoloveanu, S. Gardner, C. Jaussaud, J. Margail, A.J. Auberton-Hervé, M. Bruel: J. Appl. Phys. 62, 2793 (1987)
36. G. Papaioannou, S. Cristoloveanu, P.L.F. Hemment: J. Appl. Phys., 63, 4575 (1988)
37. G. Papaioannou, V. Ioannou-Sugleridis, S. Cristoloveanu, M. Bruel, P. Hemment: Int. Conf. on Defects in Semicond., Budapest, Hungary (1988)
38. P. Normand, D. Tsoukalas, N. Guillemot: IEEE SOS/SOI Technology Workshop, Durango, USA (1987)
39. S. Cristoloveanu, J.H. Lee, J. Pumfrey, J.R. Davis, R.P. Arrowsmith, P.L.F. Hemment: J. Appl. Phys. 60, 3199 (1986)
40. M. Levy, E. Scheid, S. Cristoloveanu, P. Hemment: Thin Solid Films 148, 127 (1987)
41. P. Dutta, G.A. Candela, D. Chandler-Horowitz, J.F. Marchiando, M.C. Peckerar: Appl. Phys. Lett., in press (1988)
42. B.Y. Mao, C.E. Chen, H.W. Lam: IEEE SOS/SOI Technology Workshop, Captiva, USA (1986)
43. R.J. Chater, J.A. Kilner, E. Scheid, S. Cristoloveanu, P.L.F. Hemment, K.J. Reeson: Appl. Surf. Sci. 30, 390 (1987)
44. F. Vettese, J. Sicart, J.L. Robert, S. Cristoloveanu, M. Bruel: J. Appl. Phys., in press (1988)
45. D.J. Foster, A.L. Butler, P.H. Bolbot, J.C. Alderman: IEEE Trans. Electron Devices ED-33, 354 (1986)
46. J.R. Davis, K.J. Reeson, P.L.F. Hemment, C.D. March: IEEE Electron Device Lett. EDL-8, 291 (1987)
47. J.P. Colinge and T.I. Kamins: IEEE SOS/SOI Technology, Durango, USA (1987) ; T.W. McElwee and I.D. Calder: ibid
48. J.P. Colinge: IEEE Trans. Electron Devices ED-34, 2173 (1987)
49. J.H. Lee, S. Cristoloveanu: IEEE Electron Device Lett. EDL-7, 537 (1986)
50. P.K. McLarty, J.W. Cole, K.F. Galloway, D.E. Ioannou, S.E. Bernaki: Appl. Phys Lett. 51, 1078 (1987)
51. P.K. McLarty, D.E. Ioannou, H.L. Hughes: Appl. Phys. Lett. (in press 1988)
52. T. Elewa, H. Haddara, S. Cristoloveanu: in The Physics and Technology of Amorphous SiO_2, Les Arcs, France (in press 1988)
53. H. Haddara, T. Elewa, S. Cristoloveanu: IEEE Electron Device Lett. EDL-9, 35 (1988)
54. D.P. Vu and J.C. Pfister: Appl. Phys. Lett. 48, 50 (1986)
55. D.J. Wouters, M.R. Tack, P.W. Mertens, H.E. Maes, C.L. Claeys: 17th European Solid State Device Research Conf., Bologna, Italy (1987)
56. T. Elewa, H. Haddara, S. Cristoloveanu: European SOI Workshop, Meylan, France (1988)
57. T. Elewa, H. Haddara, S. Cristoloveanu, M. Bruel: 18th European Solid State Research Conf., Montpellier, France (1988)
58. A. Chovet, B. Boukriss, T. Elewa, S. Cristoloveanu: 9th Int. Conf. "Noise in Physical Systems", Montreal, Canada (1987)
59. A.E. White, K.T. Shart, J.L. Batstone, D.C. Jacobson, J.M. Poate, K.W. West: Appl. Phys. Lett. 50, 19 (1987)
60. D.P. Vu, J.C. Pfister: Appl. Phys. Lett. 47, 950 (1985)
61. M. Zerbst: Z. Angew. Phys. 22, 30 (1966)
62. S. Cristoloveanu, A. Chovet, G. Kamarinos: Solid State Electron. 21, 1563 (1978)
63. T.I. Kamins, P.J. Marcoux, J.L. Moll, L.M. Roylance: J. Appl. Phys. 60, 423 (1986)
64. T. Nishimura, Y. Inoue, K. Sugahara, S. Kusunoki, T. Kumamoto, S. Nakagawa, M. Nakaya, Y. Noriba, Y. Akasaka: IEDM, Techn. Digest, 111, Washington, USA (1987)

65. A. Mohaghegh, S. Cristoloveanu, J. de Pontcharra: IEEE Trans. Electron Devices ED-28, 237 (1981)
66. G. Dimopoulos, S. Cristoloveanu, A. Chovet, T. Elewa, J. Davis: European SOI Workshop, Meylan, France (1988)
67. S. Nakashima, Y. Maeda, M. Akiya, IEEE Trans. Electron Devices ED-33, 126 (1986)
68. F. Balestra, S. Cristoloveanu, M. Benachir, J. Brini, T. Elewa: IEEE Electron Device Lett. EDL-8, 410 (1987) ; ibid, European SOI Workshop, Meylan, France (1988)

Part VI

Silicide Films

Properties of Transition Metal Silicides

C. Calandra

Dipartimento di Fisica, Università di Modena, I-41100 Modena, Italy

1. Introduction

At the beginning of the eighties solid state physicists became aware of the importance of transition metal silicides in electronics. It was already known [1] that the metallization of silicon by *d* band metals gives rise to rather stable silicon-silicide contacts, the reaction often taking place at room temperature. It had been observed that most silicides are metals and the interface properties, like the value of the Schottky barrier or the distribution of impurities and defects, depend upon the conditions under which the interface is formed. Little however was known about the nature of the chemical bond in these materials, their electrical and optical properties, their relative stability compared to that of the pure components. Indeed in many cases the observed silicide phases did not correspond to those predicted from the available binary phase diagrams.

Since then many efforts have been devoted to the characterization of the structural and electronic properties of bulk silicides and their interfaces with silicon, so that the field of silicide physics has grown up quickly. The main results obtained during its infancy were reviewed in a number of articles [2-5] published two or three years ago, to which the reader is referred for a presentation of the basic problems and the advances in the main subjects up to 1985.

In this paper I shall review the main developments that have taken place in the last few years in the field of the electronic characterization of bulk silicides and of their theoretical description.

To summarize the main results obtained up to 1985 on bulk silicides we can say that the use of photoemission had allowed a significant understanding of the distribution of filled valence states in a number of silicides formed by near-noble (Ni, Pd and Pt) and refractory (Cr, Mo, W) transition metals. In near-noble metal silicides the use of synchrotron radiation photoemission near the Cooper's minimum and of soft X-ray emission spectroscopy had made it possible to study the separate contributions of Si and metal orbitals to the density of filled states [5-7]. The systematic comparison between experimental energy distribution curves (EDC) with the available band theoretical calculations had led to a consistent picture of the general features of the chemical bond [3].

Although the bonding mechanism seemed to be understood, the work done in detail was limited. First of all no quantitative investigation of the distribution of empty states had been performed. Energy loss spectra for a number of materials and interfaces had been published, but the results were rather qualitative in character and did not permit to draw precise conclusions on the distribution of the electron states above the Fermi energy. Very few applications of optical and X-ray absorption spectroscopy had

Springer Series in Materials Science, Vol. 13
Semiconductor Silicon Editors: G. C. Harbeke and M. J. Schulz

appeared, but, due to difficulties in sample preparation, the conclusions were somewhat uncertain. Further the advanced spectroscopies for the characterization of unoccupied states, like inverse photoemission or bremsstrahlung isochromat spectroscopy (BIS), had not been used extensively to study metallic materials.

The theoretical work presented similar limitations. In the first place the complexity of the silicide crystal structures had often prevented theoreticians from performing accurate first-principle band structure calculations for the correct crystal structure. Rather a quite complete set of theoretical results was available for the density of states (DOS) of these compounds in simplified model structures. With a few exceptions the calculations performed with the real geometry were mainly semi-empirical in character, particularly for the metal rich materials. Secondly, when comparing theory with experiments no account was given for matrix element effects in photoemission or optical transitions, the agreement being limited to the correspondence of the energy location of the main structures of the calculated DOS with those appearing in the spectra. Finally it was already clear that the interpretation of the one- or two-hole spectra could be only semiquantitative if self-energy effects were not included in the theory [8,9].

Many of the above difficulties have been overcome by the work done in the last few years. On the experimental side the advent of accurate experimental equipment for inverse photoemission and the increasing application of synchrotron radiation for X-ray absorption spectroscopy have led to significant advances in the characterization of the unoccupied states. First principles approaches to the determination of the band structure and density of states for materials of complex geometry have been reported and detailed theoretical calculations of spectroscopic lineshapes have been performed.

2. Basic Features of the Chemical Bond in Silicides

To elucidate the basic mechanism, which lies behind the formation of the chemical bond in transition metal silicides, we display in Figs. 1 and 2 the theoretical DOS of calcium silicides and Pd_2Si as obtained by the linear muffin tin orbital method of band structure calculation in the atomic sphere approximation, known as the LMTO-ASA method [10,11]. The results refer to the real structure of the compounds. The choice of these systems is not casual. Calcium is at the beginning of the first transition series: its ground state configuration in the atom is $4s^2$ with a vanishing occupancy of the $3d$ levels; in the pure metal the d band lies a few eV above the Fermi energy E_F. Palladium is the last transition metal of the $4d$ series: its atomic configuration is $4d^{10}$. In the pure metal most of the d bands are located below E_F and only a minor portion is empty. Clearly these systems represent two extreme cases, which may be useful to illustrate the evolution of the electronic structure as a function of the d band filling.

We consider the Ca_2Si case first. The filled portion of the DOS is made of two main bands: a narrow band located around -7 eV and a wider and more structured part in the range between E_F and -3 eV. They are separated by a gap of more than 3 eV. An orbital analysis of their composition [12] shows that the lower band is mainly Si s derived, while the upper one arises from the joint contribution of Si p and Ca sp and d orbitals. These results lead to two important conclusions: i) the Si s states, which in pure Si give an important contribution to the bonding by hybridizing with the p orbitals [13], are not appreciably involved in the bond with Ca; ii) the presence of Ca d states below E_F implies that the d orbital occupancy increases on passing from

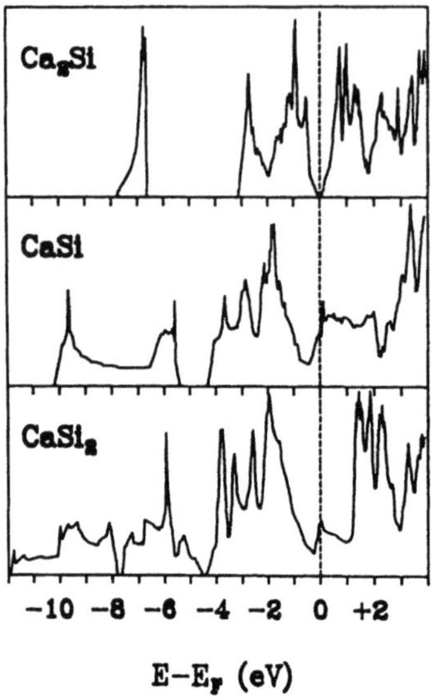

Fig 1. Theoretical single particle density of states for Ca silicides from LMTO-ASA calculations. The vertical dashed line represents the Fermi energy

pure Ca to Ca_2Si. This can occur because a significant fraction of *d* states is shifted down in energy by the coupling with Si. More precisely the structures appearing below E_F are largely due to bonding combinations of Si *p* and Ca *d* orbitals. This bonding portion of the DOS is expected to lie below the main non-bonding part of the Ca *d* band. Antibonding states are expected to produce structures above the main *d* band.

Essentially the same features are present in the DOS of Pd_2Si, except that now most of the *d* band is located below E_F and only the antibonding part is empty. This is clearly shown in Fig. 2. Notice that the Si *s* band is located at lower energies and separated by a gap of about 1.5 eV from the main part of the DOS. The high density portion of the DOS, between -0.5 and -0.2 Ryd in the figure, is mainly Pd *d* derived, while the structures at about -0.60 Ryd and 0.2 Ryd, having a more pronounced Si *p* character, correspond to the bonding and antibonding features respectively.

This description of the chemical bond in metal-rich silicides is consistent with the picture derived on the basis of the photoemission results and from calculations carried out using semi-empirical approaches or model structures [2-5]. It is however rather schematic: many details, which may be relevant in comparing theoretical results with the spectroscopical data, cannot be understood on the basis of this simple model. For example in the case of Pd_2Si it can be shown that the states located around E_F in an energy range of about 4 eV show a significant contribution from Pd *sp* orbitals. Similarly an orbital analysis of the Ca_2Si DOS gives evidence of a significant Ca *sp* orbital contribution spread throughout the conduction band. These results indicate that the *sp-d* hybridization, which is characteristic of the pure metal bond, is still somewhat present in the metal rich silicides and more correctly one should speak of a bond between Si *p* and metal hybrid orbitals. Nevertheless the simple model based on

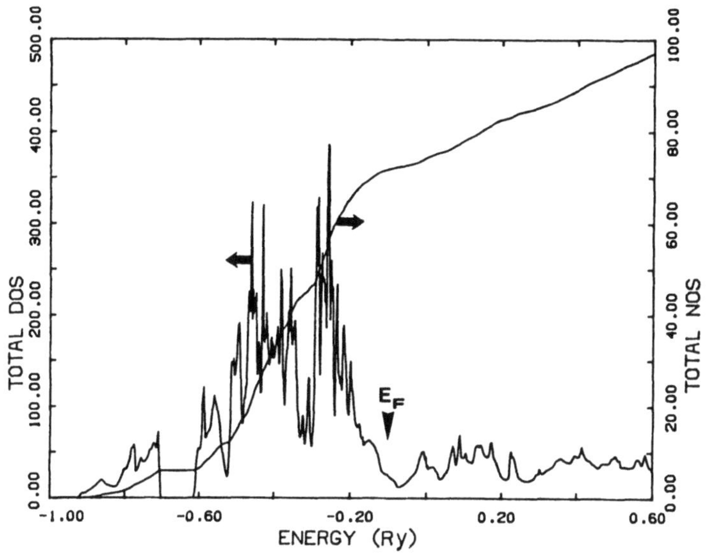

Fig 2. Theoretical LMTO-ASA density of states (DOS) for Pd_2Si and calculated total number of states (NOS) as a function of energy. Energy referred to E_F

the Si *p* - metal *d* coupling remains useful for a first description of the basic features of the bonding.

An additional important piece of information that can be drawn from the two lower panels of Fig. 1 concerns the modifications which take place on passing from metal rich to Si rich materials. In CaSi the lower band widens considerably, the band bottom shifts down in energy and the internal gap is reduced. This band is mainly Si *s* in character, but the structure around -6 eV has a significant Si *p* contribution. These results indicate that the Si states evolve towards a single *sp* band. As to the states lying above -4 eV they have a mixed Si *p* and Ca *spd* character up to -2 eV and a prominent *d* nature around E_F.

The reduction in metal character is accompanied by a downward shift of the *d* bands and an increase in their occupancy. As a consequence the Fermi energy does not fall in a sharp minimum as in Ca_2Si, but lies near a relative maximum. The density of states at E_F increases by two orders of magnitude, an effect of the greatest importance in the electrical and thermal behaviour of the system.

Turning to the disilicide $CaSi_2$ case we notice that the lower part of the band ranges from -5 to -12 eV and the internal gap is almost zero. There are several structures in the high binding energy of the curve, which are to be attributed to Si *s* orbitals and to a lesser extent to Si *p* and Ca *s*. Also the *p-d* portion of the filled DOS is widened compared to the other silicides and displays a larger number of structures.

A similar evolution has been noticed in all the transition metal silicides studied up to now. Tersoff and Hamann [14] have pointed out that it implies that the Si *s* states provide a significant contribution to the bonding in silicides. According to these authors in the case of $NiSi_2$ and $CoSi_2$ the silicon configuration is close to the sp^3 one, typical of the pure element. On the basis of this finding they questioned the validity of the Si *p* - metal *d* model for disilicides.

The problem has been considered by Lambrecht et al. [15] in a recent study of the electronic properties of $NiSi_2$ and $CoSi_2$. They compared the theoretical results for the real structure with those for the adamantane structure. $NiSi_2$ crytallizes in the CaF_2 fluorite structure with space group O_h^5. Every metal atom has eight nearest neighbour Si atoms at a distance $a\sqrt{3}/4$ and twelve second neighbour metal atoms at $a/\sqrt{2}$. In the adamantane structure a metal atom is surrounded by four nearest neighbour Si atoms at a distance $a\sqrt{3}/4$, six next-nearest neighbours Si atoms at $a/2$ and twelve metal neighbours at $a/\sqrt{2}$. The space group is T_d^2. The interest in the adamantane structure is not purely academic: it is one of the structures formed by Ni interstitials in Si that have been proposed for the early stages of silicide-silicon interface formation [16].

The calculated DOSs for $NiSi_2$ are given in Fig. 3. The comparison between the two curves shows that whereas the band bottom looks very much like that of pure silicon in the adamantane case [17], the band structure in the fluorite structure looks very similar to that of a noble metal with an almost filled *d* band cutting through a nearly free electron *sp* continuum. This means that the hybridization characteristic of the tetrahedral Si bonding is present in the adamantane structure but is lost in the real $NiSi_2$. The Ni *d* band, reported in the figure, is wider in the fluorite structure

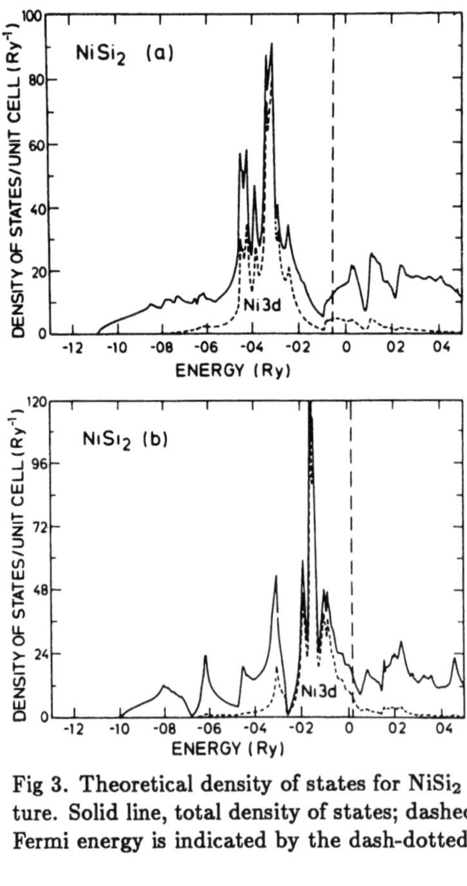

Fig 3. Theoretical density of states for $NiSi_2$ in fluorite (a) and adamantane (b) structure. Solid line, total density of states; dashed line, Ni *3d* partial density of states. The Fermi energy is indicated by the dash-dotted vertical line

and lies further below E_F, indicating an increased Si-metal bonding compared with the adamantane case. In both cases, however, the Fermi level lies above the main part of the Ni *3d* band. The difference is confirmed by a careful analysis of the charge density contours [15], showing that the valence electron distribution is mostly spherical around the atoms in the real $NiSi_2$, where the directional character of the Si covalent bond is lost, while in the adamantane structure the covalent silicon - silicon bonds are observed. In agreement with previous suggestions [3] the authors conclude that Si has a metallic character in the disilicide. The results for $CoSi_2$ are rather similar, except that the Fermi level cuts through the top of the *d* band, instead of lying just above it. This is a consequence of the reduction in the number of *d* electrons per atom.

Lambrecht et al. [15] have been able to determine the equilibrium lattice constant, the bulk modulus and the cohesive energies from total energy and pressure calculations as a function of the volume. They were able to show that the experimentally observed fluorite structure is lower in energy and to explain the stability of the compound with respect to the pure components. For details of this work, in particular for the analysis of the various contributions to the total energy, the reader is referred to the original paper.

3. Spectroscopic Results for Silicides

After so much theory it is time to turn to the experiments and to see how reliable the theoretical results are and how adequate the bonding model is for the understanding of the spectroscopic data. As I mentioned in the introduction, the most important development in the experimental characterization of silicides, that has taken place in the last few years, has been the analysis of the excited state structure through unoccupied state spectroscopies.

For the purpose of clarifying some aspects of the experimental data it is convenient to distinguish between the spectroscopies that leave a core hole in the final state, like X-ray absorption spectroscopy (XAS) and electron energy loss spectroscopy (EELS), and those that probe the empty state distribution by adding an electron to the system, like BIS or inverse photoemission. In the first case one studies the unoccupied states in presence of a more or less screened core hole. This hole may distort the distribution of final states considerably with respect to the one appropriate to the ground state so that the spectrum may not be comparable with the lineshape derived from single particle theory [18]. Also the density of excitations above the threshold may be considerably enhanced by the creation of electron-hole pairs in the screening of the hole (the so called Mahan effect) [19]. Therefore it is of crucial importance in discussing these spectra to verify the reliability of a single particle picture.

On the other hand, working with these techniques allows to use dipole selection rules to probe empty states of different orbital symmetry. For example a study of near-edge structures in the X-ray absorption for the $L_{2,3}$ edge is expected to provide information on the distribution of the empty *d* and *s* states, while the L_1 edge probes the *p* state distribution only.

Spectroscopies based on the analysis of the light emitted by injected electrons are generally interpreted in terms of density of unoccupied single particle states. A perturbation of the ground state is present in these spectroscopies too, but it is less significant than in XAS, because the Coulomb interaction between the extra-valence electron above E_F and the valence electrons below E_F is smaller than the interaction between valence electrons and a core hole. BIS and inverse photoemission therefore probe an electron distribution that is quite close to the ground state distribution. Although one cannot exclude many-body effects in these spectroscopies, they are considered the counterpart for empty states of what photoemission is for the filled ones. Indeed one can get a complete information on the distribution of single particle states simply by putting the BIS and photoemission energy distribution curves one beside the other.

A further development, that has taken place in the last few years, is the study of the structure of valence excitations through optical spectroscopy. Both ellipsometry and optical reflectivity measurements have been used to gain information on electron-hole valence excitations in silicides. A similar analysis has been performed using energy loss spectroscopy. For reasons that will be detailed later, these results are not yet well understood. However, the recent discovery that some Si-rich systems may have the appropriate spectral selectivity to be used for the realization of solar absorber is likely to stimulate further experimental and theoretical work in this field.

3.1 Inverse Photoemission Results

Both BIS and inverse photoemission are based on the same process: one varies the kinetic energy of the incident electrons and by measuring the intensity of light emitted at a given frequency one obtains information on the number and energy location of states above E_F. The relationship between the two techniques is similar to the relationship between X-ray and uv photoemission. BIS gives a spectrum that is rather close to the total density of empty states, while inverse photoemission is more sensitive to the dispersion of the electron energy bands.

The first systematic study of unoccupied states for transition metal silicides has been performed by Azizan et al. [20], who carried out both inverse and direct photoemission experiments simultaneously on three Ni silicides. Previous work by Himpsel and coworkers [21,22] had shown the possibility of probing the behaviour of empty anti-bonding states of Pd_2Si by this technique. The experimental spectra are given in Figs. 4a - 4c , where theoretically evaluated DOSs have been reported for comparison.

We shall first comment on the inverse photoemission spectra. Their main features are the following. For Ni_2Si the empty state spectrum is dominated by a large triangular structure with a peak at 0.9 eV above E_F and a well rounded shoulder at 7.5 eV. These two structures disappear in NiSi, the curve looking very much like a step function at E_F. For $NiSi_2$ the intensity at E_F decreases and one observes a linear increase of intensity at higher energies, due to continuum energy losses in the primary energy of the incident electrons, with some modulations at 3 and 5 eV.

Turning to the photoemission results we notice that the main peak of the experimental spectra is determined by emission from the filled portion of the Ni *d* band. This peak shifts to higher binding energy on increasing the Si content, while the emission from the states lying near E_F is reduced. A second structure appears between -6 and -8 eV, that is characteristic of the silicon rich materials, its intensity being higher in NiSi and $NiSi_2$ than in Ni_2Si.

The authors pointed out that some, but not all, of the observed features in the inverse photoemission curves can be interpreted on the basis of the Si *p*-metal *d* bond model and provide direct evidence of the redistribution of *d* states that takes place in silicides upon increasing the Si content. In particular the features found in Ni_2Si in a

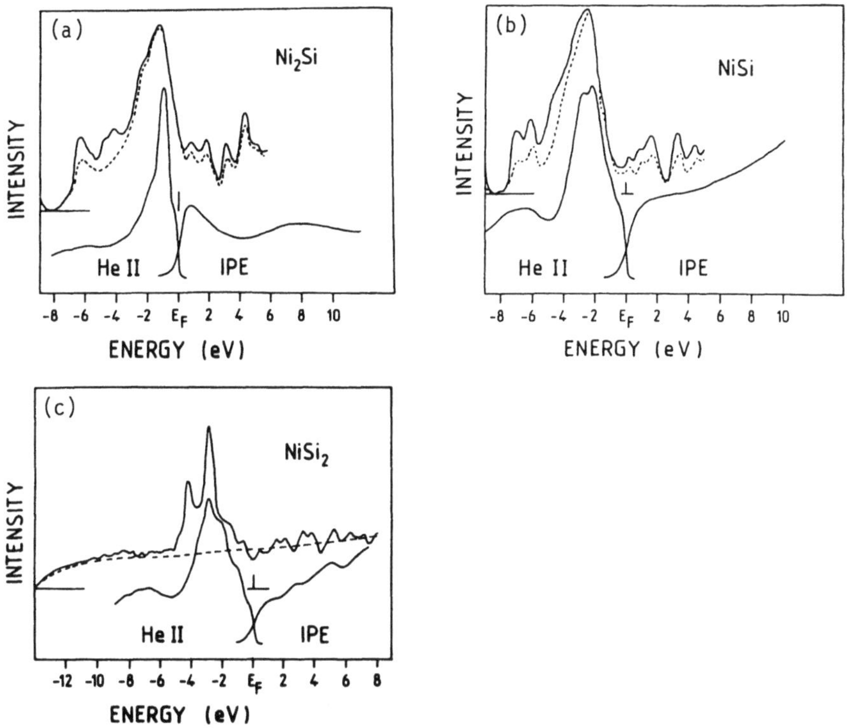

Fig 4a. Ultraviolet photoemission spectrum (He II) and inverse photoemission (IPE) spectrum of Ni_2Si compared with theoretical one particle spectrum. Upper curve, total density of states; dashed curve, Ni projected density of states

Fig 4b. Ultraviolet photoemission spectrum (He II) and inverse photoemission (IPE) spectrum of NiSi compared with theoretical one particle spectrum. Upper curve, total density of states; dashed curve, Ni projected density of states

Fig 4c. He II - IPE spectrum of $NiSi_2$ compared with a theoretical density of states. The dashed line represents a free electron background

2 eV range above E_F can be attributed to emission from the empty portion of the Ni *d* band, that is still highly concentrated in this range of energy. The gradual disappearing of significant features in the mono- and di- silicide spectra is a consequence of the fact that the coupling with Si *p* states causes the *d* character to spread over a large interval of energies, leading to a substantial reduction of the intensity above E_F. Such a model does not provide any explanation of the large shoulder at 7.5 eV which appears to be determined by some other characteristic of the electronic structure. We shall come back to this point later.

One can provide an explanation of the photoemission results on the basis of the bonding model by noticing that the experimental EDCs appear to be dominated by emission from non-bonding *d* states in the region between E_F and -3 eV, while the structure at lower energies is mainly Si *p* derived and can be associated to bonding Si-metal states. However even for the filled state spectra some important features appear unexplained. In particular it is surprising that in Ni_2Si the width of the high-density portion of the *d* band appears considerably lower than that in the other compounds and the structure at lower energies is less clearly seen. One would expect Ni_2Si to have the larger *d* band due to the higher number of Ni neighbours of every metal atom. The only explanation for such strong reduction lies in the possibility that correlation effects play a crucial role in modelling the shape of the EDC [23]. A similar remark could apply to the lower structure too.

As seen in the figures, the comparison with the results available at the time when the experiments were performed was not particularly satisfactory, partly because the loss spectrum was not subtracted from the experimental data, partly because the theoretical curves were not accurate enough and did not include matrix element effects. Notice that the comparison between the photoemission EDCs and the filled part of the theoretical DOS, that is reported in the figures, is more gratifying, since at least the change in location of the filled *d* band as a function of the Si content seems to be correctly reproduced. Also the location of the structure arising from bonding states is reasonably reproduced. The curves for Ni_2Si and NiSi reported in the figure have been obtained including some correlation effects through a perturbative approach [8]. Without this correction the bandwidth for Ni_2Si would have been about two times wider.

Very recently it has been possible to attain the end of performing a systematic comparison between highly accurate loss subtracted unoccupied state spectra and theoretical lineshapes derived from most advanced ground state band structure calculations including matrix elements effects [24]. Figures 5 and 6 show the results of such analysis for a number of silicides. Notice that both BIS and XPS spectra are reported. To emphasize the role of the matrix elements the curves are compared both with the theoretically determined lineshapes and with the broadened *d* DOSs. No correlation effect was included in the lineshape calculation.

Figure 5 allows to illustrate differences in the behaviour of metal *d* states as a function of the principal quantum number at a fixed Si content, since the three metals V, Nb and Ta have the same *d* occupancy but belong to different transition metal series. The general shape of the curves is seen to be the same, consisting of two main peaks on both sides of the Fermi energy and a deep notch at E_F. This behaviour is easily explained in terms of theoretical DOS and is somewhat characteristic of the refractory metal disilicides [25]. The structures straddling the Fermi energy are originated by the metal *d* band. Their difference in energy can be taken as a measure of the separation between bonding and antibonding states and therefore gives the intensity of the Si *p*-metal *d* orbital coupling. It is clear from the figure that the substitution of a *3d* metal by a *4d* or *5d* element leads to a larger splitting. Such a finding is in agreement with the suggestion, made by Bisi and Calandra [26] on the basis of theoretical calculations for near noble metal silicides and by Bhatacharrya et al. [27] for molybdenum compounds, that the coupling with Si increases with the weight for atoms of the same group.

The behaviour of the spectra as a function of the *d* band occupancy within the *3d* series is illustrated in Fig. 6. The most evident effect is the gradual moving of the *d* band through E_F as a result of the increasing occupancy, a fact first observed by Weaver et al.[28]. A second effect is the increased intensity due to emission from non-bonding *d* band states compared with the emission from states farther away from the Fermi energy and having a more pronounced Si *p* character.

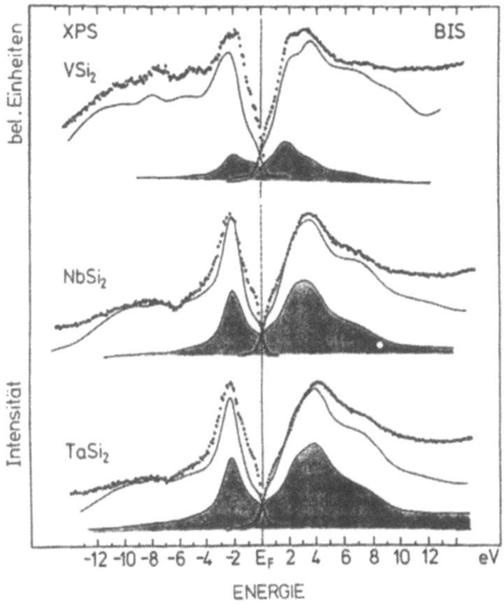

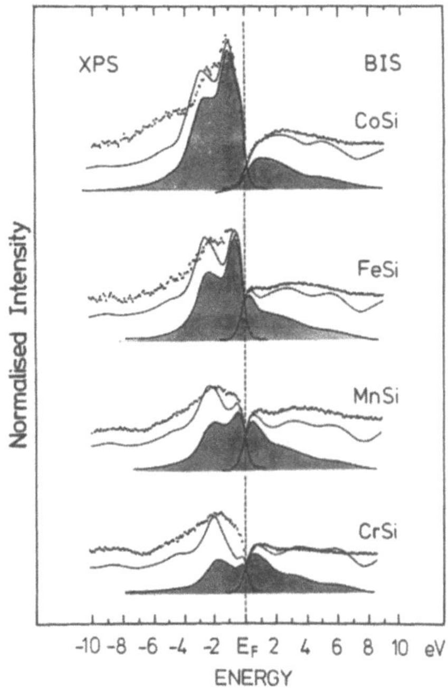

Fig 5. XPS/BIS spectra for VSi_2, $NbSi_2$ and $TaSi_2$ compared with the theoretical spectra, including matrix elements. The metal *d* contribution to the total intensity is shaded

Fig 6. XPS/BIS spectra for the *3d* transition metal monosilicides compared with the theoretical spectra, including matrix elements. The metal *d* contribution to the total intensity is shaded

As to the comparison with the theoretical curves we notice in the first place that the matrix elements are extremely important in determining the spectral lineshape, the relative intensity and the location of the main structures being often somewhat different from those appearing in the *d* DOS. The quality of the theoretical description is rather good: the experimental energy positions of the main structures generally agree with the theory and no spectral feature is found that cannot be related to the single particle DOS. There are two major discrepancies. First in the case of *3d* monosilicides the location of one of the main structures in the XPS spectra is about 1 eV higher than in the theory. Second for most of the BIS spectra the calculated lineshapes are far more structured than the experimental curves. I believe that the disagreement has to be imputed to the neglect of correlation effects in the description of single hole spectra and to limitations in the band structure approach for energies far above the Fermi level. Although there is certainly space for improvement and the need of a more advanced theoretical approach, on the whole the single particle theory appears to provide a sound description of the experimental spectra.

3.2 Absorption Edges

The first detailed study of absorption edges has been performed on palladium silicides. By using X-ray absorption spectroscopy De Crescenzi et al. [29] measured the L_1 and $L_{2,3}$ near-edge structures in Pd_2Si and PdSi. Some of their results are reported in the upper panel of Fig. 7 and compared with the Pd edge structures.

It is seen that the L_3 main edge peak moves by about 3 eV to higher energies on passing from Pd to the silicide. If one takes into account the shift in core level binding energy observed by photoemission one can estimate a shift of the structure of approximately 2.3 eV away from the edge.

In their analysis of the data the authors started from the observation that the electronic structure of Pd_2Si is somewhat similar to that of noble metals, having a nearly filled *d* band and a structureless distribution of empty *d* states above E_F mainly with antibonding character. They argued that if the final states in the absorption process were Pd *d* states the line intensity at the edge should be reduced compared to the pure metal case, in contrast with what observed in the experiment. They concluded that the behaviour of the edge could not be understood in terms of single

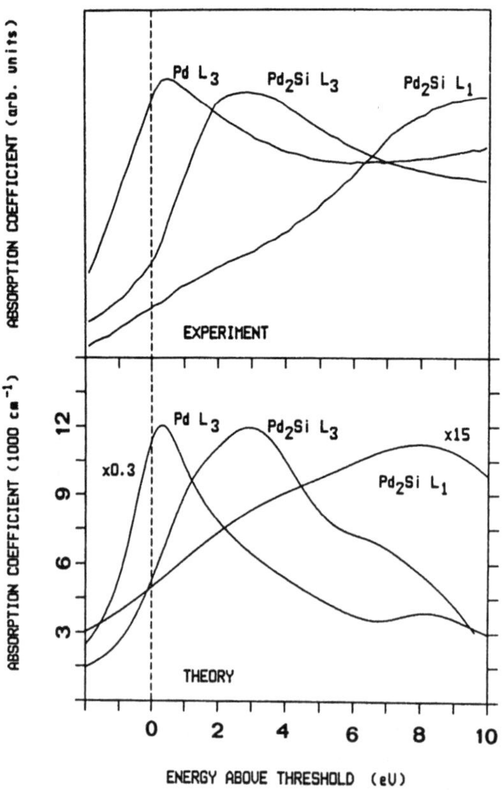

Fig 7. Comparison between the experimental (upper panel) and theoretical (lower panel) L_1, L_3 X-ray absorption spectra of Pd_2Si. The L_3 edge of pure Pd is also shown

particle spectra and proposed an interpretation based on many-body effects along the line of the Mahan-Nozieres-De Dominicis theory of singularity at simple metal edges [19]. These conclusions were somewhat surprising since it was already well known that the main features of the near edge absorption in *4d* metals could be reproduced rather satisfactorily by single particle calculations [30].

The situation was clarified in a subsequent paper by Bisi et al. [31] who pointed out that the mere comparison between experimental curves and single particle DOS is not appropriate, since matrix element effects are extremely important in weighting the various transitions occurring in the process. The calculated spectrum for Pd_2Si, given in Fig.7, compares favorably with the experiments. Bisi et al. found that its shape is essentially determined by transitions to the empty states above E_F with antibonding character. In other words the modifications of the experimental lineshape on passing from palladium to the silicides reflect the change in the *d* band due to the Si *p* - metal *d* coupling.

The possibility of describing the absorption edge in terms of the empty state distribution has been further confirmed by a similar study for Ni_2Si [32]. The observed EELS spectrum is reported in Fig. 8. It is seen that the structure near the edge shifts by approximately 0.7 eV on passing from the metal to the compound and a new structure appears about 6 eV above the edge. In view of the slow variation of the matrix element for *2p* - *3d* transition in Ni, the first structure was attributed to a maximum in the DOS of the empty *d* states 0.7 eV above the Fermi energy, in good agreement with band structure calculations [33]. The authors noticed further that the lowering in the intensity of the line near the edge could be explained by the redistribution of *d* states over a wider energy range, that takes place when the bond with Si is formed. Both observations agree with the conclusions of the inverse photoemission work by Azizan et al. [20], that we discussed previously. Indeed there is a good correspondence between the features observed in the EELS spectrum at 0.7 and 6.5 eV and those seen in inverse

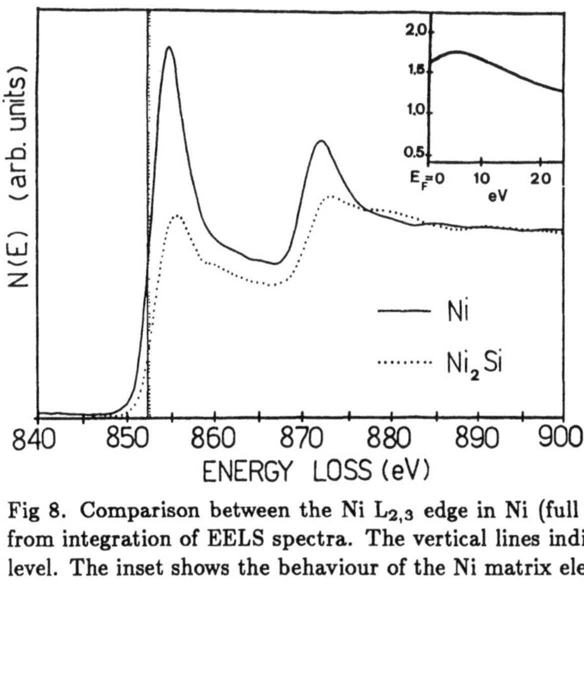

Fig 8. Comparison between the Ni $L_{2,3}$ edge in Ni (full line) and Ni_2Si (dotted line) from integration of EELS spectra. The vertical lines indicate transitions at the Fermi level. The inset shows the behaviour of the Ni matrix element for *2p*-*3d* transitions

photoemission at 0.9 and 7.5 eV. This definitely suggests that the hole creation effects are small in $L_{2,3}$ spectra of Ni_2Si.

One interesting aspect of this comparison is the possibility of obtaining information on the origin of the high-energy shoulder. It appears that its intensity is higher in the inverse photoemission spectra than in EELS. Since in the absorption mainly states with *d* symmetry are probed, this result suggests that it may have a predominant metal *p* character. The fact that both *d* and *p* metal states exist in this energy range indicates that the electronic structure for metal-rich phases is somewhat more complicated than in the simple model description and the hybridization of metal states, characteristic of the pure metal, is not entirely removed by the compound formation.

The successful interpretation of the $L_{2,3}$ spectra for near-noble metal silicides in terms of the distribution of single particle states appropriate to the ground state is somewhat deceiving. The extension of this work to other metals or to different absorption edges has shown that in general many-body effects cannot be neglected. Perhaps the best example is given by the $L_{2,3}$ spectrum of calcium and its silicides. A careful experimental study of these lines [34] has shown that the near edge structures are not modified by the chemical environment of the Ca atom and are not appreciably different from the spectrum of the pure metal. Such a finding is a strong indication that the lineshape is determined by the local density near the core hole rather than by the distribution appropriate to the ground state. A similar difficulty had been previously found in an attempt of interpreting the pure Ca spectrum in terms of the unperturbed single particle DOS [35].

Two basic mechanisms have been invoked to explain the distortions determined by the core hole in the measured empty state distribution. First the Coulomb interaction between the core hole and the excited electron can give rise to a multiplett structure, which hinder the observation of band-like states. Effects of this sort are known to occur at the $M_{2,3}$ edges of *3d* transition metals, whose interpretation is more conveniently carried out in terms of atomic states [36]. Second the attractive hole potential may pull some *d* states down below E_F and cause them to mix with the filled conduction states thus producing a substantial modification of the empty state DOS and the appearance of a bound state below the edge.

The first mechanism is rather unlikely for Ca since the Coulomb exchange interaction is expected to be rather small for a *2p-3d* system in this atom. The second effect seems to be rather likely since it provides an explanation of the occurrence of pre-edge features in the pure metal [35-37].

To explain why there is such a difference in the behaviour of the $L_{2,3}$ edges between Ca and Ni silicides we have to take resort to a more general theoretical description of the X-ray absorption process as outlined by Stern and Rehr [38]. These authors have shown that the appropriate one electron states which can be used in describing the spectrum are those for which the final state of the electron excited directly by the electromagnetic wave is orthogonal to all of the initially occupied states and the initial core state that is emptied in the transition is orthogonal to all the finally occupied states. These one electron states are not the usual single particle orbitals except in two limits, namely for nearly empty and nearly full bands, for which the many-body problem can be reduced to a single particle calculation. In the former case, which is appropriate to Ca, the local impurity DOS of the final state with the core hole is appropriate, while in the latter case, somewhat pertinent to the near noble silicides, the DOS of the unperturbed ground state is appropriate.

In conclusion we can say that, although relevant informations on the nature of the empty state distribution may be achieved by absorption spectroscopies, a full understanding of the near edge structures generally requires a consideration of the effects of the core hole.

3.3 Optical Properties

The spectrum of the valence excitations is most conveniently analyzed by using optical reflectivity and energy loss spectra, the energy of the excitations involved ranging from a few tenth of eV to values higher than the free electron plasma frequency. Both spectroscopies have been applied to silicides and results are available which allow to establish some trends as a function of the composition. However the understanding is still at a qualitative level and no model has been put forward that allow to describe at least the gross features of the spectra. In only one case (Pd_2Si) there has been an attempt to compare the experimental reflectivity with theoretical band structure calculations and the comparison is still not satisfactory. Moreover most of the EELS data have been taken in reflection mode, which leads to the additional complication of having both bulk and surface excitations in the same spectrum.

Up to now very few optical studies have been published. Because of the difficulty of preparing single crystals most of the investigations regard thin films or polycry-

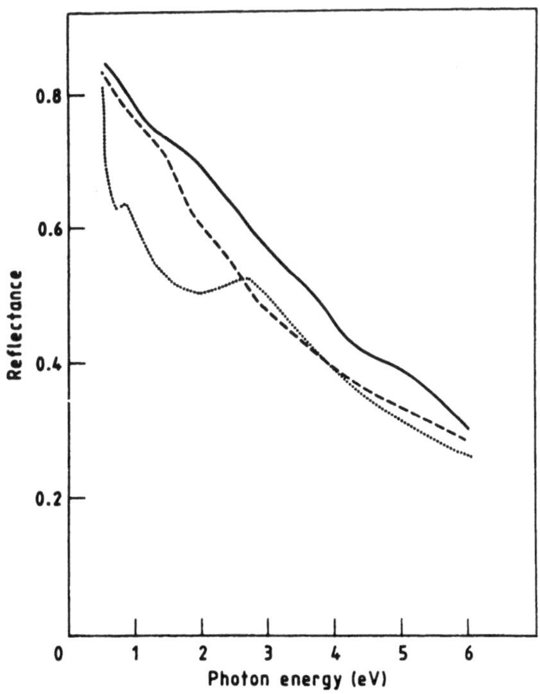

Fig 9. Room temperature reflectivity R for Pd (full line), Pd_4Si (dashed line) and Pd_2Si (dotted line)

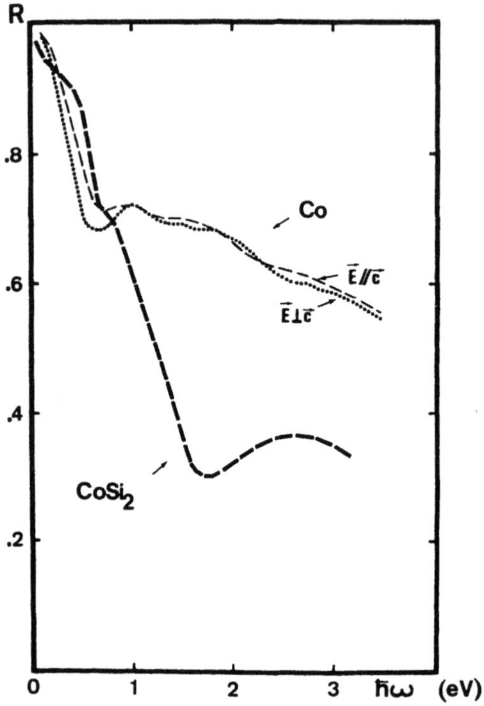

Fig 10. Room temperature reflectivity R of pure Co and $CoSi_2$

stalline samples. This is a serious limitation since in these materials thin film optical measurements by techniques such as reflectivity or ellipsometry may be affected by extrinsic factors like the grain size of the film, the interface roughness or others.

In such a situation the presentation of the results can only emphasize the observed trends and try to explain the gross features of the spectra. We shall first consider the results for silicides of late transition metals, for which a rather complete set of data has been published. Figures 9-10 display the measured room temperature reflectivity at near normal incidence for Pd silicides of various compositions and for Co disilicide [39,40]. For comparison the reflectivity of the pure metal is reported. Similar trends have been observed for Pt silicides [41,42].

The first feature one can notice is that the reflectivity for the compounds is considerably lower than for the pure metal. The difference is more pronounced in the low frequency region, where the silicide curves drop drastically, and increases as a function of the silicon content. The gradual decrease of the reflectivity of the pure metal is generally understood in terms of a superposition of Drude-like intraband absorption with the set of interband transitions due to the partially filled and high-density *d* bands lying near E_F. The onset of these transitions is in the far infrared: for Co it occurs below 0.2 eV and for Pd is lower than 0.06 eV, so that several interband features overlap in the investigated photon energy range. The behaviour of the silicide reflectivities suggests that interband transitions are still present and a clear separation between interband and intraband effects is not possible. However the reduction of reflectivity compared to the pure metal case indicates that the intensity of the interband transitions has been reduced. This may happen only as a consequence of a redistribution of metal *d* states near the Fermi energy.

A second important feature is the occurrence in Si rich materials of a maximum around 2.5 eV, which is not peculiar of the cases illustrated in the figures but has been found in Ni and Pt compounds too. This suggests that it has to be ascribed to a feature in the electronic structure that is common to all these systems. In the dielectric function of $CoSi_2$ this feature corresponds to a broad maximum in ϵ_2 and to an enhancement of ϵ_1, that causes it to cross twice the real axis [40].

The most natural assignment for this characteristic of the spectrum is a transition between the filled high density portion of the *d* band and the empty states immediately or a few tenth of eV above E_F. This assignment is consistent with the observed behaviour of the *d* band energy in near noble metal silicides: as the Si content increases the *d* band shifts down in energy and the density of states at the Fermi level decreases as a consequence of the redistribution over a wider energy range of the *d* states caused by the coupling with Si *p* orbitals. The interpretation is consistent with both photoemission and theoretical band structure results. We can therefore see the maximum in the reflectivity curve of the Si rich silicides as a fingerprint of the removal of *d* states from E_F.

The optical properties of refractory metal disilicides have been recently studied by both ellipsometry and optical reflectivity. In these cases the metal *d* band is half-filled and the Fermi energy lies approximately at its centre. Due to this fact one would expect a behaviour showing a gradual decrease of reflectivity since transitions inside the *d* bands should occur together with the usual intraband optical absorption. Instead the reflectivity shows a sharp drop at low energies and the onset of interband transitions above 0.5 eV. Results for WSi_2 and $MoSi_2$ are reported in Figs. 11 and 12 [43]. Similar behaviour has been found in silicon rich vanadium silicides [44].

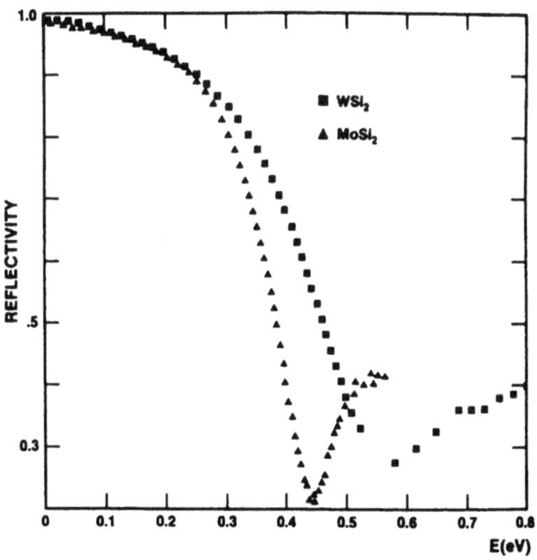

Fig 11. Reflectivities of WSi_2 and $MoSi_2$ crystals in the infrared range

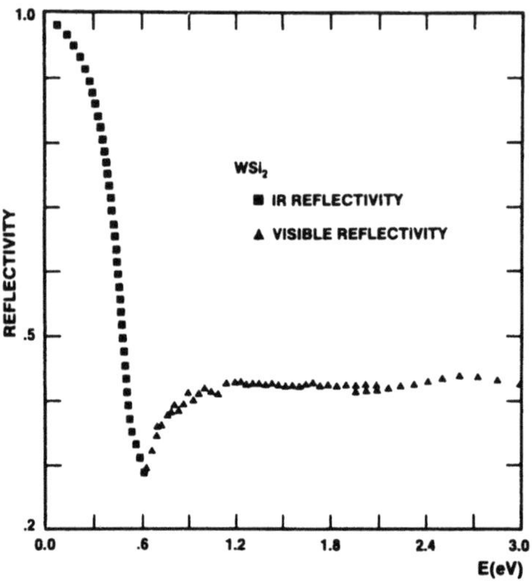

Fig 12. Room temperature reflectivity of WSi_2 under normal incidence

The explanation of these results lies in the peculiar behaviour of the *d* bands for refractory metal disilicides which, as shown in Fig. 5 for VSi_2, have a deep notch in the density of states at the Fermi energy. This pronounced minimum determines a quenching of low energy interband transitions and shifts their onset at a value of a few tenth of eV, where higher DOS values are found on both sides of E_F. An interesting feature of the reflectivity data is the displacement of the onset for interband transitions to higher energy on passing from a transition row to the subsequent one. The effect is due to the increased splitting between bonding and antibonding states, which straddle the notch at E_F and is consistent with the outcome of joint BIS and photoemission experiments given in Fig. 5.

The conclusion to be drawn from these examples is that the information provided by optical data are basically consistent with the current band model, although a detailed theoretical understanding of the spectral lineshape is still needed.

Acknowledgements

I gratefully acknowledge helpful discussions with O. Bisi and C. Mariani. The research was supported by CICAIA (University of Modena) and Ministero della Pubblica Istruzione.

5. References

1. S.P. Murarka: Silicides for VLSI Applications (Academic Press, New York 1983)
2. G.W. Rubloff: In Festkőrperprobleme, ed. by P. Grosse, Vol. 23, (Vieweg, Braunschweig 1983) p. 179
3. C. Calandra, O. Bisi and G. Ottaviani: Surface Sci. Reports **4**, 271 (1984)

4. G.W. Rubloff: In Dynamical Phenomena at Surfaces Interfaces and Superlattices eds. F. Nizzoli, K.H. Rieder and R.F. Willis (Springer-Verlag 1984), p. 220
5. G. Rossi: Surface Sci. Reports **7**, 1 (1987)
6. G. Rossi, I. Abbati, L. Braicovich, I. Lindau and W.E. Spicer: Solid State Commun. **39**, 195 (1981)
7. G. Rossi, I. Lindau, L. Braicovich and I. Abbati: Phys. Rev. **B28**, 3031 (1983)
8. O. Bisi, C. Calandra, U. del Pennino, P. Sassaroli and S. Valeri: Phys. Rev. **B30**, 5696 (1984)
9. C. Calandra, O. Bisi, U. del Pennino. S. Valeri and Y. Xi: Surface Sci. **168**, 164 (1986)
10. A. Spinsanti, O. Bisi, I. Abbati, L. Braicovich, A. Iandelli, G.L. Olcese, A. Palenzona, C. Carbone, I. Lindau: Le Vide Les Couches Minches **42**, 211 (1987)
11. O. Bisi, O. Jepsen and O.K. Andersen: Phys. Rev. **B36**, 9439 (1987)
12. L. Braicovich, C. Carbone, I. Lindau, A. Iandelli, G.L. Olcese, A. Palinzona (to be published)
13. W.A. Harrison: The Electronic Structure and the Properties of Solids (Freeman, New York 1980)
14. J. Tersoff and D.R. Hamann: Phys. Rev. **B28**, 1168 (1983)
15. W.R.L. Lambrecht, N.E. Christensen and P. Blochl: Phys. Rev. **B36**, 2493 (1987)
16. O. Bisi, L.W. Chiao and K.N. Tu: Phys. Rev. **B30**, 4664 (1984)
17. J.R. Chelikowsky and M.L. Cohen: Phys. Rev. **B14**, 556 (1976)
18. U. von Barth and G. Grossman: Solid State Commun. **32**, 645 (1979)
19. C. Almbladh and L. Hedin: In Handbook of Synchrotron Radiation, Ed. E.E. Koch, Vol. 1 (North-Holland, Amsterdam 1983) p. 607
20. M. Azizan, R. Baptist, G. Chauvet and T.A. Nguyen Tan: Solid State Commun. **57**, 1 (1986)
21. F. Himpsel and Th. Fauster: J. Vac. Sci. Technol. **A2**, 815 (1983)
22. F. Himpsel: Advances in Physics **32**, 1 (1983)
23. J.C. Fuggle, F. Ulrich Hillebrecht, R. Zeller, Z. Zoinerek, P.A. Bennett and Ch. Freiburg: Phys. Rev. **B27**, 2145 (1982)
24. W. Speier, E.V. Leuken, J.C. Fuggle, D.D. Sarma, L. Kumar, B. Dauth, K.H. Buschow, to be published; W. Speier, Thesis, unpublished.
25. C. Calandra and O. Bisi: Phys. Rev. **B31**, 8288 (1985)
26. O. Bisi and C. Calandra: J. Phys. C. **14**, 5479 (1981)
27. B.K.Bhattacharyya, D.M. Bylander and L. Kleinmann: Phys. Rev. **B32**, 7973 (1985)
28. J.H. Weaver, A. Franciosi and V.L. Moruzzi: Phys. Rev. **B29**, 3293 (1984)
29. M. De Crescenzi, E. Colavita, U. del Pennino, P. Sassaroli, S. Valeri, C. Rinaldi, L. Sorba and S. Nannarone: Phys. Rev. **B32**, 612 (1985)
30. J.E. Műller and J.W. Wilkins: Phys. Rev. **B29**, 4331 (1984)
31. O. Bisi, O. Jepsen and O.K. Andersen: Europhys. Lett. **1**, 149 (1986)
32. U. del Pennino, C. Mariani, S. Valeri, G. Ottaviani, M.G. Betti, S. Nannarone, M. De Crescenzi: Phys. Rev. **B34**, 34 (1986)
33. Xu Jian-hua and Xie Lei-ming: Solid St. Commun. **55**, 891 (1985)
34. L. Braicovich unpublished results
35. J. Fink, Th. Műller-Heinzerling, B. Scheerer, W. Speier, F.U. Hillebrecht, J.C. Fuggle, J. Zaanen and G.A. Sawatzky: Phys. Rev. **B32**, 4899 (1985)
36. H. Nakamura, K. Ichikawa, T. Watanabe and K. Tsutsumi: J. Phys. Soc. Japan **52**, 4014 (1983)

37. J. Zaanen, G.A. Sawatzky, J. Fink, W. Speier, J.C. Fuggle: Phys. Rev. **B36**, 4905 (1985)
38. E.A. Stern and J.J. Rehr: Phys. Rev. **B27**, 3351 (1983)
39. A. Borghesi, G. Guizzetti, L. Nosenzo, A. Piaggi, A. Stella and G. Majni: Semicond. Sci. Technol. **1**, 184 (1986)
40. C. Viguier, A. Cros, A. Humbert, C. Ferrieu, O. Thomas, R. Madar and J.P. Senateur: Solid St.Commun. **60**, 923 (1986)
41. A. Humbert and A. Cros: J. Physique Lettres **44**, L 929 (1983)
42. W. Henrion and H. Lange: Phys. Stat. Sol. **(b)112**, K57 (1982)
43. F. Ferrieu, C. Viguier, A. Cros, A. Humbert , O. Thomas, R. Madar andJ.P. Senateur: Solid St. Commun. **62**, 455 (1987)
44 A. Borghesi, A. Piaggi, G. Guizzetti and F. Nava, to be published.

State Density Gap in Ti-Silicide/p-Si/p $^+$Si Schottky Barriers

W. Cabanski and M. Schulz

Institut für Angewandte Physik, Universität Erlangen-Nürnberg, D-8520 Erlangen, Fed. Rep. of Germany

1. Introduction

We report on a new type of tunneling experiment which provides direct information on interface properties of metal semiconductor junctions. Modern molecular beam epitaxy (MBE) is employed to grow a tunneling layer structure of titanium on silicon as shown in Fig. 1.

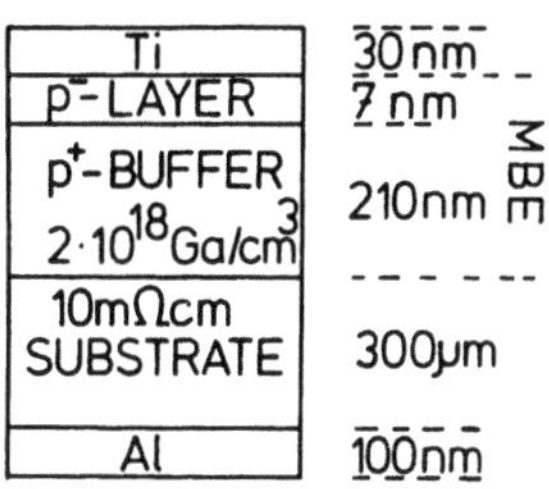

Fig. 1: Layer sequence of the tunneling structure. The buffer layer having a Ga-impurity band and the nominally undoped p⁻ spacer layer ($p<10^{16}cm^{-3}$) are grown by MBE /1/. An Al film is used as an Ohmic contact to the Si substrate. The Schottky contact is formed by the Ti film on the thin p⁻ layer.

A thin 5-10nm low-doped p⁻ Si spacer layer ($p<10^{16}cm^{-3}$) is employed to form a well-defined tunneling barrier to highly doped ($p=2x10^{18}cm^{-3}$) silicon. By using the p⁻ spacer, we avoid pileup of dopants in the metal-semiconductor interface during silicide formation which was reported to occur for highly doped substrates. The high doping in the p^+ buffer layer results in impurity band conductivity /1/. The impurity band provides us with a sharp density of state peak for a tunneling study of the metal semiconductor interface state density.

2. Experimental

Two sets of tunneling diodes are prepared on portions of the same MBE grown wafer. In the first set, titanium dots with a diameter of 0.3 mm are evaporated without removal of the native oxide from the surface. In the second set, the titanium dots are evaporated on the Si substrate immediately after an HF-dip to remove the native oxide.

Some of the samples of both sets are heated by rapid isothermal annealing (RIA) at 550 °C for 20s. The temperature chosen is the threshold temperature reported for Ti silicide

Springer Series in Materials Science, Vol. 13
Semiconductor Silicon Editors: G. C. Harbeke and M. J. Schulz

formation /2/. The RIA is chosen to avoid a complete consumption of the spacer layer by the silicide formation.

Rutherford backscattering (RBS) data indicate that no or only a marginal (<5nm) interdiffusion and reaction has taken place during the RIA process. We assume that at least at the interface there is a Ti silicide film even when the silicide formation has not completely penetrated the evaporated Ti film.

3. Tunneling current characteristics

Tunneling current-voltage characteristics measured at a temperature of 11K are shown (continuous line) in Figs. 2a and 2b for forward and reverse bias, respectively. In forward direction, the current increases continuously with increasing bias. In reverse direction, we observe an increasing current with a kink in the characteristics.

The basic features of the characteristics measured are in accordance with the basic models shown in Figs. 3a and 3b. The monotonous increase of the tunneling current in forward direction is caused by the reduction of the tunneling barrier height and the shortening of the tunneling depth with increasing forward bias. In reverse direction, the tunneling current remains low up to the bias threshold for the onset of tunneling into the band. This onset is visible by the kink in the I/V characteristics Fig. 2b.

The threshold bias voltage of the kink is a direct observation of the activation energy $eV_{th} = E_{Ga}-E_{v}$ from the gallium impurity band into the valence band (see Fig. 3b).

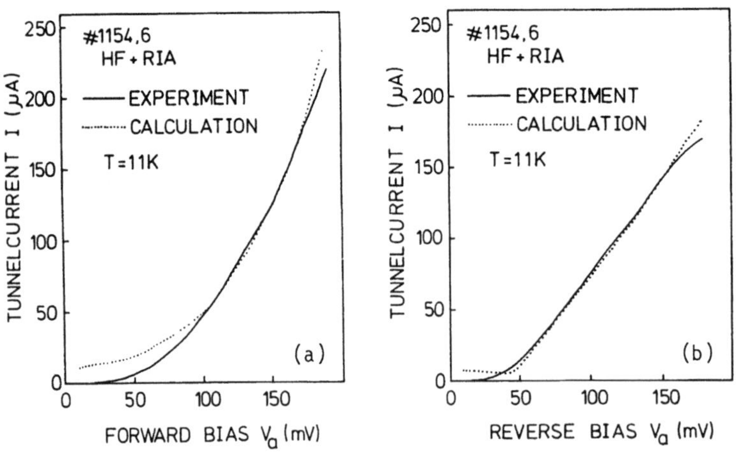

Fig. 2: Current-voltage characteristics for the HF etched and RIA annealed sample # 1154,6. The continuous line is measured, the dotted line is calculated. (a) Forward and (b) reverse characteristics.

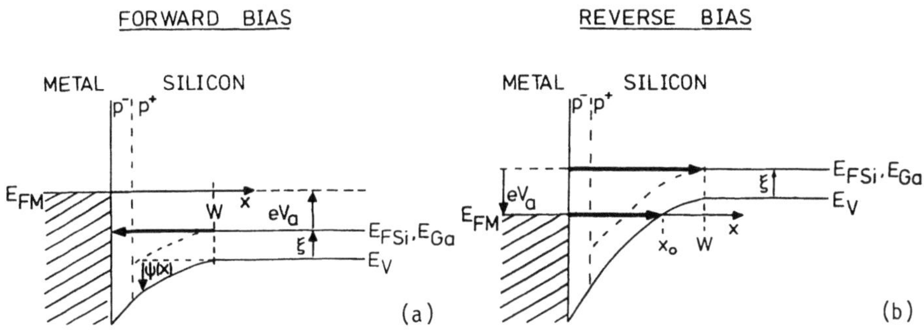

Fig. 3: Band diagram of the tunneling structure in a) forward and b) reverse bias direction. The Fermi energy in Si is pinned to the Ga impurity band. The bold arrows indicate tunneling paths for holes. Two tunneling paths are possible in reverse direction (b) when the applied bias voltage exceeds the gap to the impurity band $eV_a > \xi$.

This threshold voltage V_{th}=45mV is less than the activation energy $E_{Ga}-E_v$ = 74 meV reported in the literature for Ga-activation in silicon. The reduced barrier observed may be due to the high doping of the MBE material which leads to a band tail and a finite width of the impurity band /3/.

4. Discussion

The tunneling current into and out of an impurity band is directly proportional /3/ to the density of states on the metal side of the junction. Since the tunneling probability for a constant state density as a function of the applied bias voltage can be calculated by standard models /3-6/, we can evaluate a relative state density distribution $D_{rel}(V_a)$ as a function of the applied voltage V_a by simply dividing measured and calculated currents.

The ratio of the two measured and calculated curves (Fig. 2) is replotted to yield the relative density of state distribution shown in Fig. 4. The plot extends from approximately 40mV in reverse bias, i.e. the onset for tunneling into the valence band beyond which the evaluation is not valid, to 180mV in forward bias direction. The arrow in the graph indicates the bias voltage V_{ao}=100mV for which the constant density states for normalization is taken.

A significant drop in the density of states is visible for sample # 1154,6 in the bias voltage range -40mV ≤ V_a ≤ + 80mV around zero bias voltage. A less pronounced density of state drop in a narrow range -40mV ≤ V_a ≤ + 40mV is visible for sample # 1154,2 having the native oxide. We interpret this drop as a gap in the density of state distribution which appears on Ti-silicide Schottky barriers. The gap is well developed and >100meV wide for direct reaction of Ti with a HF-cleaned Si surface. In this case, the density of state

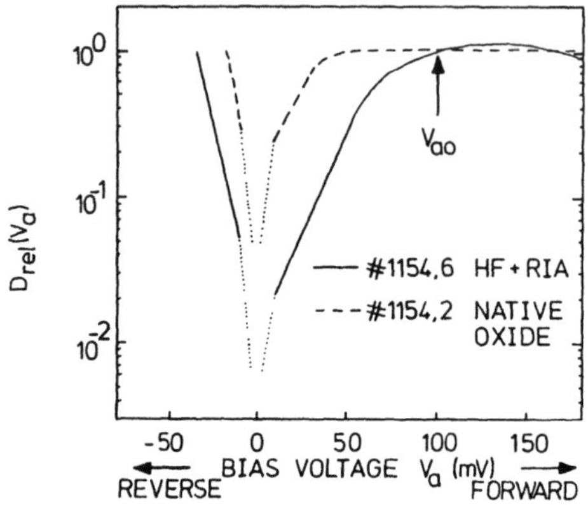

Fig. 4: Relative metal state density D_{rel} as a function of applied bias voltage V_a. The state density is normalized to the value at V_{ao}=100 mV (arrow). Full and dashed lines are calculated for two different samples as indicated. The dots indicate the voltage range V_a << ± 10mV, where the evaluation is not valid due to limitations by the finite temperature (11K) and the width of the impurity band (≤ 15 meV)

drops in the gap to a value approx. 2% of the magnitude outside the gap in forward bias where the density of states is roughly constant. In reverse bias this range is not accessible by the measurement structure using the gallium impurity band.

The sample # 1154,2 having the native oxide which inhibits the Ti silicide formation only shows a weak reduction of state density to a value of 35%. This gap is probably due to a patchy inhomogeneous Ti silicide formation in areas where Ti has penetrated the thin native oxide. The bias voltage range ± 10mV around zero bias which is drawn by dotted lines in Fig. 4 does not represent a real density of state distribution. The model calculation is not valid in this range, because it is a zero temperature approximation and assumes an infinitely sharp impurity band.

For the incomplete silicide formation in a limited depth range we expect that Ti-rich silicides, e.g. TiSi, are formed. These expectations are consistent with recently reported XPS observations /7/.

The state density gap observed is probably that predicted by the computed band structures for TiSi /8/. Our tunneling measurement is the first observation of a state density gap for a Ti-rich silicide. The Fermi energy is located in the center of the gap.

5. Conclusions

We have studied the Ti-silicide Schottky barrier on p-type Si by analyzing tunneling currents. The special feature of our tunneling structure is a p^- Si film grown by MBE on a highly Ga-doped Si substrate. The impurity band of the Ga-doping is used as a sharply peaked density of state probe for the tunneling. The p^- spacer layer ensures a well-defined tunneling barrier.

The tunneling characteristics measured is well described by model calculations. We can determine the activation energy of the Ga-impurity band in silicon to $E_{Ga}-E_v$=45meV and estimate the width to $\Delta E \approx$ 10-15meV. For the Ti-Si Schottky barrier, we find a gap of $\Delta E \approx$ 100meV in the density of states. We interpret this gap by an interfacial layer of a Ti-rich silicide formed by the rapid thermal annealing applied. The Schottky barrier is pinned in the center of this state density gap.

Acknowledgements

The authors are grateful to Drs. E.Kasper and H.J.Herzog at AEG Research Lab, Ulm for preparing silicon wafers having the MBE-grown p^- spacer layer which is crucial for these tunneling experiments. The RBS test measurements on the interfacial reaction by Thomas J. Pollock, Electrostatics International, USA are gratefully acknowledged.

References

1. A.Casel, H.Jorke, E.Kasper and H.Kibbel, Appl.Phys.Lett. 48, 922(1987)
2. M.A.Taubenblatt and C.R.Helms, J.Appl.Phys. 53, 6308(1982)
   see also M.A. Taubenblatt, D.Thomson and C.R.Helms, Appl.Phys.Lett. 44, 895 (1984)
3. W.Cabanski, M.Schulz, to be published in Appl.Phys.A (1988)
4. E.L.Wolf, Principle of Electron Tunneling Spectroscopy, International Series of Monographs on Physics 71, Oxford Science Publication, 1985, p. 19ff
5. J.W.Conley, C.B.Duke, G.D.Mahan and J.J.Tiemann, Phys.Rev.150, 466 (1966)
6. J.W.Conley and G.D.Mahan, Phys.Rev. 161, 681 (1967)
7. S.A.Chambers, D.M.Hill, F.Xu and J.H.Weaver, Phys.Rev.B 35. 634 (1987)
8. J.H.Weaver, A.Franciosi and V.L.Moruzzi, Phys.Rev.B 29, 3293 (1984)

Morphology and Structure of Thin $TiSi_2$ Films on Silicon

A. Catana[1], *M. Heintze*[1], *F. Lévy*[1], *P.E. Schmid*[1], *and P. Stadelmann*[2]

[1]Institute of Applied Physics, Swiss Federal Institute of Technology, CH-1015 Lausanne, Switzerland

[2]Institute of Electron Microscopy, Swiss Federal Institute of Technology, CH-1015 Lausanne, Switzerland

INTRODUCTION

In the past few years, there has been a considerable interest in the study of transition metal silicides because of their importance in the silicon device technology. Among all silicides, $TiSi_2$ occupies an interesting position, mainly because of its high conductivity and its ability to grow selectively on silicon (SALICIDE process). The electrical characteristics of the contact are strongly related to the structural properties of the silicide which, in turn, depend on the preparation conditions. Structural knowledge is very important for the optimization of the technological process parameters and for a better understanding of the complex solid phase reactions leading to the silicide formation /1,2/. At the same time, this system offers the opportunity of studying a well defined solid/solid reaction on the atomic scale and the matching conditions of different crystal structures.

SAMPLE PREPARATION

Si (111) and (001) oriented substrates were first thermally cleaned at 1000 °C under UHV conditions ($7 * 10^{-8}$ Pa). Subsequently, about 25 nm titanium was deposited by e-gun evaporation. The sample was then annealed by Joule heating of the substrate. The chemical composition and contaminants were controlled by electron spectroscopy. Another preparation technique is the magnetron, dc-sputter deposition of 40 nm Ti on Si (001) wafers lightly doped, followed by rapid thermal processing in an argon ambient. The HRTEM investigations were performed on a Phillips 430 ST microscope with a point resolution of less than 0.2 nm. Both flat-on and cross-sectional thin samples were prepared for the transmission microscope studies using standard ion-thinning techniques.

FILM MORPHOLOGY

The film morphology as a function of the annealing temperature has been investigated on a 25 nm Ti layer, UHV deposited on a Si (001) heat cleaned sample. In a first step, the sample was annealed at 500 °C for 10 minutes. A second annealing was then carried out at 800 °C. The chemical surface composition after each step was checked by AES. An in-situ light scattering experiment was used to observe changes in the surface roughness and morphology during the heat treatments. After the first annealing, a $TiSi_2$ layer entirely covering the Si substrate was formed. During the second annealing, a significant rise in light scattering intensity was observed, as well as a large excess of Si with respect to

Springer Series in Materials Science, Vol. 13
Semiconductor Silicon Editors: G. C. Harbeke and M. J. Schulz

Fig. 1. SEM (a) and TEM (b) micrographs of $TiSi_2$ islands after a two step anneal (500+800 °C)

the $TiSi_2$ stoichiometry. These facts suggested a breakdown of the silicide layer into islands. The island structure is illustrated by the top-view SEM and cross-sectional TEM micrographs in Fig. 1. The TEM image shows that the silicide islands are partially buried in the silicon substrate. Moreover, epitaxial regrowth of Si between the islands was observed, which seems to indicate that the initial $TiSi_2$/Si interface is below the final Si surface level. The mechanisms explaining this phenomena are not well understood yet. However, it is believed that Si transport and precipitation are combined in a complex process /3/.

The island formation is due to a morphological transformation of the $TiSi_2$ layer, consisting of silicide agglomeration and movement into Si /4/. From the light-scattering data we found that the transformation was completed after about 250s. The morphological degradation of $TiSi_2$ films induced by high temperature annealing was also observed by Revesz et al. but only after a 7 hour annealing at 900 °C /3/. This difference might be explained by the presence of contaminants which are supposed to increase the stability of the silicide films.

The initial conditions at the Ti/Si interface also influence the film growth and morphology /5/. For example, the deposition of 25 nm Ti on an argon bombarded Si surface and annealing at 700 °C results in a silicide layer which completely covers the substrate. However, the interface is highly corrugated. This sample also displays a high density of stacking faults which belong to the $TiSi_2$-C49 metastable structure. The same preparation process performed on an unroughened substrate leads to the formation of the normal bulk $TiSi_2$-C54 structure. Thus, the initial Si surface does not only influence the final interface roughness, but also the silicide microstructure and the C49-C54 phase transformation temperature.

HRTEM ON $TISI_2$/SI (001) AND (111) INTERFACES

Atomic scale investigations reveal very interesting interface features which depend on the substrate orientation, silicide structure and preparation techniques. Samples prepared using production techniques (magnetron dc-sputter and annealing in an Ar ambient) show two main interfacial microstructures. The first case is characterized by monoatomic steps. It occurs when silicide planes are parallel to the (002) Si planes and the interface is slightly inclined. An example is shown in Fig. 2. The interfacial correspondence at steps requires a high density of misfit dislocations, due to the large difference (0.12 nm) between the interplanar spacings of the (002) Si and (111) $TiSi_2$ planes. In the second case, the interface is not parallel to the Si (002) planes but rather follows the close packed (111) planes, producing a faceted interface as illustrated in Fig. 3. The micrograph shows that the (131) planes of the

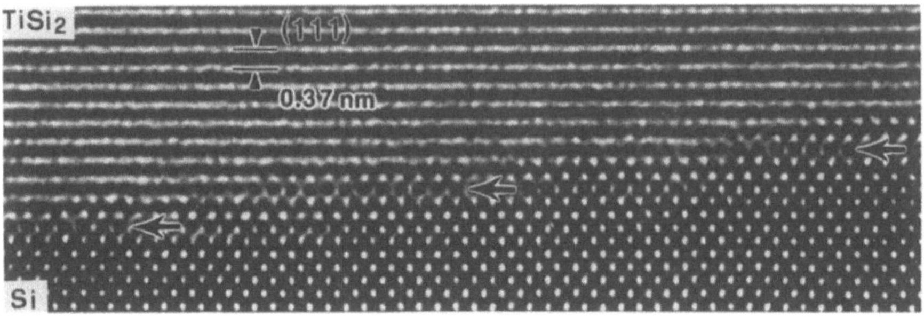

Fig. 2. HREM micrograph of a stepped $TiSi_2$/Si (001) interface structure

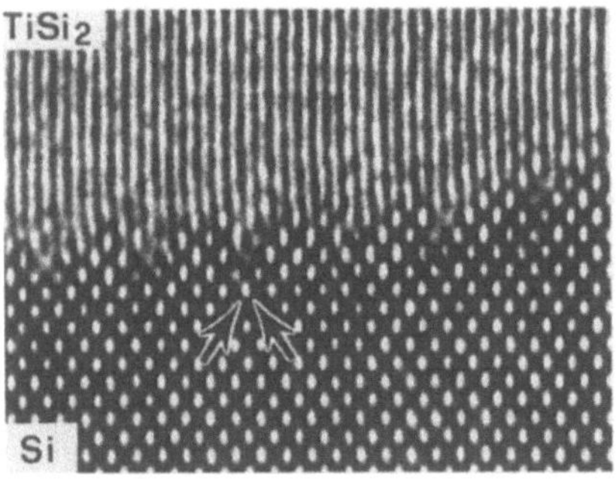

Fig. 3. HREM micrograph of a faceted $TiSi_2$/Si (001) interface

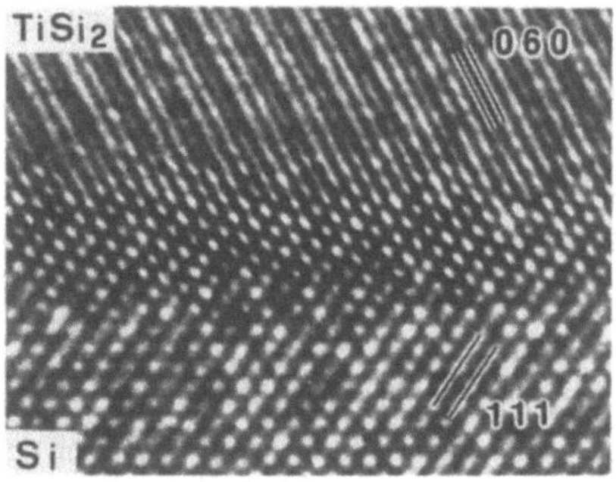

Fig. 4. HREM showing an interface layer accommodating the (060) $TiSi_2$ and (111) Si planes

metastable $TiSi_2$ structure (C49) are parallel to the (220) Si planes. The interplanar mismatch is 12.5%. Good match is achieved by the introduction of misfit dislocations every 8 (220) Si or 7 (131) $TiSi_2$ planes. The observation of the micrograph at grazing incidence indicates that the continuity of these planes across the interface requires considerable elastic distortion. The preservation of the continuity of atomic planes as they cross the interface appears to be an important feature of this silicide / silicon system. The micrograph in Fig. 4 emphasizes this property. It shows that, when the continuity is not directly realized, an interfacial layer is formed, which accommodates the $TiSi_2$ planes on one side and the Si planes on the other.

UHV titanium deposition and annealing on Si (111) result in flat and abrupt interfaces. No interfacial layers are observed. A HRTEM micrograph of a remarkably sharp interface is shown in Fig. 5. In this case, the (111) Si planes are parallel to the (202) $TiSi_2$ planes with <110> Si parallel to <161> $TiSi_2$. Other orientations were reported by Fung et al. /6/ and Wu et al. /7/. The (202) silicide plane was found to be a preferential interface plane. The epitaxial orientations occuring in this plane are reported and discussed on the basis of a fit-misfit model elsewhere /8/. In order to study the atomic arrangement of $TiSi_2$ in this particular orientation with respect to silicon, the atomic positions were projected along the corresponding directions and compared with calculated and experimental images. The image calculations were carried out using a program developed by one of us /9/. Figure 6 shows that, for an underfocus of 90 nm and a thickness of 6 nm, the computed images are in good agreement with the

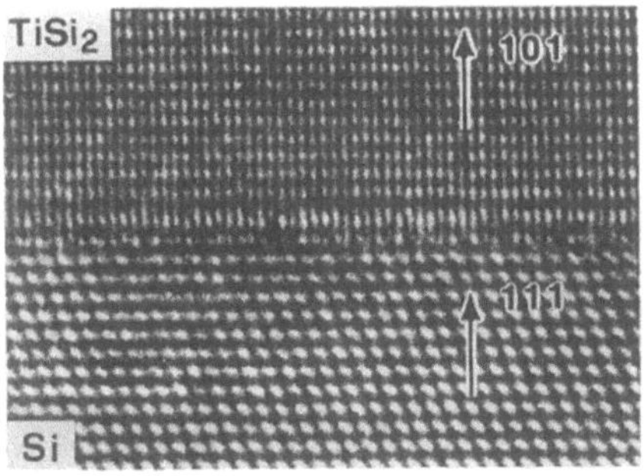

Fig. 5. HREM micrograph of a flat and abrupt $TiSi_2$/Si (111) interface

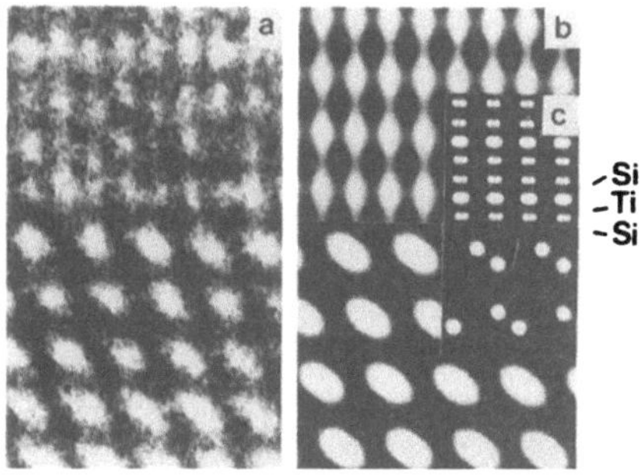

Fig. 6. (a) HREM, (b) computed image and (c) projected atomic potential used for the image calculations

experimental image. A further comparison with the projected atomic positions shows that the brightness maxima normal to the interface are due to the presence of Ti atoms and that the plane stacking order normal to the interface is expressed by pure Ti and Si layers with a repeating Si-Ti-Si sequence (Fig. 6c). Since the terminating (111) Si plane is not adjacent to the bright lines, we conclude that the interface is essentially represented by Si-Si bonding.

CONCLUSION

Morphological transformations leading to the island formation of $TiSi_2$ films were observed after annealing above 700 °C, suggesting a rather complex phenomena of plastic deformation. TEM micrographs show that the islands are partially buried in the substrate. HRTEM was used to study the influence of the substrate orientation on the interfacial microstructure. The results show that $TiSi_2$/Si (001) interfaces are mainly characterized by monoatomic steps parallel to the interface or by (111) facets indicating that the close packed planes are favourable planes for the epitaxy of $TiSi_2$. Silicide films prepared in UHV on <111> oriented substrates show flat and abrupt interfaces in spite of the large crystallographic differences between both crystal structures. The study of the atomic positions in one of the preferential epitaxial relationships indicates that the terminating silicide plane at the interface is a pure Si plane.

REFERENCES

1. G.W. Rubloff, R.M. Tromp and E.J. van Loenen: Appl. Phys. Lett. 48, 1600 (1986)
2. R. Butz, G.W. Rubloff, T.Y. Tan and P.S. Ho: Phys. Rev. B30, 5421 (1984)
3. P. Revesz, L.R. Zheng, L.S. Hung and J.W. Mayer: Appl. Phys. Lett. 48, 1591(1986)
4. C.Y. Wong, L.K. Wang, P.A. McFarland and C.Y. Ting: J. Appl. Phys. 60, 243 (1986)
5. M. Heintze, A. Catana, P.E. Schmid, P. Stadelmann: In "Le Vide, Les Couches Minces" 21, 236 (1987)
6. M.S. Fung, H.C. Cheng and L.J. Chen: Appl. Phys. Lett. 47, 1312 (1985)
7. I.C. Wu, J.J. Chu and L.J. Chen: J. Appl. Phys. 60, 3172 (1986)
8. A. Catana, M. Heintze, P.E. Schmid, P. Stadelmann: In Inst. of Phys. Conf. Ser. 87, 259 (1987)
9. P. Stadelmann: Ultramicroscopy 21, 131 (1987)

Modelling Diffusion in Silicides

A.A. Brown, P.B. Moynagh, and P.J. Rosser
STC Technology Ltd., Harlow, Essex, CM17 9NA, UK

INTRODUCTION

As device geometries are reduced, so the maximum speed of a circuit is governed increasingly by the RC constant of the interconnect. Local interconnects of polysilicon, with a resistivity greater than 100 μΩcm, are a particular problem, but one which can be overcome by using low resistivity silicides. One promising approach is to use a self-aligned silicide process (1) in which the silicide is formed by reacting a metal (for example titanium) with the exposed silicon. Unreacted metal can subsequently be removed in a selective wash. The silicon consumed during silicide formation is generally highly doped and it is important to be able to understand and model the redistribution of the dopant in this layer. Also, the formation of titanium disilicide has been shown to inject vacancies (2) which will interact with the dopant present in the substrate. It is therefore desirable to have a model for this process.

Silicides can also be used as a diffusion source to form shallow junctions. Such shallow junctions are formed by implantation into the silicide and subsequently driving in to form the junctions (3). This technique allows the formation of very shallow junctions with high surface concentrations and no implantation damage in the substrate. Clearly a model is required if this is to be implemented in future VLSI processes.

STATIC MODEL

The growth of metal silicides is a complex process to model since it involves the consumption of both metal and silicon while the silicide layer grows. A simple approach is to consider the silicide to have been deposited and only model the subsequent diffusion in the silicide and silicon. This method is particularly appropriate for modelling the

Springer Series in Materials Science, Vol. 13
Semiconductor Silicon Editors: G. C. Harbeke and M. J. Schulz

diffusion of dopant implanted into the silicide layer and diffused to form shallow junctions in the underlying silicon. Cobalt disilicide has been modelled in this way by modifying the diffusion parameters for one of the user defined materials in SUPREM 3 (4). Figure 1 shows the experimental profile of a boron implanted cobalt silicide film after annealing at 1000°C for 30 seconds determined by Secondary Ion Mass Spectroscopy (SIMS), Spreading Resistance Analysis (SRA) and Bevel and Stain. Figure 2 shows the simulated results using the default diffusion coefficients of boron in silicon and for the diffusivity enhanced by a factor of 8. In both cases, the diffusivity in the silicide is 0.445 μm^2/minute. This is a factor of 5000 greater than the default boron diffusivity in silicon. The higher diffusivity in silicon gives a better fit to the experimental data and suggests that the presence of the silicide may enhance the diffusivity of boron in the silicon. In the case of arsenic diffusion a 100 fold enhancement was observed. These diffusivity enhancements beneath the cobalt silicide are consistent with considerable vacancy injection by the silicide.

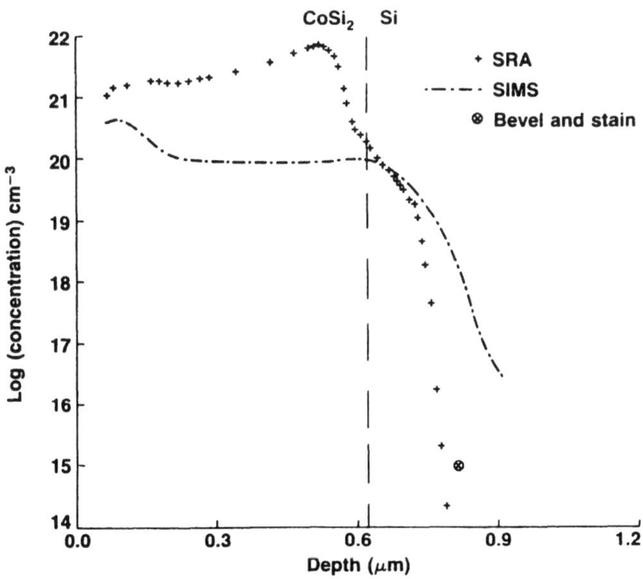

Fig. 1. Experimental boron diffusion from a $CoSi_2$ source anneal: 1000°C, 30 secs

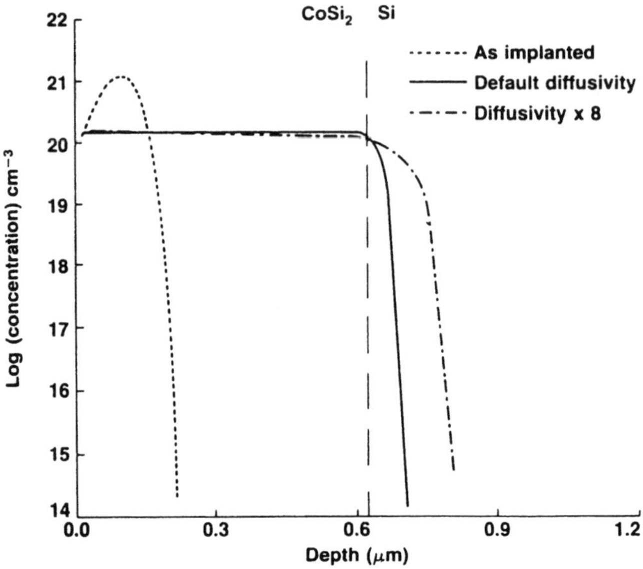

Fig. 2. Simulated boron diffusion from a $CoSi_2$ source anneal: 1000°C, 30 secs

DYNAMIC MODEL

The static model described above is not, however, suitable for modelling the more complex situation of silicide formation for which a moving boundary model is required. This can be seen from the Auger Electron Spectroscopy(AES) analysis of titanium sputtered over arsenic doped silicon and the same structure after silicide formation (figure 3). Layer evolution during the silicide formation is apparent as is considerable dopant redistribution. A qualitative model has been developed, to model the formation of titanium disilicide and the diffusion of dopants within the structure. The four layers considered are silicon, silicide, titanium with a constant oxygen concentration, and titanium nitride.

Figure 4 shows a typical sequence of process steps as they are represented by this model for the formation of titanium disilicide in a self aligned process. After titanium deposition, the silicide is formed by annealing at 650°C for 100 seconds in nitrogen. During this anneal, the silicide and nitride layers grow, consuming silicon and metal. Any oxygen in the titanium is segregated into the unreacted layer remaining, leading to a high oxygen concentration in this layer. Figure 4 shows the

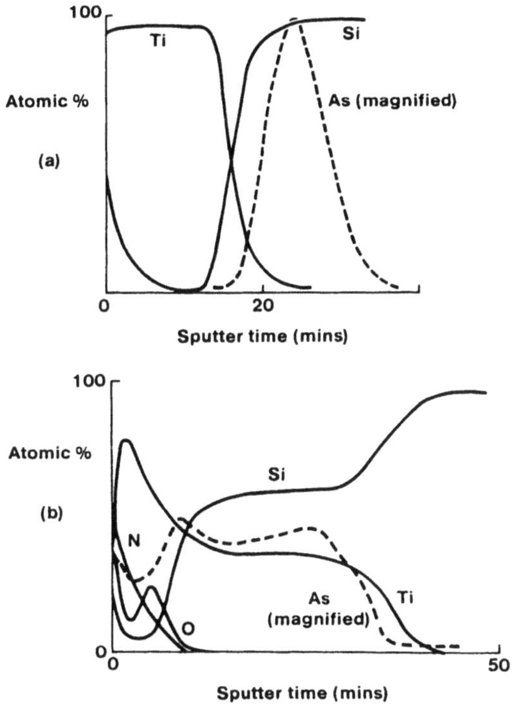

Fig. 3. AES of arsenic.
(a) before $TiSi_2$ formation,
(b) after $TiSi_2$ formation

situation at the end of this process and the modelled redistribution of dopant within the layers. Subsequent processing involves removing the two capping layers and annealing to reduce the resistivity of the silicide. Dopant is lost both when the impurity layers are removed and due to evaporation during the subsequent anneal. Finally, annealing at 1100°C for 20 seconds in nitrogen converts the silicide to titanium nitride and this step redistributes the dopant back into the substrate.

This four layer moving boundary model is complex and it will need to be integrated in a full process model if fully qualitative results are to be produced.

CONCLUSIONS

As silicides are incorporated into VLSI processes, it will become increasingly important to model their effect on dopant redistribution. Two models have been presented here which address this problem. The static

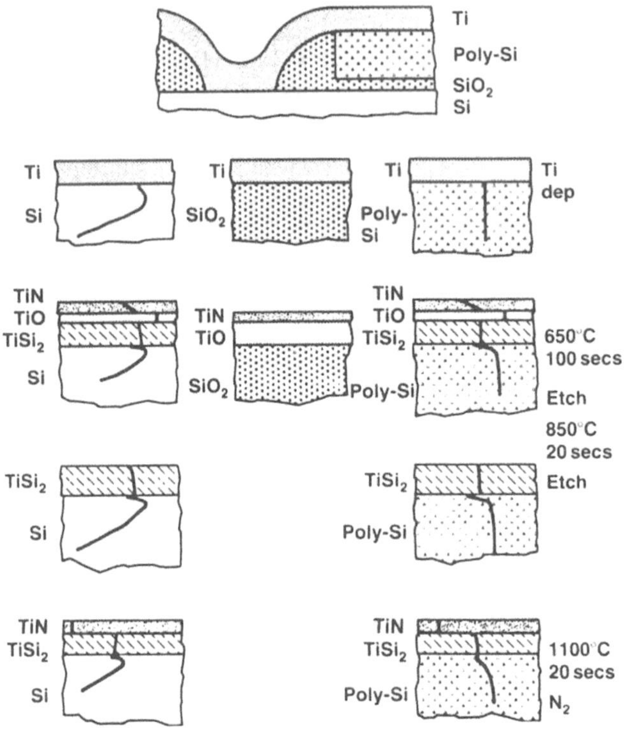

Fig. 4. Structure of silicide during formation

model has been shown to give results which compare favourably with experimental data. From this model, the diffusivity of dopants in silicides has been estimated to be 3 or 4 orders of magnitude greater than in silicon.

The dynamic model is at present purely a qualitative model. However, if this model is implemented in a process model, it will provide a useful tool for determining the effect of silicides on VLSI processes.

Both models are at an early stage of development and further experimental work to determine the diffusion and segregation coefficients of dopants in silicides is required. Nevertheless, promising initial results have been achieved and, with further development, these models will provide a useful extension to existing process models in simulating advances in VLSI processes.

ACKNOWLEDGEMENTS

This work was sponsored by the Alvey Directorate under contracts ALV/APP/VLSI/005, 016, 066.

This paper is based on a paper presented at the Alvey Conference, Swansea, July 4-7 1988.

REFERENCES

1. Rosser, P., and Tomkins, G., 1985, Vacuum, 35, 419-434.

2. Wen, D.S., Smith, P.L., Osburn, CM.M, and Rozgonyi, G.A., 1987, Appl. Phys. Lett., 51, 1182-1184.

3. Kwong, D.L., Ku, Y.H., Lee, S.K., Louis, E., Alvi, N.S., and Chu, P., 1987, J. Appl. Phys., 61, 5084-5088.

4. Stanford University/Technology Modelling Associates, SUPREM 3, process model.

Part VII

Devices

4 Mbit Technology

S. Röhl

Siemens AG, Postfach 830857, D-8000 München 83, Fed. Rep. of Germany

1. Introduction

The Siemens MEGA Project is a major effort to speed up the development of microelectronics as a key technology for many system applications. The project comprises three sections: the 1M DRAM, the 4M DRAM and the MEGA Logic part. Memories play a fundamental role in the basic project strategy, both as high volume standard products and as snow plough for a whole technolgy generation.

The ideal semiconductor memory would have the following properties: high performance, high density, low power dissipation, random access, nonvolatility, high reliability and common use throughout the industry [1]. Up to now, no single device having all these characteristics has been developed, although the various available MOS memories fulfil some of the above requirements (Fig. 1). Dynamic random access memories have the largest market share of the microelectronics market (in sales 43% in 1987) with an average annual growth rate in the eighties of 22% [2].

	DRAM	SRAM	EPROM	E^2PROM	ROM
percentage of total MOS memory market [expected growth rate]	43% [14%]	19% [13%]	26% [7%]	3% [32%]	9% [4%]
Major applications	mainframes, PC, video, graphics	mainframes, MC-applica-tion	MC-programs printers, terminals	telephone machine control	voice synthesis digital imaging
volatile/nonvolatile	v	v	n	n	n
power consumption [W]	0.3	0.1...0.3	0.3	0.5	0.3
access time [nsec]	100	50	200	150	150
integration density (production 1987)	1M	256k	256k	256k	1M
type of memory cell	one tran-sistor	poly load; 6 transistor	one tran-sistor	two tran-sistor	one tran-sistor
supply voltage [V]	5	5	5	5	5
special characteristics	refresh; high densi-ty; low price per bit	high perfor-mance, low power consump-tion, ease of use		"ideal memory"	

Fig. 1 Overview over the different classes of semiconductor memories. Market information according to ICE [2]

Springer Series in Materials Science, Vol. 13
Semiconductor Silicon Editors: G. C. Harbeke and M. J. Schulz

Apart from ROMs they reach the highest integration densities with the onset of the 1Mbit market in 1987. The main reason for the dramatic growth of the DRAM market has been the ability to continuously decrease the price per bit (the so called learning curve) down to presently 1 millicent/bit. This inflation of cheap memory capacity has greatly influenced the evolution of mainframe computers, workstations and personal computers.

2. Evolution of DRAM Memory Technology

Product integration density has increased by a factor of four so far in fairly constant time intervals of three years per generation. This rapid progress in memory density has been accomplished by two factors illustrated in Fig. 2.

Comparing the 4k and 4M generations the thousandfold increase in integration density can be attributed to miniaturization (two orders of magnitude) and chip size (about one order of magnitude). To realize a 10 μm^2 sized 4Mbit DRAM cell which basically consists of one MOS transistor and one MOS capacitor a minimum feature size of 0.8 μm is necessary.

The evolution leading to these integration densities with submicron feature sizes started in 1970 with the three transitor memory cell. The main innovations on the way from 1K to 4M are illustrated by the memory cell cross sections of fig. 3. The simple one transistor cell was already introduced for 4K memories. This cell type uses a pass transistor to read and store the electrical charge in a MOS capacitor. The storage capacity is fixed by a certain signal to noise ratio necessary for reliable sensing of charge on the bitline. The available capacitor area is however decreasing from generation to generation and has been compensated for by reducing oxide dielectric thickness down to a limit of 8nm for 1Mbit DRAMs. Further shrinkage requires new three-dimensional storage capacitor concepts, like the trench capacitor and the stacked capacitor shown in

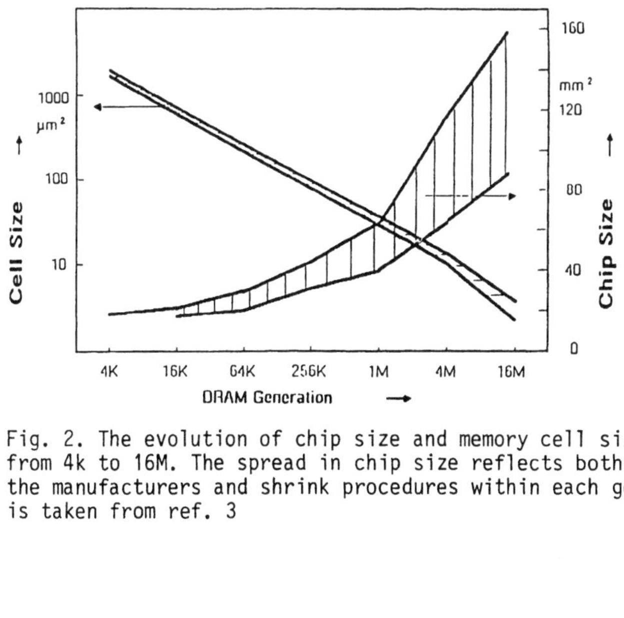

Fig. 2. The evolution of chip size and memory cell size for dynamic RAMs from 4k to 16M. The spread in chip size reflects both variations between the manufacturers and shrink procedures within each generation. The figure is taken from ref. 3

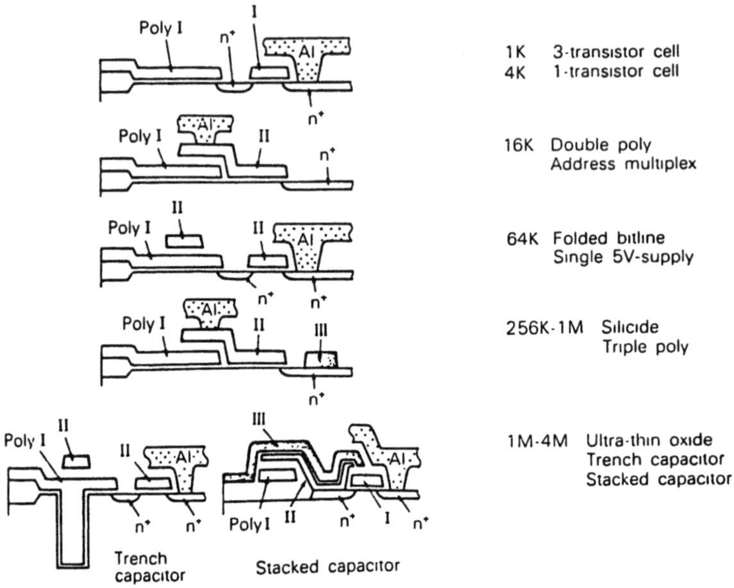

Fig. 3 Cross sections of typical 1k to 4M DRAM cells together with major innovations. The figure is taken from ref 4.

fig. 3. A variety of more sophisticated 3D cells have been suggested for 4M memories [4].

To illustrate the necessity of 3D cells the noise contributions to the read signal are estimated in fig. 4 for a trench cell with mid level concept [5]. The read signal is determined by the ratio of the storage capacitance to the sum of storage and bitline capacitance which typically reduces the available read signal for the sense amplifier to 1/10 of the cell voltage. The main noise contribution is the "α-margin" which accounts for the charge loss due to the collection of charges generated by α- particles (from chip materials, package and environment) in the depletion region around the trench (logical <1>) or in the bit line node (logical <0>). Measurements of the α-margin indicate that a cell capacitance of 40 fF is required for acceptable performance [5]. The capacitor formula $C = \varepsilon \varepsilon_o F/d = 40$ fF illustrates how this could be achieved:

1) decrease in thermal SiO_2 dielectric thickness d:
   less than 6 nm of SiO_2 would be required, which is so close to the onset of electron tunneling through the oxide, that major leakage and reliability problems will be encountered.

2) new dielectrics with higher dielectric constant ε:
   Tantalum oxide (Ta_2O_5) with $\varepsilon = 22$ is being investigated for example [6] but the development status is still far from being capable of integration into a DRAM process. Chemical vapor deposited (CVD) Si_3N_4 with $\varepsilon = 7$ is compatible with current processes and high quality films of several nm thickness can be homogeneously deposited especially on trench corners, where thermal oxide becomes

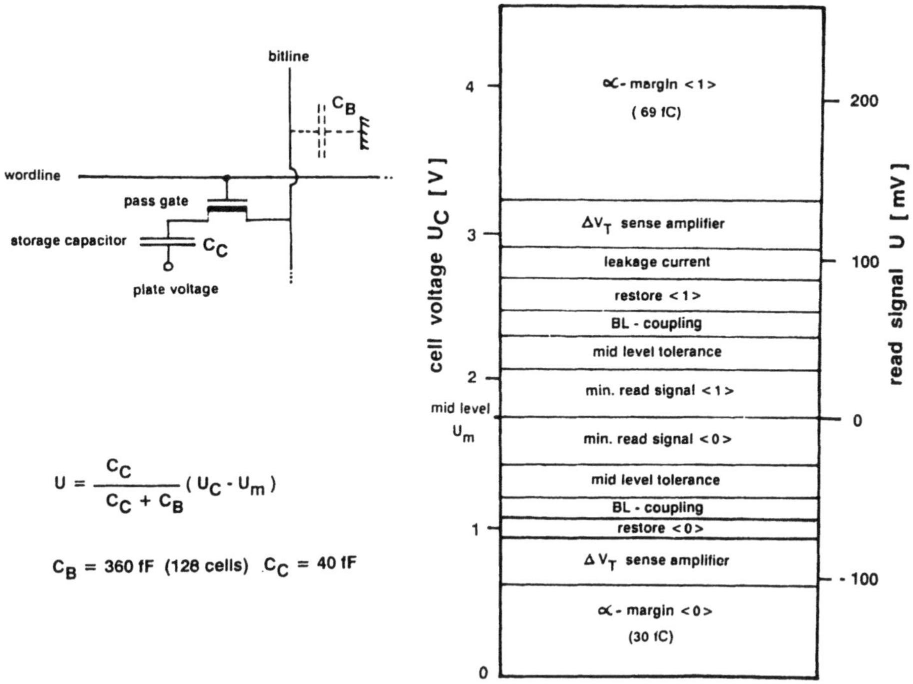

Fig. 4 Noise contribution to the read signal of a 4Mbit DRAM cell [5]. Cell plate is on mid level and operating voltage is 5V.(The worst case is drawn in the figure)

thinner. It suffers, however, from low breakdown voltages, high surface state densities and high electron trapping probability. These drawbacks can be compensated for by combining it with top and bottom oxide in a triple dielectric (ONO) sandwich. Scaling of this composite dielectric is limited to 10 nm equivalent oxide thickness [7].

Due to these limitations, the folding of the capacitor into the third dimension (trench or stacked capacitor) is the only possibility for further area reduction. These structures increase process complexity and introduce several new process steps. 3D cells are therefore one of the main development areas of submicron technology. A second paper concentrates on specific trench cell technology problems. Other major development areas for cell shrinkage are microlithography, reduction of nesting tolerances by self aligned contact techniques and improvement of contact hole reliability. They are outlined in the next two sections.

3. Microlithograpy

Optical lithography using wafer stepper equipment has become the dominant microlithographical method in semiconductor technology for feature sizes below 2 µm [8]. Whilst the capability of optical lithographic equipment has been improved continuously, the "safety margin" for mass production

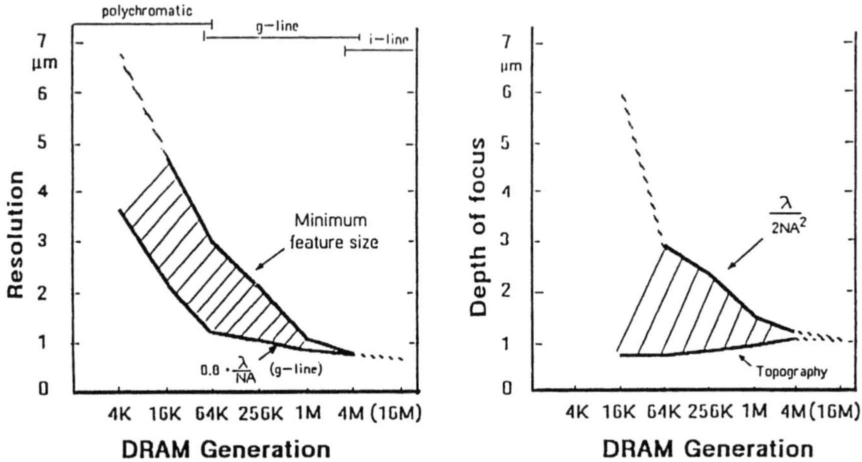

Fig. 5 Left: The resolution capability achievable with the lithography equipment available at the time period corresponding to each memory generation and minimum feature size for the same generation. Right: Depth of focus of lithography equipment and device topography versus DRAM generation (taken from ref.3)

requirements has nevertheless become critically small with regard to resolution and depth of focus requirements, as shown in Fig. 5. The resolution of wafer steppers has been primarily improved in the past by increasing the numerical aperture (NA) of the lenses, leading presently to a practical resolution limit $0.8\,\lambda/NA$ i.e. about 0.8 µm. However, the depth of focus varies as λ/NA^2 which leads to problems of line definition due to device topography, defined as the maximum step height on the wafer surface.

Topography has increased over the last generations primarily because of increasing process complexity. That is feature sizes have been reduced, whilst the number of patterned layers has increased. Furthermore, the thickness of the various layers could not be decreased correspondingly for electrical and reliability reasons. The effect of this small safety margin in lithography on routine mass production is still not fully known and presents one of the major challenges in this sub µm generation of IC's. The total amount of defocus contains several contributions (Fig. 6).

One important factor is the local thickness variation (LTV) of the silicon wafer. The relevant area to be taken into account is the field size of the wafer stepper(typically 15mm x 15mm). The percentage of 6"-wafers within a specified LTV-range (90% usable area) varies between 10% and 40% depending on the detailed requirements. Improvements in the polishing procedures during silicon substrate manufacturing are proceeding presently.
In order to ensure stable process conditions, the continuous measurement of many small features is necessary immediately after a specific process step has been carried out. In this field of metrology the intensive usage of scanning electron microscopy (SEM) for online inspection is the most important innovation. Since insulating layers cover the substrate in many cases (e.g. resist or SiO_2), low voltage SEM is necessary to avoid

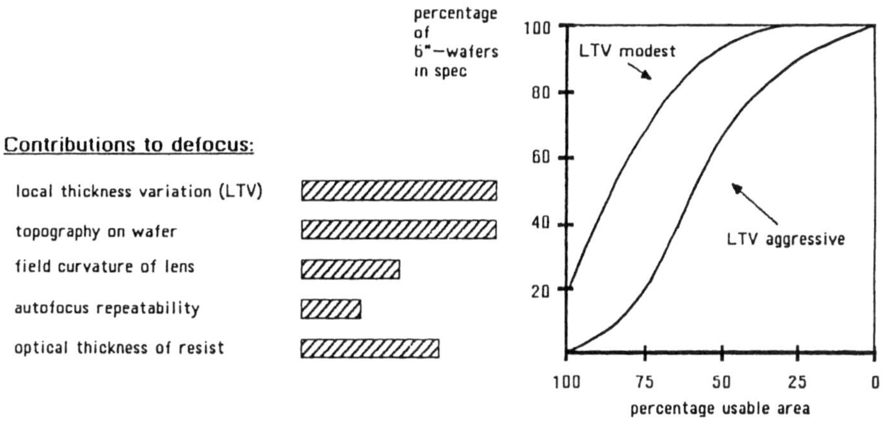

Fig. 6 Left: Contribution to the total defocus budget of a wafer stepper with state of the art technology. Right: Typical yield distribution of 6" wafers for aggressive and modest requirements to LTV (taken from ref. 3)

Fig. 7 0.8 µm resist line over poly-Si 3 topography. Resist profile appears adequate in the cross section but oblique view reveals resist bridges due to topography

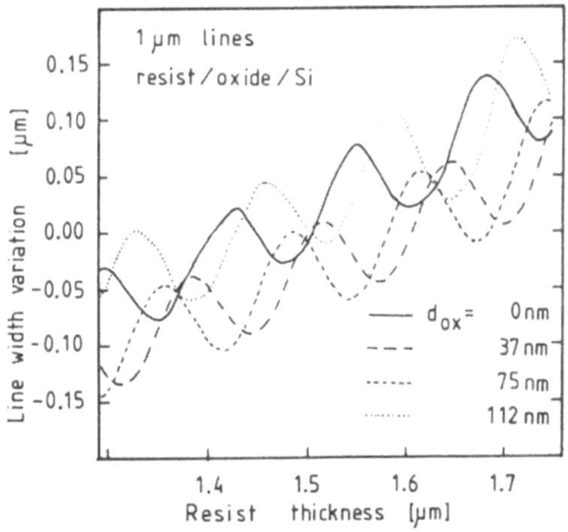

Fig. 8: Calculated variations of critical dimension (line width) with resist thickness and thickness of underlying SiO_2 layer (interference effect)

charging effects. Relative accuracies of 20 nm can be achieved by linear regression of the intensity signal across the measured structure [9].

Besides the limited focus depth, light reflection from topographic steps and from interfaces of underlying layers also disturb fine line patterning. Fig. 7 demonstrates that resist bridges on otherwise perfect 0.8 µm resist lines still exist due to the poly-Si 3 topography. Fig. 8 illustrates that already in a standard case of resist over SiO_2 on Si critical dimension (line width) variations of 0.3 µm occur due to interference effects in the resist and the underlying layers.

4. 4M DRAM Concept and Major Process Innovations

The process flow of the Siemens 4Mbit DRAM is given in table 1 together with the key performance data. A SEM cross section of the memory cell along the bitline (fig. 9) illustrates the trench cell concept with n^+-

Table 1: Process flow and main features of 4Mbit DRAM

Substrate CZ p, $\varrho = 5\ \Omega$cm	**Organization**	4M x 1 and 1M x 4
p-well	**Technology**	Twin tub CMOS 0.9 µm, FOBIC-trench cell
n-well	**Cell size**	2.3 x 4.6 = 10.58 µm²
Advanced LOCOS isolation	**Chip size**	6.5 x 14.05 = 91.3 mm²
Trench etching	**Power supply**	5 V ± 10% ext. and int.
Trench doping (As)	**Active/Standby current**	<70 mA / < 1 mA
1st gate oxide ($T_{eff} = 13.5$ nm)	**Access/Cycle time**	70 / 150 ns typical @ 5V
1st poly-Si and trench refilling	**Static column cycle**	45 ns typical @ 5V
Channel implantation	**Refresh**	1024 / 16 ms (CAS before RAS, hidden, RAS only)
2nd gate oxide ($T_{ox} = 20$ nm)	**Redundancy**	4 rows / 4 columns for each 512 k block
2nd poly-Si and CVD SiO_2 gate definition	**Package**	SOJ 26/20, 350 mil
LDD implantation (P)	**Operating modes**	Read, Write, Read-Modify -Write, Fast Page Mode, Static Column (Fuse Option)
Spacer definition		
S/D implantation (As, BF_2)	**Test mode**	8 bit parallel according to JEDEC
CVD oxide deposition and flow		
Direct contact definition		
3rd poly-Si/$MoSi_2$		
BPSG deposition and flow		
Contact hole definition		
AlSi/TiN/Ti metallization		
Passivation		
Pad opening		

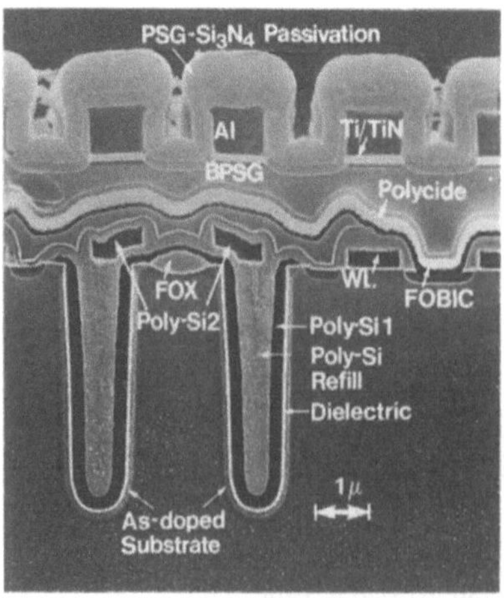

Fig. 9 SEM cross section of 4M DRAM Cell

storage node, capacitor dielectric, poly-Si 1 capacitor plate, trench refill and the pass gate transistor (word line). The aluminum lines with a diffusion barrier shunt the word lines. The bitline with the fully overlapping bitline contact (FOBIC) is perpendicular to the word lines.

FOBIC and diffusion barrier are discussed in this section while a separate paper concentrates on the trench cell.

The 25% area saving capability of the selfaligned contact (FOBIC) can be seen from the layout of fig. 10. The FOBIC overlaps the poly-Si gates (word lines) and the field oxide isolation. The important aspect in the cross section of fig. 11 is the nitride etch stop layer which allows an uncritical isotropic etch of the contact window. A short careful etch of the etch stop layer leaves the lateral gate encapsulation intact. Nitride layer thickness and deposition temperature had to be optimized for low stress to prevent defect generation in the Si substrate [10,11].

The contact resistance of Al to source/drain transistor regions is another important topic as in submicron contacts the Si precipitation at the interface from the Al/Si metallization increases the contact resistance by a large factor (fig. 12). This is of special concern at the present time because the BF_2 implants used for shallow p^+ source/drain regions, result in higher substrate surface damage which acts as a seed for epitaxial Si regrowth [13] during metallization sintering. Another problem is Al spiking by diffusion of Si into Al and vice versa due to

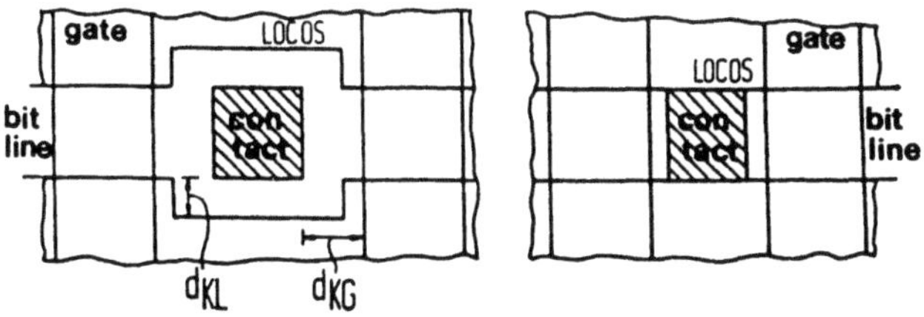

Fig. 10 Area reduction in cell layout due to the fully overlapping bitline contact (FOBIC) technology (right). Registration tolerances d_{KG} and d_{KL} are indicated (taken from ref. 11).

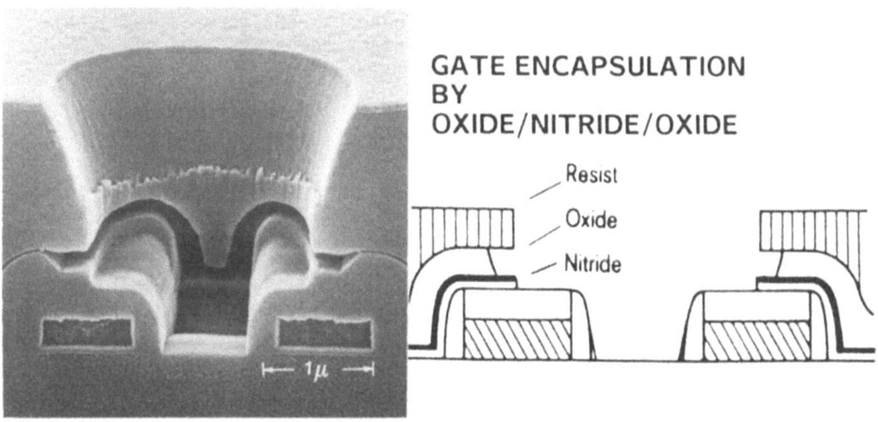

Fig. 11 Cross section of FOBIC contact

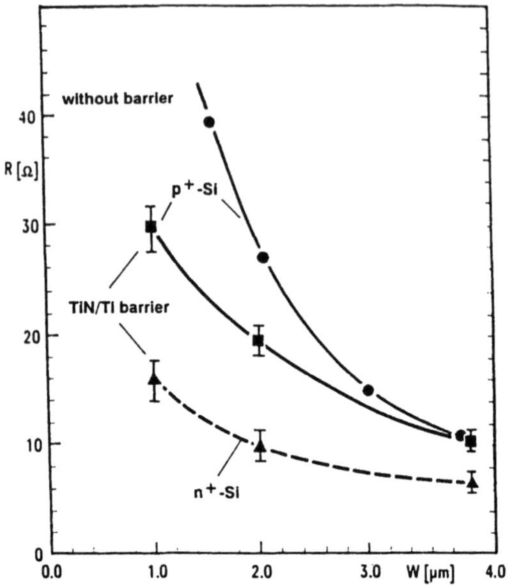

Fig. 12 Contact hole resistances with and without diffusion barrier for different contact hole sizes

temperature and/or electric field stress. These spikes introduce p-doped regions within the n^+ diffusion thus shorting the junction. These effects make a diffusion barrier for submicron Al contacts absolutely necessary. A sputtered Ti/TiN barrier has therefore been introduced in our process [12]. $TiSi_2$ is formed with the bulk Si and improves contact resistance (fig. 12) while TiN prevents Al and Si diffusion.

Summary

The present status of the semiconductor memory market and the evolution of DRAM cells from 1Kbit to 4Mbit is briefly reviewed.The 4Mbit technology with chip area of 100 mm^2, memory cell size of 10 μm^2 and minimum feature size of 0.8 µm is discussed.The microlithography requirements for fabrication of 0.8 µm features, the three dimensional trench capacitor cell, the selfaligned contact technology and the metallization diffusion barrier are addressed.

Acknowledgment

The author is indebted to the 4Mbit engineering teams, especially L. Risch, Ch. Nölscher, K.H. Küsters and G. Higelin for their contributions. I thank W. Beinvogl and E. Hopf whose work was partly used for this paper.
The work reported here has been supported by the Federal Department of Research and Technology of the Federal Republic of Germany (contract NT 2696).

References

/1/ B. Prince, G. Due-Gundersen; Semiconductor Memories , Wiley, New York, 1983

/2/ "Status 1988", A Report on the Integrated Circuit Industry, 1988, Integrated Circuit Engineering Corporation

/3/ W. Beinvogl, E. Hopf, in: Festkoerperprobleme (Advances in Solid State Physics), U. Rössler (ed.), Vieweg, Braunschweig 1988,Vol.28,87

/4/ H. Sunami, Japan Semiconductor Technology Reports, JST Reports, Vol. 3, No. 1; 1987, p. 21.

/5/ L. Risch, W. Müller, R. Tielert, ntz 40 (1987) Nr. 3, S. 192.

/6/ B.W. Shen, I.C. Chen, S. Banerjee, G.A. Brown, J. Bohlman; P.H. Chang, and R.R. Doering, IEDM 1987, Digest of Technical Papers 1987, p. 582.

/7/ K.K. Joung and W.G. Oldham in "Silicon Nitride and Silicon Dioxide Thin Insulating Films "eds. V. Kapoor and J. Haukins, The Electrochemical Society, 1987, p. 471.

/8/ W. Arden, K.H. Müller; Microelectronic Engineering 6 (1987) 53

/9/ M. Miyoshi,Japan Semiconductor Technology Reports, JST Reports, Vol. 3, No.1, 1987, p.56

/10/ K.H. Küsters, G. Enders, W. Meyberg, H. Benzinger, B. Hasler, G. Higelin, S. Röhl, H.M. Mühlhoff, W. Müller; Proc. of the 1987 Symposium on VLSI Technology, Nagano (1987) p. 93.

/11/ K.H. Küsters, H.M. Mühlhoff, G. Enders, E.G. Mohr and W. Müller; Proc. of the First International Symposium on ULSI Science and Technology 1987, p. 640

/12/ G. Higelin, U. Fritsch, K.H. Küsters, G. Enders and W. Müller; 1988 Proc. Fifth Int. IEEE VLSI Multilevel Interconnection Conference Santa Clara, 1988, p. 29

/13/ T. Maeda, S. Shima, T. Nakayama, M. Kakumu, K. Mori, S. Iwabuchi, R. Aoki and J. Matsunage; IEDM 1985, Digest of Technical Paers 1985, p. 610

Technology and Reliability Problems of Trench Cell Capacitors

S. Röhl

Siemens AG, Postfach 830857, D-8000 München 83, Fed. Rep. of Germany

1. Introduction

From the 4Mbit DRAM generation onwards the folding of the storage capacitor into the third dimension has become necessary [1]. Apart from the stacked capacitor, the trench cell is the most frequent choice of memory chip makers. Before the introduction of a trench cell into the DRAM fabrication process several technology and reliability problems have however to be solved to guarantee high production yield and dielectric lifetime of 10 years. The main items are the reproducible etching of deep trenches, the doping of three dimensional structures and the quality and reliability of the dielectric inside the trench. A further important topic is trench cell leakage caused by p-n junction and dielectric leakage currents , cell to cell cross talk and α-particle induced charge loss. Leakage levels have to be below a fraction of a picoamp per cell in order to keep charge loss during the memory refresh cycle of 16 msec within 5%.

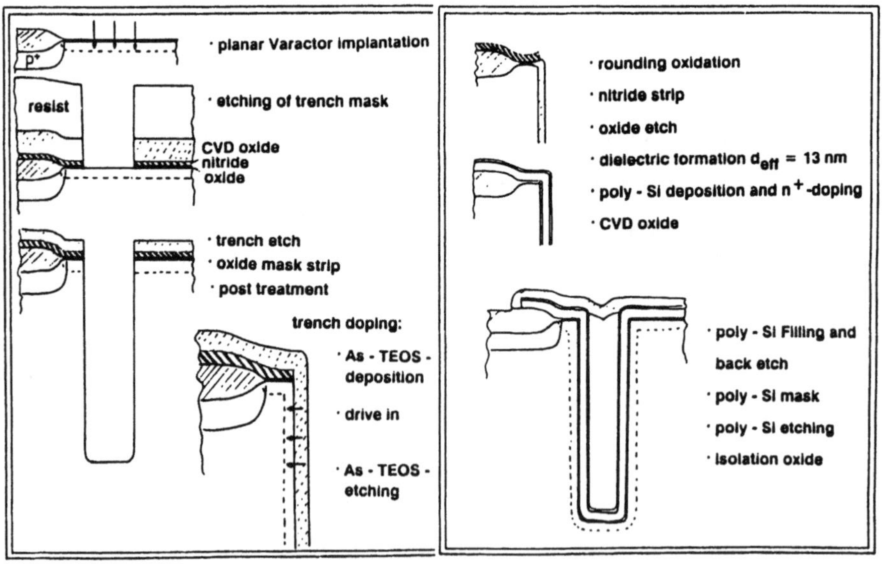

Fig. 1: Process flow for trench storage capacitor of 4Mbit DRAM [2]

Springer Series in Materials Science, Vol. 13
Semiconductor Silicon Editors: G. C. Harbeke and M. J. Schulz

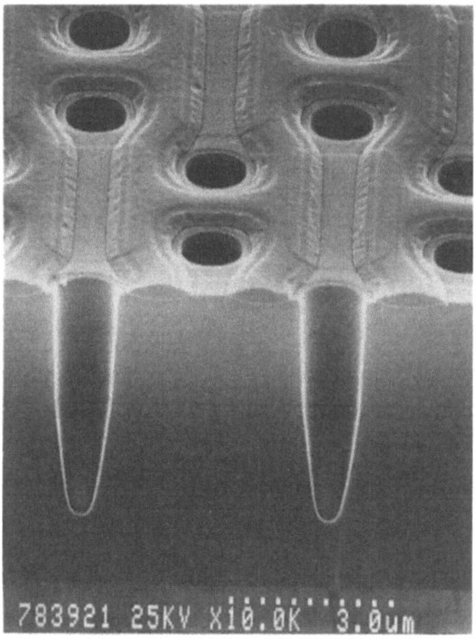

Fig. 2: Oblique view of trench cells after trench mask stripping

Table 1: 4M memory cell parameters and basic requirements

Trench Cross Section	0.9 x 1.0 μm^2
Trench Depth	4.5 μm
Trench Doping (As)	1 E 19 cm^{-3}
Gate Oxide (T_{eff})	13 nm
Cell Capacitance	40 fF
P Well Concentration (N_s)	4 E 16 cm^{-3}
P Well Depth (Cell Array)	8 μm
Trench Profile:	round bottom, smooth sidewall no undercut and trenching
Breakdown field strength:	± 10 MV/cm (safety margin of 3 over operating field strength)
Defect Density: (early gate oxide failures)	< <.1/cm^2 (active area 1 cm^2)
Long term stability:	100 Fit in first year (1 Fit = 1 failure in 10^9 device hours) 10 years lifetime
Leakage Current:	<.1 $pA/_{cell}$ (<5% charge loss for 16ms refresh time)

2. Trench Cell and Basic Requirements

A typical process flow for a trench storage capacitor is shown in fig. 1. The cell is formed in a deep p-well which is necessary to reduce trench to trench leakage currents, and a planar varactor implantation which is necessary for contact to the source/drain region of the pass gate transistor. For the trench etch process, an oxide etch mask is used in combination with a nitride etch stop layer. This latter layer is used to prevent etch-back of the field oxide isolation during the trench oxide strip process. The SEM cross section of fig. 2 shows an oblique view at this stage. Trench post treatment is important for optimization of dielectric quality, which can be either a thermal oxide or a triple oxide-nitride-oxide layer. The poly-Si electrode is doped conventionally by a gas source and the trench is then filled by an oxide and an undoped poly-Si layer which are etched back. After this processing the poly-Si electrode is patterned and encapsulated by an oxide isolation. A SEM cross section of the completed cell has already been given in the preceeding paper. Table 1 gives the Siemens 4M memory cell parameters and lists the basic requirements which will be addressed in the next three sections.

3. Trench Technology and Dielectric Quality

The trench etch process has been sucessfully developed as is illustrated by the SEM of fig. 3 which shows a 9.5 µm deep trench with 0.7 µm diameter and perfect geometric profile [3].

A variety of problems have, however, occured before such good results could be obtained (see table 2). The reproducibility and stability of trench etch processes concerning profile, etch rate, selectivity and etch machine cleaning cycles were and are still of major concern. The gate oxide quality especially, which inside the trench depends on the trench post treatment, is a main topic. Examples of 6 different trench etch processes (single wafer etcher (# 1-4) and batch etcher(#5,6) are given in fig. 4 (after etching of

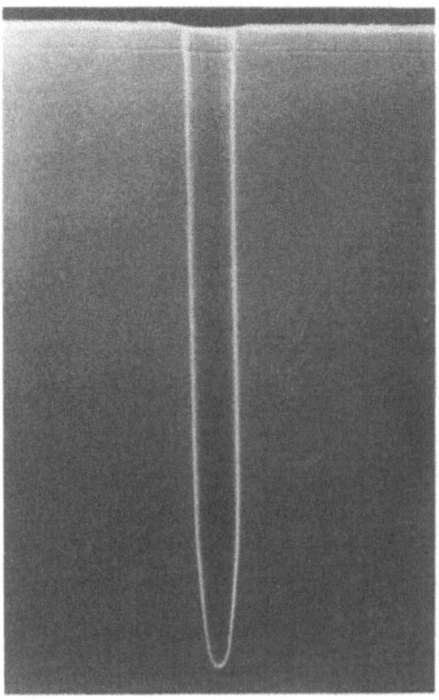

Fig. 3: 9.5 µm deep trench of 0.7 µm diameter etched with $CBrF_3$ in a single wafer triode etcher. Selectivity to oxide mask was 18 : 1

Table 2: Summary of main trench etch problems

- sidewall films	- sidewall contaminations or etch residues
- undercut or bowing	- long trenches etch easier than circular holes
- trenching at trench bottom	- problems increase with depth
- reproducibility of trench profile and etch rates	- gate oxide quality
- homogeneity of profile and etch rate across wafer	

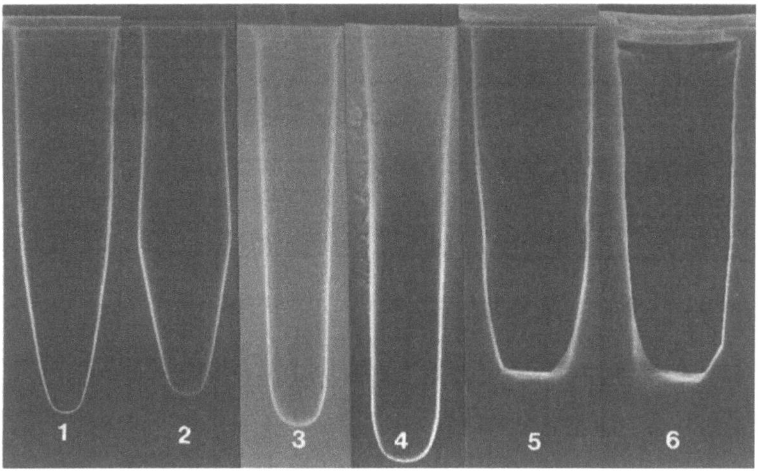

Fig. 4: SEM micrographs of the trench profile of six different etch processes after removal of sidewall redepositions.

a) Single wafer processes with $CBrF_3$
1 low
2 medium gas flow
3 high
4 N2 diluted

b) Batch processes
5 $CBrF_3/BCl_3/Cl_2$
6 BCl_3/Cl_2 (including wet anisotropic Si etch)

the sidewall redepositions).The redepositions consist mainly of SiO_2 and are necessary to obtain good profiles. Fig. 5 illustrates the status before redeposition etching. Redepositions also shield the Si surface from etch products and Si damage yielding high dielectric quality. However, processes # 1 and # 2, with high redeposition rates, are not reproducible. The stable processes, # 3 and # 4 with higher gas flow and less redeposition show reduced gate oxide quality, which can only be improved by Si consuming post treatments (fig. 6, [4]). Triple dielectrics appear much less sensitive to

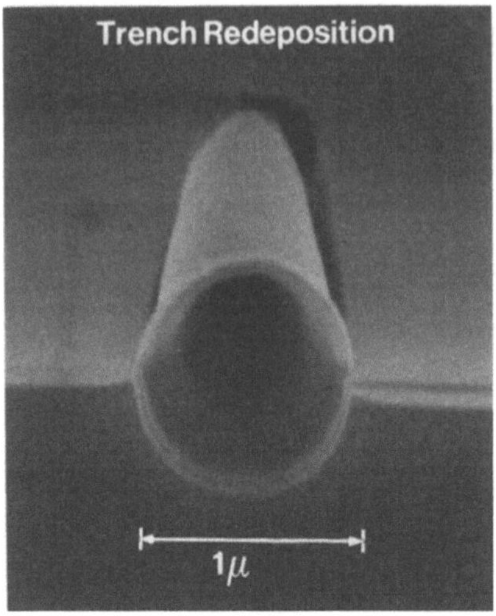

Fig. 5: Oblique view of trench with SiO_2 redepositions on trench sidewall

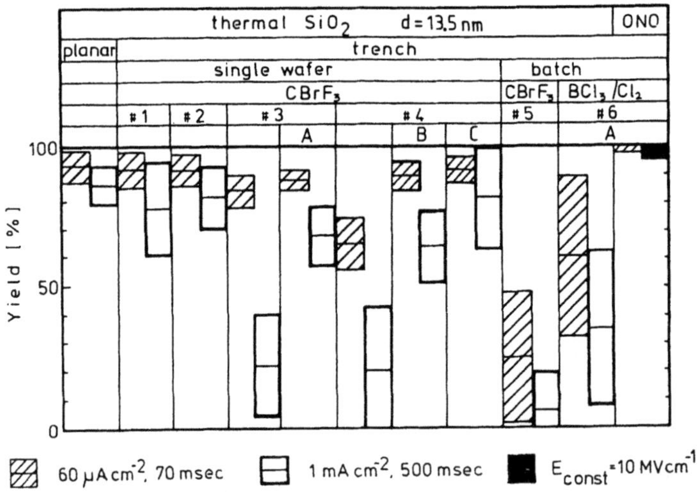

Fig. 6 Gate oxide results for single wafer and batch etch processes. Hatched boxes indicate means and standard deviations. For # 3 and # 4 results for the following post treatments are shown: A = wet anisotropic Si etch, B = O_2 plasma + BHF C = O_2 plasma + BHF + dry isotropic Si etch. Triple trench dielectric (ONO) and planar reference are shown for comparison.

the particular trench etch process (fig. 6) and are a good choice for 4Mbit DRAMs. The advantages of a triple dielectric are also demonstrated by the tunneling currents of planar and trench capacitors (fig. 7) which show a strong increase for SiO_2 due to gate oxide thinning at the upper trench corner while ONO shows no difference planar to trench: Scalability is, however, limited to approximately 10 nm oxide equivalent thickness.

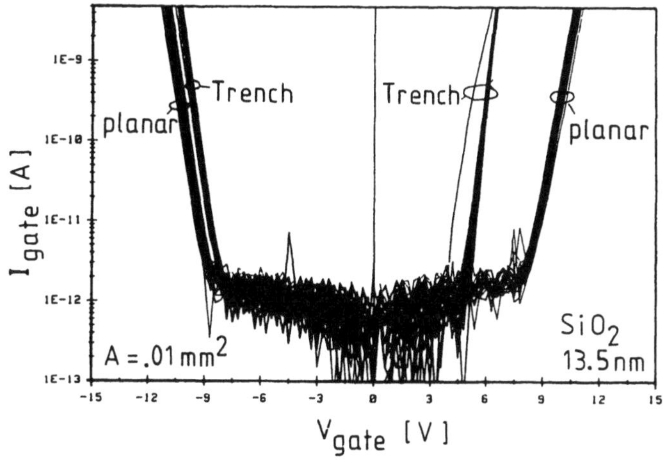

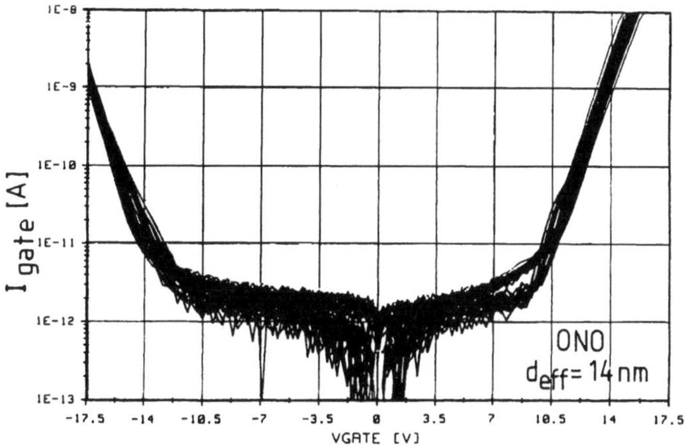

Fig. 7: Tunneling currents of planar and trench capacitors for SiO_2 and ONO dielectric. Curves are from 24 dies on one wafer

4. Doping Techniques for Trench Capacitors and Trench to Trench Leakage

The trench sidewall has to be n-doped to a level of $10^{19}cm^{-3}$ in order to achieve a conductive capacitor plate with sufficient capacity in the operating range. The voltage and doping concentration dependence of the trench capacity is shown in fig. 8. As the n^+ junction has to be shallow to avoid trench to trench punchthrough [5, 6], the slowly diffusing As is the dopant of choice.

Different methods for doping of trench structures are summarized in table 3 all of which have been recently applied [5-10]. Ion implantation is a very elegant doping method because of its flexibility concerning ion dose, ion energy and incident angle. Expensive hardware optimization is, however,

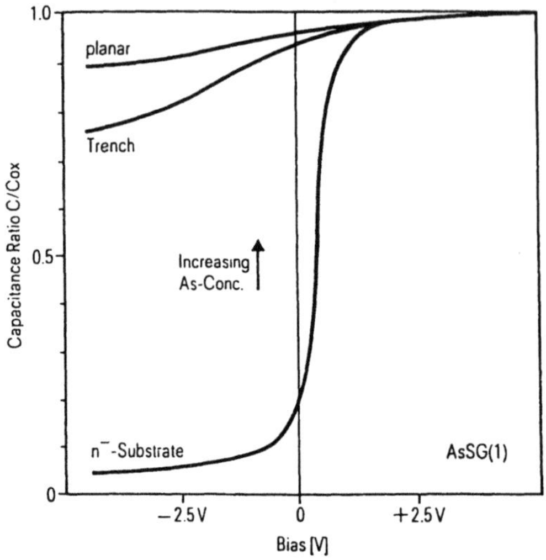

Fig. 8:
C(V) curves for trench and planar capacitors on As TEOS doped Si in comparison to capacitor on lightly doped n-substrate. The DRAM operating range is indicated with +/- 2.5V

Table 3: As doping methods for trench capacitors (for details see ref. 5)

Method	Advantages	Disadvantages
Spin on glass (SOG)	Safe and low cost High As content possible	Delicate Spin on and curing process Repeatability and uniformity of layer thickness in trench (removal !) Cracks possible
Ion implantation	Ion, -dose, -energy, -incident angle easily changeable Simple process	Expensive hardware (rotation, angle) Sensitive to trench shape(asperities,taper, undercut) Substrate damage Enhanced doping at trench bottom
Selective epitaxial growth	Shallow and abrupt junction Good oxide quality Rounding capability	Lateral undergrowth and crystal defects at upper trench corner Facetting Expensive
Solid sources	Simple process	Repeatability Source aging effects Throughput Surface quality ?
Poly Si source*	Good step coverage Effective dopant source	No etch stop for source removal Surface quality
AP CVD oxide* source	Well studied process	Poor step coverage Surface damage
As TEOS source	Good step coverage Simple process No dangerous gases	New process High As concentrations difficult

( )*: No data for trench application

necessary to minimize angular spread and rotation of wafers. It was found that adequate doping can also be achieved on the shadowed side of the trench if implantation is performed at an oblique angle and that forward scattering enhances trench bottom doping [8]. The trench sidewall angle must be well controlled and conditions such that slight bowing or

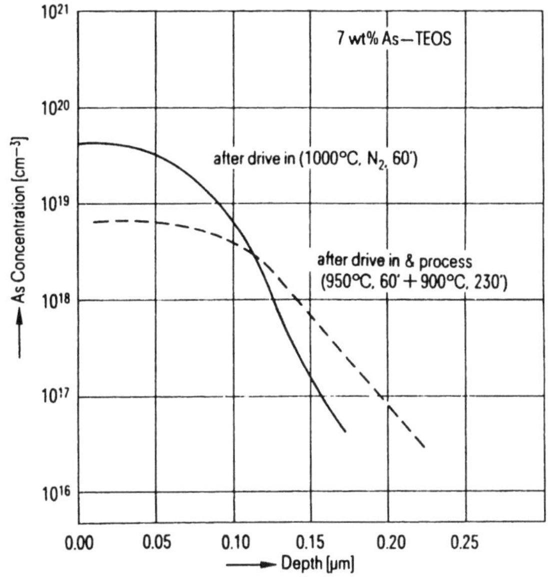

Fig. 9: SIMS profile of As in Si (planar) after drive-in of As from As TEOS source and after full DRAM process

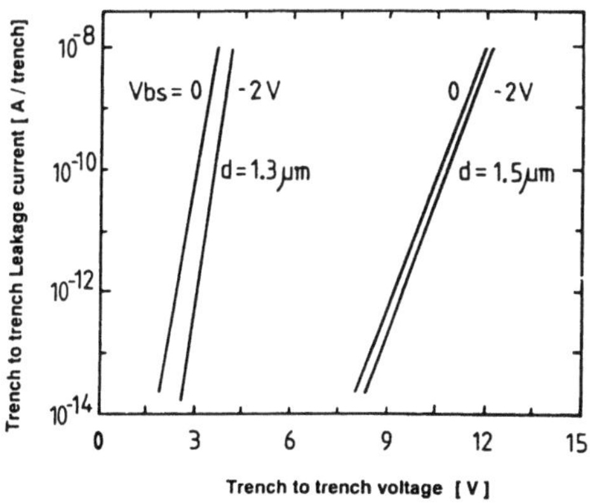

Fig. 10:
Trench to trench current versus voltage for two different trench to trench distances d and two different substrate biases V_{bs}. The figure is taken from ref.11

asperities in the sidewall are prohibited. New 16M DRAM concepts nevertheless make extensive use of ion implantation.

An innovative method is the selective epitaxial growth of in-situ As doped silicon. The cleaning process used before this epitaxy rounds the trench bottom and the upper corners: homogeneous epitaxial layers of 120 nm could be obtained on the trench sidewalls (see ref. 6). Futher optimization has to be done for the epitaxial Si to Si interface at the upper trench edge. Spin on glass is a cheap means for doping the trench. We found, however, reproducibility problems concerning sidewall coverage and we observed cracks at the upper trench corner, so that doped CVD oxide (As TEOS) was chosen as a very reproducible and reliable doping source [5]. An As SIMS profile is shown in fig 9 after drive-in and after processing which fulfills the above mentioned requirements.

The trench to trench leakage current results obtained with this doping method are shown in fig. 10. For 1.5 μm trench to trench separation the critical leakage current level for 5 % loss of the stored charge (ca. 1pA per trench) is reached only at voltages well above the 5 Volts which occur during operation.

5. Composite Dielectrics and their Reliability

The reason for using composite dielectrics is to combine the advantages of high dielectric films (e.g. Si_3N_4) with the good properties of the SiO_2/Si interface. Si_3N_4 has higher interface state density, lower breakdown voltage, higher leakage current and higher electron trapping probability than SiO_2 and is therefore not useful as single layer dielectric. Chemical vapor deposited nitride has twice the dielectic constant of SiO_2 and shows a good step coverage. It is therefore advantageous to combine oxide and nitride in an oxide-nitride-oxide stack. The oxide layer (tunnelling barrier) prevents electon trapping within the nitride. It is used on both sides of the nitride because of electric field reversal during operation in most cell concepts.

A TEM cross section of an upper trench corner shows a triple dielectric with 6 nm of bottom oxide, 8 nm CVD nitride and 3 nm top-oxide (fig. 11). The nitride shows no thinning at upper trench corners. Tunneling curves for the ONO layer have already been shown in fig. 7. Planar and trench structures show no difference in current voltage characteristics due to the self-healing effect of trapped electrons. If, for example, within a thinned oxide region a current is flowing, electrons will be trapped in the nitride thus reducing the electric field across the thin oxide. ONO layers also show a dramatic decrease in defect density even though three different process steps are required for fabrication [12]. This is probably due to self-healing and the effect that defects in one layer are covered by the other two layers. For the ONO dielectric much higher voltages than for a

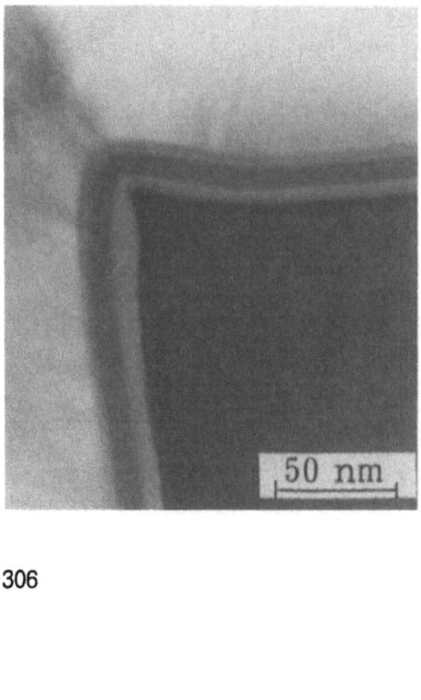

Fig. 11: TEM cross section of upper trench corner with triple oxide-nitride-oxide dielectric

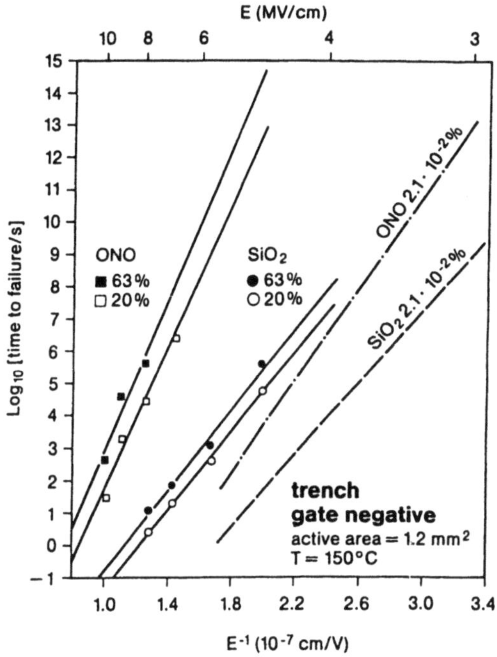

Fig. 12: Breakdown times of trench cell capacitors for two cumulative failures versus reciprocal field strength. The dotted and dashed lines are extrapolated from the 20% and 63% fractiles The fractile 2.1×10^{-2}% corresponds to a mean failure rate of 100 Fit during one year for a 5 mm^2 device. For details see reference [13]

monolayer are needed to drive a certain current. Therefore constant-voltage rather than constant-current stress measurements are used to determine the intrinsic lifetime of the oxide and the ONO in trench capacitors [13].

The 20% and 63% failure rates for three different field strengths can be used to extrapolate to the operating field strength of 3MV/cm (fig. 11). Further extrapolations from test to device area and to a reliability criterion of 100 Fit in the first year (1 Fit = 1 failure per 10^9 device hours) lead to the conclusion that oxide trench dielectrics just meet the reliability criterion while ONO is 6 orders of magnitude better.

6. Summary:

A typical trench cell process is described and the basic requirements for the use in a 4Mbit DRAM are given. The main technology and reliability problems are adressed which arise when a novel process like that for the trench cell capacitor has to be introduced into a fabrication process. It is shown that the trench etch process has considerable influence on the dielectric quality. Different doping techniques are compared and the trench to trench leakage is discussed. It is further shown how the long term stability of the dielectric can be ensured by introduction of a triple oxide-nitride-oxide dielectric. All the given examples illustrate the kind of problems, which have to be considered before a new process step can be integrated into an existing VLSI fabrication process.

Acknowledgements:

I thank my collegues from the group for electrical characterization and the analytical and process engineering teams for making available their material for this paper. The work reported here has been supported by the Federal Department of Research and Technology of the FRG (contract NT 2696).

References:

/ 1/ H.Sunami, IEDM Technol. Dig. 1985, 696 (1985)

/ 2/ K.H. Küsters, G. Enders, W. Meyberg, H. Benzinger, B. Hasler, G. Higelin, S. Röhl, H.M. Mühlhoff, W. Müller; Proc. of the 1987 Symposium on VLSI Technology, Nagano (1987) p. 93

/ 3/ M. Engelhardt, S. Schwarzl, Proc. of the Symp. on Dry Process, J. Nishizawa, Y. Horiike, M. Hirose, K. Suto (eds.) The Electrochemical Soc. (1988) p. 48

/ 4/ S. Röhl, M. Engelhardt, W.K. Kellner, A. Schlemm, Conference Proc. ESSDERC 1988, p. 283

/ 5/ F.S. Becker, H. Treichel, S. Röhl and D. Pawlik ECS Extended Abstracts Proc. Volume 86-2, 1986, p. 396 and F.S. Becker, M. Treichel and S. Röhl, to be published in J. Electrochem. Soc.

/ 6/ H. Binder, H. Geiger, R. Kakoschke, H.M. Mühlhoff, and S. Röhl, Extended Abstracts of 18th Conference Solid State Devices and Materials, (SSDM), p. 299 (1986)

/ 7/ Y. Morita, K. Ishikawa, K. Yoneda, H. Oishi, S. Terekawa, M. Yoshida, and M. Inoue, Ext. Abstr. 19th Conf. on Solid State Devices and Materials, Tokyo, p. 311 (1987)

/ 8/ K. Kakoschke, H. Binder, S. Röhl, K. Masseli, I.W. Rangelow, S. Saler, and R. Kassing, Nucl. Instr. and Meth. in Phys. Research B 21, 142 (1987)

/ 9/ M. Nagatamo, K. Mashiko, M. Yoneda, N. Kotani, S. Woya, S. Osaki, M. Hirayama and T. Matsu Kawa, IEDM Technol. Dig. 1986, 144 (1986)

/10/ T. Biwa, H. Miyatake, Y. Nogami, H. Shimizu, Ext. Abstr. 19th Conf. on Solid State Devices and Materials, Tokyo p. 315 (1987)

/11/ P.A. Murkin, H.M. Mühlhoff, S. Röhl, W. Meyberg, W. Müller, W. Bergner and R. Kircher, Conf. Proc. ESSDERC 1987, p. 761

/12/ P. Hiergeist, M. Kerber, R. Baunach and A. Spitzer, Conference proceedings ESSDERC 1987, p. 829

/13/ P. Hiergeist, A. Spitzer, S. Röhl, to be published in IEEE Transactions on Electron Devices

Heavy Doping Effects and Their Influence on Silicon Bipolar Transistors

R. Mertens

IMEC, Kapeldreef 75, B-3030 Leuven, Belgium

1. INTRODUCTION

In recent years silicon bipolar technology is receiving a renewed interest and an excellent speed performance has been demonstrated with VLSI compatible processes.It is therefore important to define the limitations and the ultimate speed performance that one can expect if the dimensions of the transistors are scaled to their minimum values. The purpose of this presentation is to discuss the different physical phenomena determining the performance of scaled silicon bipolar transistors. Emphasis is put on the effects of heavy doping. Advanced silicon bipolar technology has advanced to the point that both the emitter and the base are extremely shallow as schematically indicated on figure 1.

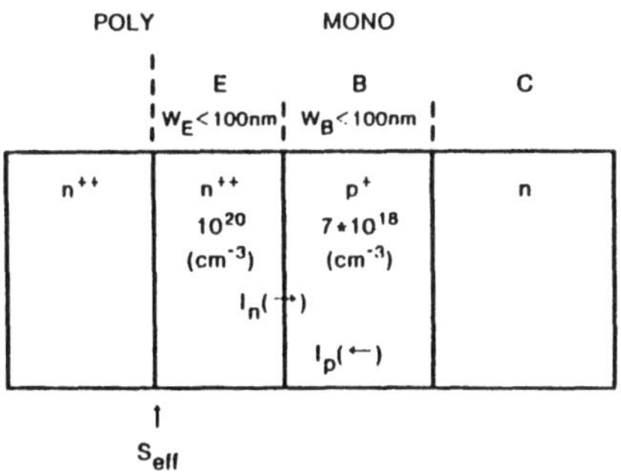

Fig. 1: Schematic representation of an advanced silicon bipolar transistor

Both W_E (emitter junction depth) and W_B (base junction depth) are in the sub-100 nm range. Both the emitter (10^{20} cm^{-3}) and the base doping (> 5 x 10^{18} cm^{-3}) levels are high. As also indicated on figure 1 the emitter is a "poly" emitter indicating that the emitter dopant is implanted in the poly and is subsequently diffused in the monocrystalline substrate.To describe injection of minority carriers into the poly-emitter the poly/mono interface is characterized by an effective interface recombination velocity (s_{eff}). Important parameters of the bipolar transistor are the current gain,the cut-off frequency f_T,characterizing its dynamic performance and the three parasitic resistances r_E,r_B and r_C.The current gain ß is the ratio of the collector to base current and equals the ratio of I_n/I_p;In being the electron current injected from the n-emitter through the p-base into the n-collector and I_p the back injected hole current,injected from the base into the

Springer Series in Materials Science, Vol. 13
Semiconductor Silicon Editors: G. C. Harbeke and M. J. Schulz

emitter.In practice ß is around 100.A crucial parameter and an important figure of merit of the bipolar transistor is the cut-off frequency f_T for which ß equals 1.f_T is inversely proportional to the emitter-collector transit time τ: the time needed for the electrons to travel from the emitter contact to the collector.In a charge control approximation τ can be modeled as the ratio of the excess charge to the collector current.It is obvious that for high speed operation f_T must be as large as possible.
Moreover the three parasitic resistances must be kept as small as possible.

2. HEAVY DOPING EFFECTS IN SILICON

As in advanced bipolar transistors both the emitter current and the base current are injected in heavily doped regions. Heavy doping effects will strongly influence the device performance. One heavy doping effect is of particular importance in bipolar devices: the fact that if the doping level exceeds $10^{18} cm^{-3}$ the bandgap starts narrowing and becomes position dependent.

2.1 Physics underlying bandgap narrowing

At high doping concentrations both many-body effects and carrier-impurity interactions give rise to bandgap narrowing. The term many body effects also refers to the jellium model in which impurity scattering is ignored. The effect of carrier-carrier interaction on the position of the bandedges has been calculated by Mahan (1) and Berggren and Sernelius (2). A downward shift ΔE_{cj} of the conduction bandedge and upward shift ΔE_{vj} of the valence band result in an effective gap narrowing ΔE_{gj}; where j refers to the jellium model. The theoretical predictions of ΔE_{vj}, ΔE_{cj} and ΔE_{gj} as a function of the doping level in n-type material are given by curves 1,2 and 3 of figure 2.

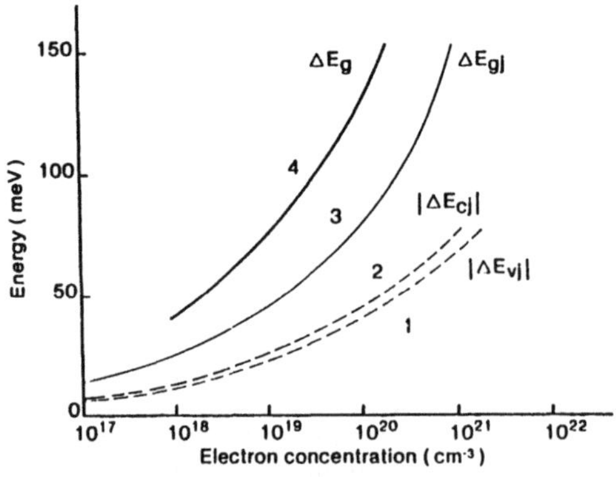

Fig. 2: Bandgap narrowing versus doping concentration in n-type Si
Curve 1: upward shift of the valence band edge in the jellium model
Curve 2: downward shift of the conduction band edge in the jellium model
Curve 3: total bandgap narrowing in the jellium model
Curve 4: total bandgap narrowing including carrier-impurity interaction

The effect of impurity scattering on ΔE_g was calculated using perturbation theory by Berggren and Sernelius (2). The effect of impurity-carrier interaction is two fold: it creates rigid shifts ΔE_{ci} and ΔE_{vi} of the conduction and valence band edges ($\Delta E_{gi} = \Delta E_{ci} + \Delta E_{vi}$) and gives also rise to bandtails. Curve 4 of figure 2 represents the total bandgap narrowing $\Delta E_g = \Delta E_{gj} + \Delta E_{gi}$ as a function of the dopant concentration.

The theoretical values of ΔE_g agree reasonably well with the experimentally measured values as can be seen on figure 3.

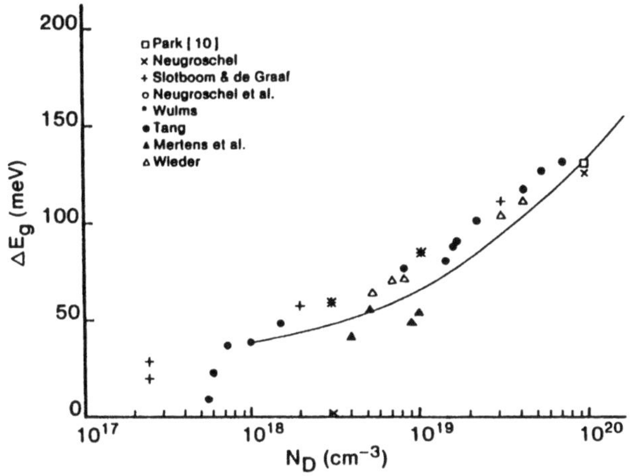

Fig. 3: Experimental values for bandgap narrowing versus doping in n-Si

The different experimental data points have been obtained using electrical techniques. The solid line corresponds to the theoretical curve (curve 4 of figure 2). It should be noted (figure 3) that bandgap narrowing can be quite substantial; e.g. at 10^{20} cm $^{-3}$ the electrical bandgap narrowing is around 120 meV; this implies that the minority carrier concentration is two orders of magnitude larger than expected on the basis of a model that neglects bandgap narrowing

2.2 Inclusion of bandgap narrowing in the transport equations

The position dependence of the bandgap Eg creates quasi-fields. For the minority carrier current in a neutral region the quasi-field can be modeled by a single parameter $\Delta Eg(x)$. If ΔEg is known as a function of the dopant concentration the quasi-field can be easily computed:

$$E\ quasi = \frac{d}{dx}(\Delta E_g) \qquad (1)$$

This field term has to be added to conventional electrostatic field to obtain the total field acting on the minority carriers.

2.3 Effects of bandgap narrowing on device performance

A first important effect of bandgap narrowing is that it decreases the current gain ß. In a shallow transistor structure, where bulk recombination is unimportant, ß can, under the condition of a homogeneously doped emitter and base, be written as:

$$ß = \frac{(N_D/D_p)\exp(-\Delta E_{g,E}/kT)W_E + (N_D/S_{eff})\exp(-\Delta E_{g,E}/kT)}{(N_A/D_n)\exp(-\Delta E_{g,B}/kT)} W_B \quad (2)$$

In (2) N_D and N_A, $\Delta E_{g,E}$ and $\Delta E_{g,B}$ are the doping levels and the bandgap narrowing in the emitter and the base respectively. D_n and D_p are the minority carrier diffusion constants and S_{eff} the recombination velocity at the interface between the poly and the monocrystalline part of the emitter. Bandgap narrowing has two important consequences for ß: 1) it strongly reduces ß bccause $\Delta E_{g,E} >> \Delta E_{g,B}$ and 2) it makes the emitter transparent for the holes injected from the base into the emitter causing the base current to be very sensitive to the value of S_{eff}. Therefore the use of a polysilicon emitter has a strong influence on the ß value since it reduces the value of S_{eff}.

The second important effect of bandgap narrowing is the higher emitter charge storage,resulting in a lower f_T.Whereas in a conventional "deepdiffused" transistor the total delay is dominated by the base,the delay in a narrow-base transistor consists three equally important delays, τ_E, τ_B and τ_C related to the emitter, base and collector respectively. The importance of the emitter delay comes about by the bandgap narrowing,making the effective emitter doping of the same order of magnitude as that of the base.This effect is enhanced by the presence of a polyemitter, reducing s_{eff}.The fact that τ_E and τ_B are about equal in an advanced transistor is shown on figure 4 (3).

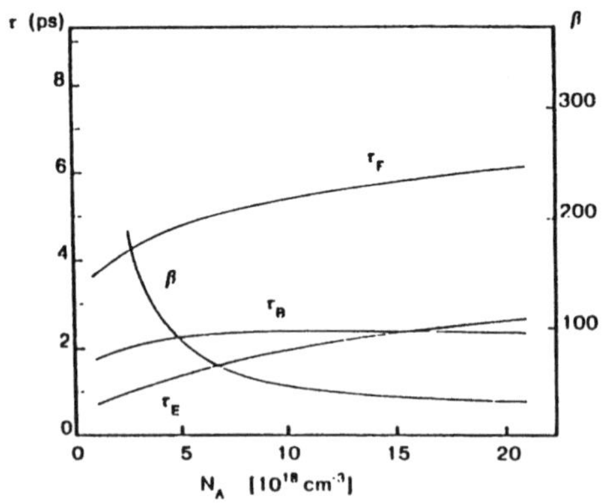

Fig. 4: Transit time as a function of base doping level.
(W_B = 50 nm, W_E = 40 nm)

3. SILICON BIPOLAR TRANSISTORS OPERATING AT LIQUID NITROGEN TEMPERATURES

In recent years there has been an increasing interest in the operation of Si chips at liquid nitrogen temperatures.Whereas most interest has gone into CMOS technologies operating at 77 K,the growing importance of merged BICMOS technologies has created a need for Si bipolar transistors operating at

LN.Conventional bipolar transistors operate very poorly at cryogenic temperatures for three reasons:a reduction of ß with decreasing temperature due to bandgap narrowing,the potential freeze-out of the majority carriers in the base,increasing the base resistance, and finally minority carrier freeze-out in the base yielding a larger τ_B.The majority carrier freeze-out problem in the base has been studied by Dumke (4) who calculated the dependence of the base resistance on the temperature.The problem of minority carrier freeze-out, caused by freeze-out of the minority carrier electrons,trapped in shallow compensating donor levels in the p-type base,has been studied by Woo and Plummer(5). The result of the effect is an important increase in base transit time.It has been suggested that in order to avoid minority carrier freeze-out the presence in the base of compensating donors,always present in the epi-layer,should be avoided.This can be done by using a high resistivity epi-layer or,more drastically,by using an epitaxial process for the base formation thereby avoiding all n-type compensation coming from the collector.

4. MAXIMUM f_T OF A SILICON HOMOJUNCTION BIPOLAR TRANSISTOR, INCLUDING BANDGAP NARROWING

Bandgap narrowing introduces an important emitter delay time influencing the maximum f_T that one can expect from a homojunction silicon bipolar transistor. In order to determine the maximum f_T one first has to determine the maximum allowable doping level in the base
It is well known that the leakage current of a pn junction increases if both sides of the junction are heavily doped. This effect, caused by direct

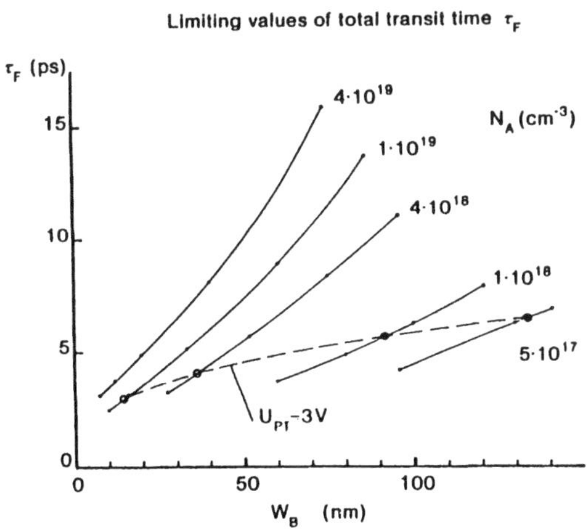

Fig. 5: Forward transit time as a function of base width with the base doping level as parameter. The dashed line gives the punch-through limit (after (3)) The dashed line on the figure indicates the punch-through limit;all points below the curve are in the punchthrough regime and must be avoided.As the maximum allowable value for N_A,set by tunneling, is ~ 10 $^{19}cm^{-3}$ the figure clearly shows that a minimum value for τ_F is around 4 ps,obtained for a 20 nm basewidth.This corresponds to an f_T of 40 Ghz.To go beyond that limit heterojunctions are necessary

interband tunneling between the base and the emitter,puts an upper limit on the practical base doping level in silicon homojunction bipolar transistors. The practical limit on base doping level is around $10^{19} cm^{-3}$. A second important constraint on the base doping and base width is emitter to collector punchthrough that has to be avoided in the voltage range in which the transistor is operational.For a given base width W_B a minimum base doping level N_A can be defined for which punch-through does not occur. The optimum base profile has therefore to take these two constraints(tunneling and punchthrough) into account. Figure 5, taken from work by Schaber et al (3), shows the minimum values for the total (forward) transit time τ_F as a function of the base width with the base doping level as parameter;a realistic shape of the base profile has been assumed.

5. HETEROJUNCTIONS IN SILICON BIPOLAR DEVICES

As homojunction silicon bipolar transistors are strongly limited by bandgap narrowing, heterojunction devices offer an interesting alternative.
The potential advantages of heterotype emitter-base junctions are well known. They include a very large emitter efficiency, the elimination of emitter stored charge, and therefore of the emitter delay τ_E, and the possiblility of using a very heavily doped base without having the limitation imposed by tunneling. To obtain an emitter-base heterojunction one can either use a wide-gap emitter or a narrow-gap base. Considerable work has been done by the author (6) on widegap materials compatible with Si technology. Amorphous and microcrystalline silicon are such widegap materials. The banddiagram of the amorphous-crystalline silicon heterojunction is shown in figure 6;theoretically high emitter efficiency can be expected since most of the bandgap difference between a-Si(EG=1.7 eV) and c-Si(EG=1.1 eV) is present as a difference in valence band edge.

We have made several transistors using these emitters (figure 7) and achieved considerable higher current gains than with conventional homojunctions.

Amorphous silicon emitters however have a major problem;their doping efficiency is poor and therefore ohmic contacts with low enough contact

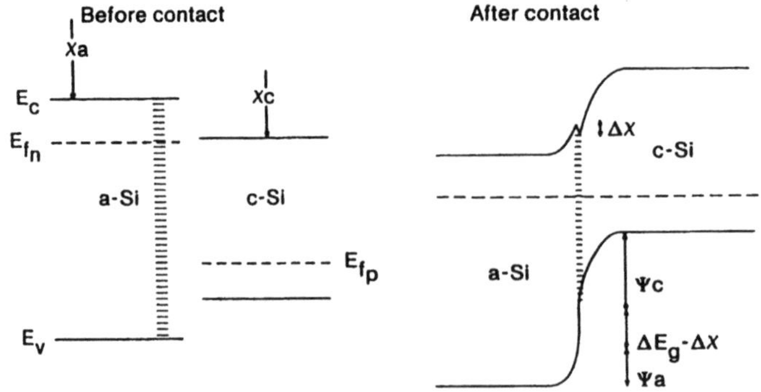

Fig. 6: Banddiagram of the amorphous silicon/crystalline silicon heterojunction

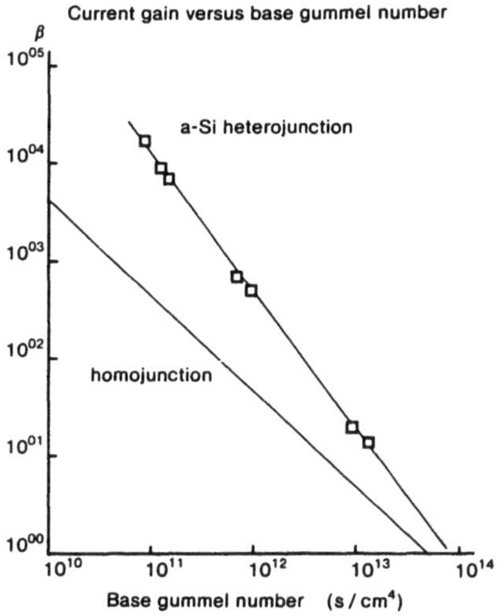

Fig. 7

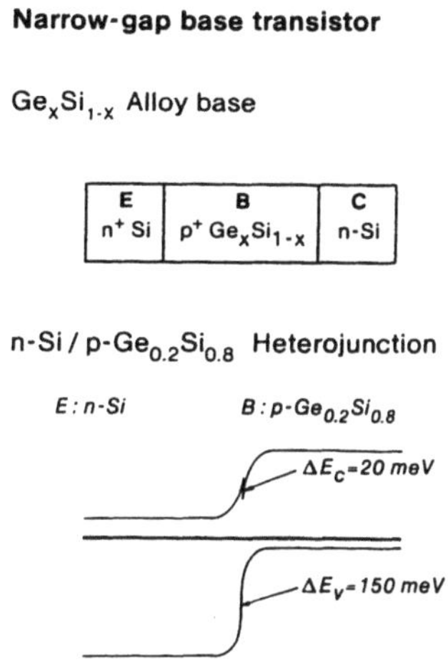

Fig. 8

Fig. 7: Current gain versus base gummel number for amorphous silicon emitter transistors

Fig. 8: Band diagram of a silicon-germanium heterojunction bipolar transistor

resistance as required for VLSI applications are impossible to achieve.Microcrystalline silicon is an interesting alternative since it has much better doping efficiency and still maintains the larger bandgap although it is doubtful wether the contact resistance will meet the $10^{-6}\Omega cm^2$ value as requested in advanced bipolar transistors.

Most attention in the field of silicon heterojunction bipolar transistors is now focused on the germanium alloyed base transistor(figure 8). This idea was first proposed by J.C.Bean (7).

Pseudomorphic growth of such a structure is possible if the base is sufficiently thin. In practice it turns out that with a Ge fraction of 20% the critical thickness for pseudomorphic growth is 200nm which is sufficiently large to make a punch-through-free base.For a 20 % Ge fraction a bandgap difference of 170 meV can be expected.Almost the entire bandgap difference is seen as ΔE_V at the valence band edge,which is an almost ideal case for the heterojunction application.Recently the first working Ge alloy base transistor has been presented (8.) Although the actual performance is still limited one can expect that these structures will yield extremely high f_T and low base resistance.

REFERENCES

1. G.D. Mahan: J. Appl. Phys., 51,p. 2634
2. K.F. Berggren and B.E. Sernelius: Phys. Rev., B 24, 1971 (1981); Phys. Rev., B 29, p. 5575 (1984), Solid State Electronics, 28, 11 (1985)

3. J. Schaber, J. Bieger, B. Benna and T. Meister: Proceedings of the ESSDERC 87, p. 365 (1987)
4. W.P. Dumke: IEEE Tr. on El. Dev., ED-28, 5, p. 494, (1981)
5. J.C. Woo and J.D. Plummer: IEDM 87 Technical Digest, p. 401 (1987)
6. R. Mertens, J. Nijs, J. Symons, K. Baert and M. Ghannam: Proceedings of the ESSDERC-87, p. 377 (1987)
7. J.C. Bean: Proceedings of the 6th American Conference on Crystal growth (Atlantic City, N.J.) (1987)
8. G.L. Patton, S.S. Iyer, S.L. Delage, S. Tiwain and J.M.C. Stork: El. Dev. Let., 9, p. 165 (1988)

Ionizing Radiation Effects in MOS Devices

J.M. McGarrity, F.B. McLean, T.R. Oldham, and H.E. Boesch, jr.
US Army Laboratory Command, Harry Diamond Laboratories,
Adelphi, MD 20783-1197, USA

Introduction

One of the major effects associated with radiation-induced ionization in electronic materials is that of trapped-charge buildup, which induces internal space-charge fields that interfere with the normal (designed) operation and control of devices. This problem is primarily associated with the insulating films used in modern IC technologies, and it is the major type of effect encompassed in the term "total-dose ionization." The major focus of work in this area over the past 20 years or so [1] has been on the MOS technologies because of the charge-trapping effects in the thin silicon dioxide films employed in these technologies, both as gate oxides over the active semiconductor channel region and as field, isolation, and passivation oxides. However, as device dimensions shrink and hardness requirements increase, the total-dose problem is increasing in concern as well for the bipolar technologies in connection with trapped-charge-induced leakage paths near field and passivation oxides. Here, for the purposes of discussing the physics, we focus on the total-dose charging problem in the SiO_2 gate oxides of MOS structures. The basic mechanisms underlying the charge trapping is the same in both gate and field/passivation oxides; only the circuit manifestations differ.

Figure 1 shows a simple schematic of a MOSFET, in this case an n-channel into the Si surface region immediately below the gate region. If the gate bias is sufficiently large and positive (for n-channel operation), the majority carriers (holes in p-type Si) will be repelled from or depleted in this surface region, and minority carriers (electrons) will be attracted to this region, forming what is called an inversion layer. If now also a potential difference is applied between the source and drain contacts (n^+-doped regions in fig. 1), the inversion layer provides a low-resistance current channel for electrons to flow from the source to the drain. The device is then said to be turned on (fig. 1a), and the control gate bias potential at which the channel just begins to conduct appreciable current is called the turn-on voltage or threshold voltage of the device.

The total-dose ionization problem that occurs in this structure is then due to the radiation-induced charging (normally positive) of the thin gate oxide region, which generates additional space-charge fields at the Si surface. These additional induced fields result in voltage offsets or shifts in the turn-on voltages of the devices, which lead to circuit degradation and failure. For example, for sufficiently large amounts of trapped positive charge for the device schematically shown in figure 1, the device may be turned on even for zero applied gate bias (fig. 1b). In this case, the device is said to have failed by "going depletion mode."

Springer Series in Materials Science, Vol. 13
Semiconductor Silicon Editors: G. C. Harbeke and M. J. Schulz

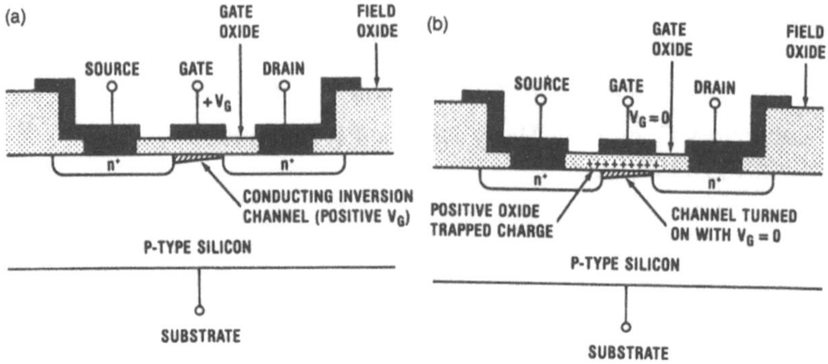

Figure 1. Schematic of n-channel MOSFET illustrating basic effect of total-dose-ionization-induced charging of gate oxide: (a) normal operation and (b) postirradiation

1. Overview of Radiation Response of MOS Structures

Figure 2 shows a schematic energy band diagram for an MOS structure where positive bias is applied to the gate, so that free electrons in the oxide layer will be swept toward the gate and holes will be attracted to the Si substrate. (In an energy band diagram--which is essentially equivalent to a potential energy diagram--electrons want to fall downhill, whereas holes will float upwards.) Also indicated in figure 2 are the four major basic processes contributing to the radiation response of such a system, which we now discuss: (1) electron-hole creation and recombination (2) hopping transport of holes (3) hole trapping at the Si/SiO_2 interface and (4) radiation induced interface traps.

A major electrical consequence of the radiation-induced charging of the SiO_2 film (including transporting holes, trapped holes, and interface traps) is a shift in pertinent voltage operating points for devices, such as in the threshold voltage V_t for a MOSFET. Let us write the threshold voltage as

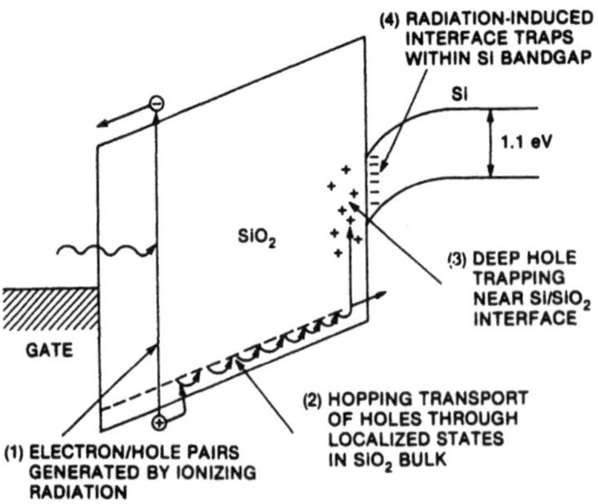

Figure 2. Schematic energy band diagram of SiO_2 MOS structure for positive gate bias, indicating major physical processes underlying radiation response

$$V_T(t) = V_T^o + \Delta V_T(t) \quad (1)$$

where V_T^o is the threshold voltage before irradiation and $\Delta V_T(t)$ is the voltage shift following radiation exposure, and for reasons already discussed, $\Delta V_T(t)$ is in general time dependent. The preirradiation threshold voltage is defined and its explicit expression is given in any textbook discussing MOS transistors. It depends upon temperature, gate-semiconductor work function difference, substrate doping density, oxide layer thickness, and substrate bias. There are also contributions from any oxide charges and interface traps existing before irradiation (and there are always some small amounts of these, but usually for good devices their areal densities are $<10^{10}$ cm^{-2}.

Based on figure 2 and its discussion, we break the radiation-induced threshold voltage shift into three components:

$$\Delta V_T(t) = \Delta V_{st}(t)+\Delta V_{ot}(t)+\Delta V_{it}(t), \quad (2)$$

where $\Delta V_{st}(t)$ is the (short-term) contribution from the radiation-generated mobile holes transporting in the oxide bulk, $\Delta V_{ot}(t)$ is due to the deep trapped holes near the interface, and $\Delta V_{it}(t)$ is the contribution from the charged interface traps. Note that all three components are in general time dependent. They are given explicitly as

$$\Delta V_{st}(t) = -q/C_{ox}\int_o^{d_{ox}} dx(x/d_{ox})\, n_h(x,t), \quad (3a)$$

$$\Delta V_{ot}(t) = -q/C_{ox}\Delta N_{ot}(t), \quad (3b)$$

$$\Delta V_{it}(t) = -\Delta Q_{it}(t)/C_{ox}, \quad (3c)$$

where q is the electronic charge, d_{ox} is the oxide thickness, and C_{ox} is the oxide capacitance per unit area ($C_{ox} = \varepsilon_{ox}/d_{ox}$) In equation (3a), $n_h(t)$ is the space- and time-dependent density of free (mobile) holes and distance x in the oxide is measured relative to the gate/SiO_2 interface. In equation (3b) $\Delta N_{ot}(t)$ is the radiation-induced areal density of deep trapped holes near the SiO_2/Si interface; it is time dependent both because of its time-dependent buildup as the transporting holes reach the interface and because of its long-term annealing, which can extend out to years. In equation (3c), we leave the sign of the radiation-induced interface trapped charge $\Delta Q_{it}t)$ unspecified, as it can contribute either a net negative or net positive charge depending on the position of the Fermi level at the Si surface at inversion. (Usually ΔQ_{it} is negative for n-channel and positive for p-channel devices at the threshold voltage.) We note that in general a voltage offset is simply proportional to the first moment of the induced oxide charge relative to the gate interface. For the mobile hole distribution, equation (3a) retains this definition explicitly, since the transporting holes can be distributed through the bulk of the oxide. For both the long-term trapped holes and interface traps, $\langle x\rangle \equiv d_{ox}$, with the resulting simplifications exhibited in equations (3b) and (3c).

In any case, because of the several different physical processes involved, each having different characteristic times, it is apparent that the overall time history of the recovery of threshold voltage can be fairly complex, with important implications for sensible testing procedures, hardness assurance, and prediction. In simplest terms, one must be able to extract from test data--usually done over a very limited time regime and for a limited range of experimental conditions--the expected performance of an IC in a particular radiation environment which may involve a completely different time regime (perhaps orders of magnitude shorter or longer). In this paper we discuss in further detail the basic physical processes responsible for this complex time history.

2. Descriptions of Physical Processes Underlying Radiation Response of MOS Devices

2.1 Initial Hole Yield

The initial hole yield in the oxide determines the initial (maximum) voltage shifts of MOS devices and sets the scale for the amount of damage (at any time) due to ionizing radiation. The two major factors that determine the initial hole density are the electron/hole pair creation energy and the field-dependent fraction of holes which escapes the initial recombination processes. The initial value of the threshold voltage shifts is then simply related to the initial hole density via the dose and geometric (oxide thickness) factors.

The electron/hole pair creation energy E_p was determined by Ausman and McLean [2] to be 18± 3 eV, based on an analysis of experimental data of Curtis et al [3]. This result has been confirmed independently by others since [4,5], including a recent more accurate set of measurements and analysis by Benedetto and Boesch [6] which establishes E_p to be 17±1eV.

From the value of E_p, the initial electron/hole pair density per unit dose is easily determined to be 8.1×10^{12} cm^{-3} $rad^{-1}(SiO_2)$ (see fig. 2). But this initial density is quickly reduced by the initial recombination processes occurring in picoseconds, before the electrons are swept out of the oxide and collected. The fraction of holes escaping initial recombination, $f_y(E_{ox})$, which determines the final hole yield, is determined mainly by two factors: the magnitude of the oxide electric field E_{ox}, which is acting to separate the charge pairs and the initial line density of electron/hole pairs created by the incident radiation particle. The pair line density, which is determined by the linear energy transfer (LET) and therefore a function of the incident particle type and energy, is inversely proportional to the average separation distance between electron/hole pairs. Obviously, the closer the average spacing of the pairs, the more recombination that occurs for a given field, and the less will be the final yield of holes.

The initial recombination problem has not been solved analytically for arbitrary pair line density. However, analytic solutions do exist in the limiting cases, namely, where the charge pairs are far apart (geminate model) and the opposite, where the electron/hole pairs are very close together (columnar model). The characteristic distance scale which distinguishes these cases is the thermalization radius of a single electron/hole pair, i.e., the average separation distance between an electron and hole of the same pair after they have dissipated their excess kinetic energy and reached thermal equilibrium energies. For SiO_2 the average thermalization distance is about 8 nm [2,7].

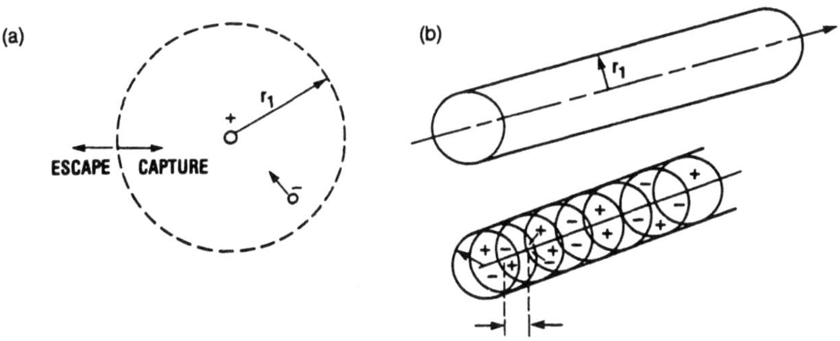

Figure 3. Schematic diagrams indicating limiting pair separation distances for two recombination models: (a) geminate (separate electron/hole pairs) and (b) columnar (over lapping electron/hole pairs)

Figure 3 illustrates schematically the two limiting cases for which recombination models have been solved. For the geminate case (fig. 3(a)) the separation distance between pairs is much larger than the thermalization distance. Then one can treat only the recombination between the individual members of the same pair. That is, one need only calculate the recombination probability of a single isolated charge pair whose members are attracted to each other via the mutual Coulomb attraction while, at the same time, they undergo both a drift motion (in opposite directions) in response to the local electric field and a random diffusive motion driven by the thermal fluctuations of the system. The geminate recombination model was first formulated by Smoluchowski [8] and later solved by Onsager [9], originally for recombination of electrons and positive ions in gases.

The other limiting case (fig. 3(b)) occurs when the average pair separation distance is much less than the thermalization distance. In this case--columnar recombination--the individual electron/hole pairs lose their identity, and recombination must be considered between many electrons and holes lying in a cylindrical distribution around the track of the incident particle. Obviously, the probability of recombination of an individual carrier in this case is significantly greater than in the geminate model because of the enhanced probability of recombination encounters between opposite charges. The columnar model was originally solved by Jaffe [10], expanding on previous work by Langevin [11]; more recently, a more accurate, numerical treatment has been done by Oldham [7] which also extends the range of applicability of the model.

These models have been applied to the initial recombination process in thermally grown SiO_2 films with good success in recent years. For example, low-LET particles such as high-energy electrons (including the high-energy secondary Compton electrons from Co^{60} gamma interactions) generate a sparse density of charge pairs along their tracks (for a 1-MeV electron, the average pair separation distance is ~50nm), and the geminate model fits the results very well. On the other hand, very-high-LET particles, such as protons, alpha particles, and other heavier ions generate a high density of pairs along their tracks (for example, a 1-MeV proton generates an electron/hole pair on the average about every 0.3 nm), and the columnar model is successful in describing the experimental results. As one might expect, many experiments of practical interest fall in the transition region between these two models for the limiting cases of pair separation distance. However, the experimental results reflect a smooth continuous transition between the limiting cases. That is, intermediate cases show increasing columnar behavior (stronger recombination) as the density of electron/hole pairs is increased. Figure 4 is a compilation of a number of experimental results [2,3,5,6,12] of the fractional hole yield f_y versus electric field for a number of particles spanning the range from low to high LET. For these particle sources at a field of 1 MV/cm, the yield varies from almost 90 percent for low-LET particles (12-MeV electrons and Co^{60}) to only about 6 percent for the high-LET 2-MeV alpha particles. This figure clearly shows that recombination is a real and often important effect when MOS responses to different radiation sources are compared.

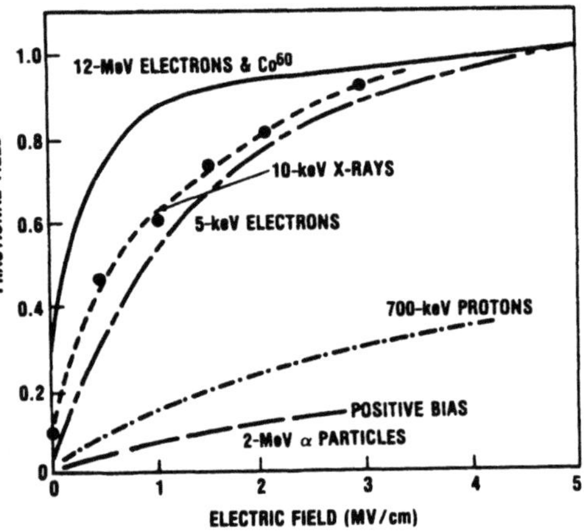

Figure 4. Experimentally measured fractional hole yield versus electric field in SiO_2 for a number of incident particles

Finally, knowing the initial pair volume density per rad (g_o = 8.1 x 10^{12} cm^{-3} $rad^{-1}(SiO_2)$ and the fractional hole yield after recombination $f_y(E_{ox})$ (from fig. 4), the initial threshold (or flatband) voltage shift is easily obtained. Assuming a uniform generation density across the oxide layer, the total initial areal charge density of holes which escape recombination is $\Delta Q_h = qg_o d_{ox} f_y(E_{ox})D$, where d_{ox}, is the oxide thickness and D is the dose in $rad(SiO_2)$. The initial threshold shift is related to ΔQ_h as

$$-\Delta V_T(0^+) = \Delta Q_h/2C_{ox} = [qg_o d_{ox} F_y(E_{ox})D]/2C_{ox} \tag{4}$$

where $C_{ox} = \varepsilon_{ox}/d_{ox}$, and the factor of 2 comes from the fact that the centroid of a uniform charge density is $d_{ox}/2$. Finally, substituting for the constant factors in equation (4), we have

$$-\Delta V_T(0^+) \text{ it= } 1.9 \times 10^{-8}\, d_{ox}{}^2 f_y(E_{ox})\, D \tag{5}$$

Here, the units of ΔV_T are volts if d_{ox} is expressed in nanometers and D in $rad(SiO_2)$. Equation (5) explicitly shows the dependence of the initial voltage shifts on oxide thickness squared, indicating a significant improvement in the radiation susceptibility of devices having thinner gate oxides.

2.2 Hole Transport

The transport of the holes through the SiO_2 oxide layer, which is responsible for the early-time recovery of V_T, has been studied extensively by several groups [13-24]. It has been found to exhibit some rather unusual properties, the principal ones being the following: (1) It is highly dispersive in time, taking place over many decades in time following a radiation pulse. (2) It is universal in nature, meaning that changes in temperature, field, and thickness do not affect the shape or overall dispersion of the recovery curves, when plotted in terms of log-time. Changes in these variables affect only the time scale for the recovery. (3) The transport is field activated. (4) At temperatures above about 140 K, the transport has an Arrhenius-type temperature activation dependence, but below 140 K, the transport becomes essentially not thermally activated. (5) The characteristic recovery time, or hole transit time, has a strong superlinear power-law dependence on oxide thickness.

The best overall description of the experimental data seems to be provided by the hopping transport mechanism. Specifically, the available transport data on clean, hardened oxides, in which the short-term recovery is not masked by significant amounts of deep hole trapping near the SiO_2/Si interface, has been successfully described by a continuous-time-random-walk (CTRW) formalism originally developed by Montroll, Weiss, Scher, and others [25-28] and applied to hole transport in SiO_2 by McLean et al [13,16,17,21-24,29] and Hughes [14,15,20]. In essence, a response function (e.g., a radiation-induced voltage shift) may be characterized by the functional form $F(\alpha;t/t_s)$, where α is a disorder parameter describing the amount of dispersion (or shape) of the response and is independent of T, E_{ox}, and d_{ox}. These parameters enter only in the characteristic time scale, $t_s = t_s(T, E_{ox}, d_{ox})$, for the response. For further discussion of the transport the reader is referred to the references given above.

One of the salient features of hole transport is illustrated by the data sets shown in Figure 5. Figure 5 shows the effect of temperature

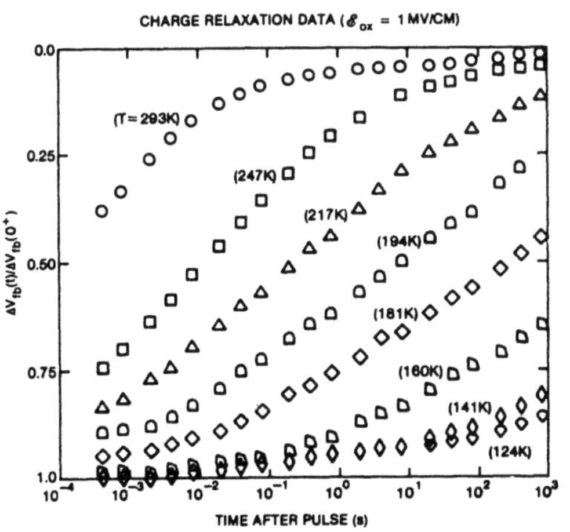

Figure 5. Normalized flatband voltage recovery data following pulsed 12-MeV LINAC electron irradiation of 96.5-nm oxide MOS capacitor under 1-MV/cm oxide field for series of temperatures between 124 and 293 K [21]

[21]. These data are measurements - the flatband voltage shift versus log-time from 10^{-4} to 10^{3} s following pulsed 12-MeV electron (LINAC) irradiation of MOS capacitors. The data in these figures are all normalized to the initial shifts, $\Delta V_{FB}(0+)$ immediately after the radiation pulse. The initial shifts are not those at the earliest measurement (10^{-4}s), but rather are the calculated shifts before any transport occurs. The actual values of $\Delta V_{fb}(0^{+})$ vary with dose, field, and thickness as described in the previous subsection. The data are plotted in the negative direction because the voltage shifts are negative, indicative of net positive charge induced in the oxide layer. Figure 5 shows the response for a series of temperatures between 125 and 293 K at a single oxide field of 1 MV/cm. The strong temperature activation above 141 K is apparent; in fact, plotting the time, say, at which half-recovery occurs versus 1/T (Arrhenius-type plot) yields a straight line, the slope of which yields an activation energy of ~0.60 eV.

2.3 Deep Hole Trapping and Annealing

After the radiation-generated holes have had time to complete their transport through the oxide, MOS structures typically exhibit a negative voltage shift component, ΔV_{ot}, in their electrical characteristics (e.g., V_T, V_{fb}) that is not sensitive to silicon surface potential and that persists for hours to years. This long-lived radiation effect component is the most commonly observed form of radiation damage in MOS devices and is of greatest concern in space environments. It is attributed to the long-term trapping of some fraction of the radiation-generated holes in the oxide layer within about 5-10 nm of the SiO_2/Si interface. This effect generally dominates other radiation damage processes in MOS structures, including negative charge (electron) trapping and interface-state buildup effects, unless specific device processing changes are made to alter the oxide and consequently reduce the hole trapping or enhance the other effects.

An important point that is worth emphasizing is the following: radiation effects involving interface phenomena in general (hole trapping and long-term annealing (removal) of trapped holes, interface trap buildup) are all highly dependent upon oxide processing. There can be orders of magnitude variation in the different effects among oxides of varying processing histories.

The effect of processing on the fraction of holes which undergoes deep, long-lived trapping is illustrated by the data shown in Figure 6 [48]. Here, the threshold voltage shifts versus total dose are plotted for two n-channel devices where the radiation hardness was varied by changing the high-temperature processing. The dose rate was 8 x 10^{4} rad per minute and the measurements were obtained within 10 minutes of the exposure at each dose. Clearly, at each dose there is roughly an order of magnitude difference in the threshold voltage shift, which in turn is most likely due to an order of magnitude difference in the number of hole traps present in the two oxides. In general, the fraction of radiation-generated holes that undergoes long-term deep trapping has been found to vary from as little as 1 or 2 percent for a good, improved oxide, to the order of 20 to 40 percent for good-quality commercial oxides, and to as much as 50 to 70 percent in very soft commercial oxides.

Evidence concerning the location of the trapped holes is contained in figure 7 which shows long-term ΔV_{fb} measured in MOS capacitors on n-type

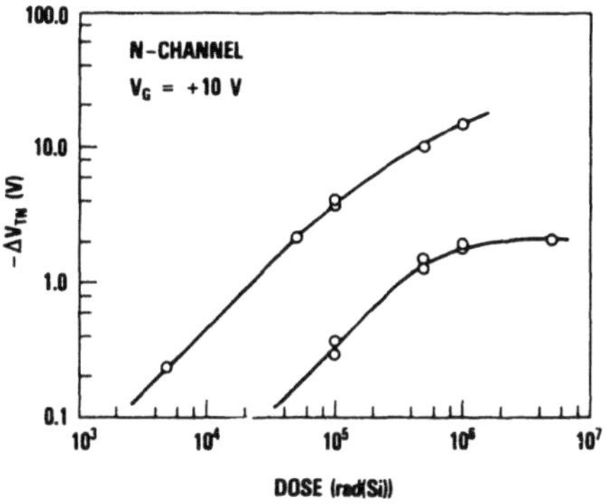

Fig. 6

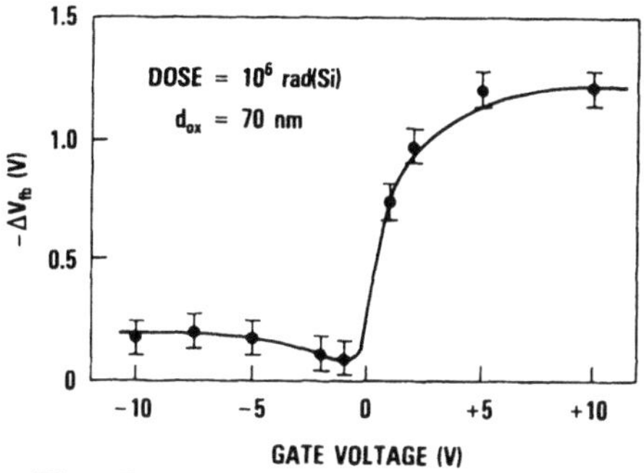

Fig. 7

Figure 6. Effect of processing variation on hole trapping: threshold voltage shift versus total dose for two n-channel MOSFET's which received different high-temperature processing [15]

Figure 7. Bias dependence of radiation-induced voltage shift: radiation-induced flatband voltage shift versus gate voltage for MOS capacitor following 1 Mrad(SiO_2) exposure [31]

silicon irradiated to 1 Mrad(SiO_2) at gate voltages from -10 to 10 V [31]. The negative ΔV_{fb} includes contributions from both the oxide trapped charge and radiation-generated interface states, but the response is dominated by the trapped hole component. The shift under positive V_g is much greater than that observed at equivalent negative V_g. From similar data, Zaininger [32] first in 1966 inferred that the positive oxide charge must be fairly mobile since it apparently moves rapidly to and is trapped near the SiO_2/Si interface under positive bias and moves toward the gate under negative bias. If bulk trapping was involved rather than trapping near the interfaces, the shifts for positive and negative polarity for the same field magnitude would not differ nearly as much as data such as Figure 7 indicates. A great amount of other work, including oxide etch-back analysis [33-36], internal photoemission spectroscopy [37,38], and inferences from the postirradiation buildup [39-41] and annealing [42] of the trapped charge all indicate that the location of the long-term trapped holes under positive gate bias is generally within ~10 nm of the SiO_2/Si interface.

Besides the location of the trapped holes in the oxide within ~5-10 nm of the SiO_2/Si interface and the fact that the hole trapping fraction is highly dependent on processing, other pertinent pieces of information concerning the characteristics of the hole traps--obtained from the work of many investigators over the years--include the following:
(1) The number density of the hole traps typically lies in the range from 10^{18} to 10^{19} cm^{-3} [33,43-45], with the corresponding areal density range from being 10^{12} to 10^{13} cm^{-2}. (2) The cross section for hole trapping is usually found to be around 5×10^{-14} cm^2 for a 1 MV/cm oxide field [44,45]. (3) The cross section decreases as $E_{ox}^{-1/2}$ for fields above 1 MV/cm in magnitude (46-48). For the data shown in Figure 7, the increase in ΔV_{fb} as the bias is increased from zero is attributed to the increased yield of holes with increasing field, but the flattening of the curve for larger positive bias (and sometimes a turn-around has been observed) is probably due to the decreasing cross section for hole capture as the field

increases above 1 MV/cm. (4) It is commonly observed [5,40,41,43,45] that the total number of trapped holes, N_{ot}, tends to a saturation level in the 1 to 10 Mrad dose range. Depending on specific circumstances, this is due to one or more operating factors, including trap filling (hard saturation), large space charge effects to the point of field reversal in some regions of the oxide layer, and recombination of the trapped holes with radiation-generated electrons moving through the trapped hole distribution. (5) There are several different experimental indications [36,42,49,50] that the energy level associated with some (but perhaps not all) of the deep hole traps is about 3 eV above the valence band edge of SiO_2. (6) Based upon an increasing accumulation of electron spin resonance (ESR) spectroscopy data [49,51,52], the microscopic structure of the trapped holes seems to be the so-called E' center, which is a trivalent silicon defect associated with an oxygen vacancy in the SiO_2 structure [53-55]. In essence, present evidence suggests the following model for the hole trapping process. The SiO_2/Si interface region of MOS structures is characterized by high local strain [56] and a deficiency of oxygen atoms which result in a number of strained Si-Si bonds (instead of normal Si-O-Si bonding configurations). A hole encountering such a strained bond may break the bond and recombine with one of the bonding electrons. The resulting positively charged structure relaxes to the E' center configuration, with one of the Si atoms retaining the remaining electron from the broken bond and the positive charge residing with the other trivalent Si atom. Figure 8 shows a simple schematic of the trapping process.

Let us now address the question of the long-term stability of the deeply trapped holes. In fact, the holes in deep traps in the SiO_2 layer of an MOS structure after irradiation are not truly "permanently" trapped. Instead, they are observed to disappear from the oxide over times from milliseconds to years. This discharge of the hole traps, as commonly observed at or near room temperature, is the major contributor to the so-called "long-term annealing" of radiation damage in MOS devices. The annealing of the trapped holes has two manifestations that may reflect different hole removal processes. The first is the slow bias-dependent recovery of ΔV_{ot} typically observed at normal device operating temperatures (-55° to 125°C, for instance). Aspects of this process have been described through a tunneling model. The second is the relatively rapid and strongly temperature-dependent thermal detrapping or recombination of the holes observed when MOS structures are deliberately subjected to thermal annealing cycles at elevated temperatures (150° to 350°C). This process has been described through a thermal detrapping model. We will restrict our attention here to just the "tunnel anneal" process important at normal operating temperatures.

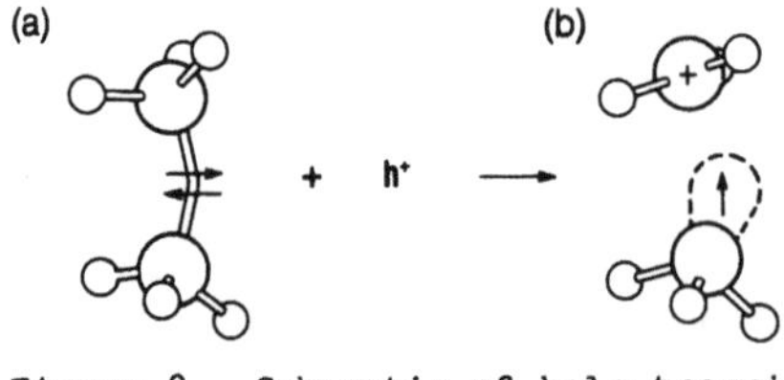

Figure 8. Schematic of hole trapping process indicating (a) initial strained Si-Si bond (oxygen vacancy) and (b) relaxed E_1'center configuration after hole capture

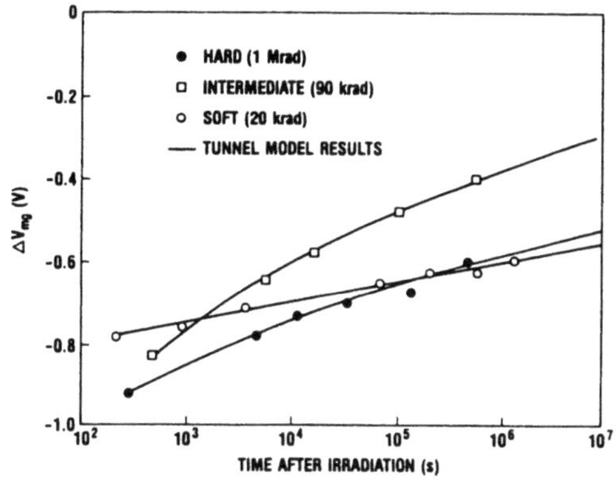

Figure 9. Long-term annealing data for three MOSFET's of varying radiation hardness. Devices received indicated doses to produce comparable initial shifts [42]

Figure 9 presents typical results for the time dependence of hole annealing at room temperature. Midgap voltage shift, ΔV_{mg} is plotted as a function of time after irradiation for three n-channel MOSFET samples with different dry oxide processing [42]. The discharge of the trapped holes is roughly linear in log(t). This behavior is a hallmark of the hole anneal process and has been used in convolution schemes for predicting the response of MOS devices under low-dose-rate irradiation conditions [30,57,58]. We note that the three devices shown in Figure 9 have been irradiated to different doses to achieve roughly comparable shifts for the first measurements.

Several investigators have suggested a tunneling process to explain the log(t) recovery and other aspects of hole annealing in MOS structures [23,42,50,59]. These models assume that electrons from the silicon substrate tunnel to, and recombine with, the trapped holes in the distribution of traps near the SiO_2/Si interface, as illustrated by the schematic diagrams in Figure 10. (Equivalently, one can think of the holes tunneling from the traps to the silicon valence band.) As a consequence of the exponential decay of the tunneling probability with distance into the SiO_2, at a given time t the hole traps are emptying at a depth $X_m(t)$ from the silicon that increases logarithmically with time [60,61]:

$$X_m(t) = (1/2\beta)\ln(t/t_o) \qquad (8)$$

where β is the tunneling barrier height parameter and t_o is the time scale parameter. For distances only slightly greater than X_m, essentially all the traps remain full at time t; for distances only slightly less than X_m, essentially all the traps have been emptied by tunneling. Thus, hole removal proceeds via a "tunneling front" that moves into the oxide with a "velocity" $\Delta X_m = 1.15/\beta$ per decade in time. (ΔX_m has been found to be about 0.2 nm per decade [50,59]). For $X_m \ll d_{ox}$ and a uniform distribution of traps in the oxide within tunneling distance of the silicon, the resulting trapped charge loss (and hence the decrease in ΔV_{ot}) at a given time due to the tunnel anneal is then proportional to X_m; i.e., ΔV_{ot} decreases as ln(t). Thus the tunneling model explains qualitatively the observed approximate log(t) anneal of the trapped holes.

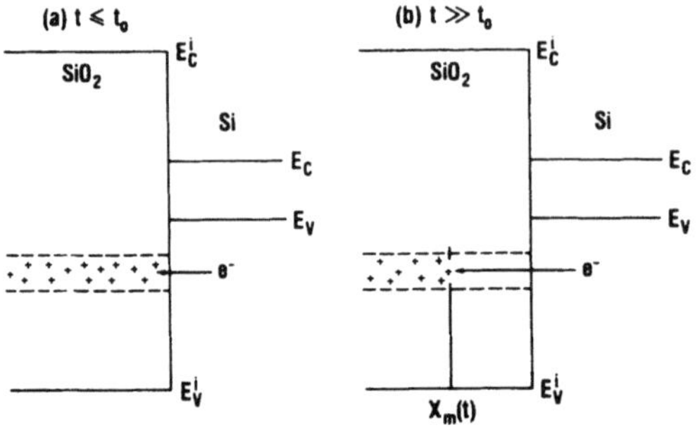

Figure 10. Schematic of trapped hole removal by electron tunneling from Si substrate

This simple model fails to describe the observed responses in detail. As shown in Figure 9, the annealing curves are not completely linear; the rate of annealing tends to decrease with time. This is clearly to be expected, since the rate of annealing should approach zero as the amount of charge left to be removed (remaining ΔV_{ot}) approaches zero. The simple log(t) anneal arises from the assumption that the trapped holes are distributed uniformly into the oxide from the silicon interface. In reality, the hole trap density generally falls off with distance from the silicon. Oldham et al [42] assumed an exponential falloff in occupied hole trap density from the interface and incorporated this form into the appropriate expressions from tunneling theory; this yielded the solid line fits to the data as shown in Figure 9. An important conclusion from this analysis was that the trapped hole distribution lies close to the SiO_2/Si interface in hard oxides (i.e., the distribution falls off relatively sharply from the interface), whereas it extends deeper into the oxide bulk for the soft oxides. This is apparently related to the size of the strained, oxygen-deficient region near the interface, a property highly dependent upon processing conditions. The fact that the tunneling rate depends exponentially on tunneling distance then explains the many orders variation in long-term annealing rates among different oxides [62].

Additional aspects of the tunnel anneal process are its dependencies upon temperature and oxide field. The temperature effect seems to be adequately explained [50] by a linear temperature dependence of the trap energy levels, $E_{t(T)} = E_t^o - bT$, where $E_t^o = 3.6$ eV and $b = 2.0$ meV/K. The electric field dependence arises from the field modification of the tunneling potential barrier height; a positive electric field has the effect of lowering the barrier to tunneling and thereby increases the rate of annealing.

2.4 Radiation-Induced Interface Traps

Interface traps N_{it} are localized electronic states located at or very near the SiO_2/Si interface, having their energy levels distributed within the Si bandgap. They can exchange charge with the Si conduction and valence bands. Their occupancy, or charge state, depends upon the position of the Fermi level at the interface, i.e., upon the value of the surface potential. The major effects of N in MOS systems are to (1) produce distortions in device characteristics (e.g., C-V_g or I-V_g) as the gate bias is varied; (2) shift the threshold voltage due to the net interface trapped charge at the turn-on or inversion point; and (3) introduce additional Coulomb scattering centers for carriers moving in the surface channel of a MOSFET which degrade the carrier mobilities. Before irradiation, the areal density of interface traps in good modern devices is $<10^{10} cm^{-2}$, a density such that N_{it} effects normally are not much of a problem.

Upon irradiation, however, interface traps can build up to a significant level, resulting in discernible effects in devices. In general two components of radiation-induced interface traps (ΔN_{it}) have been observed, one a prompt component present at the earliest measurements following an irradiation and the other a time-dependent component which can continue building up for thousands of seconds. The relative ratio of the two components can vary greatly, with one or the other dominating in a particular system. As a rather loose general statement, the delayed time-dependent ΔN_{it} seems to dominate in metal (aluminum) gate devices [63-69], prompt ΔN_{it} dominates in thick, steam oxides (as used as field oxides) [70-71], and both components seem to be present in more or less comparable amounts in the polysilicon gate devices [72] that are more representative of present-day technology.

Figure 11 shows the time-dependent buildup of interface trap density (for states with energies between midgap and inversion) for Al-gate wet oxide capacitors for a series of oxide fields between 1 and 6 MV/cm, from 2 to almost 10^5 s following pulsed LINAC irradiation [64]. Note that the final N_{it} levels in this case are almost solely due to the time-dependent buildup process. There is obviously a strong field dependence on both the rate of buildup and the final value of N_{it}. For all fields, the buildup apparently begins on the order of seconds (the point at 1 s on figure 11 is also the preirradiation N_{it} value), continuing for several hundred seconds at the highest field (6 MV/cm) before leveling off. For the lower field values (1 and 2 MV/cm) the generation rate is much lower, but N_{it} is still seen to be increasing even at the latest measurement times.

In figure 12, the field dependence of the radiation-induced interface trap buildup is compared for three types of MOS capacitors [73]: aluminum gate, polysilicon-gate samples receiving "typical Si-gate processing," and special "hardened" Si-gate capacitors. Plotted in the figure is the

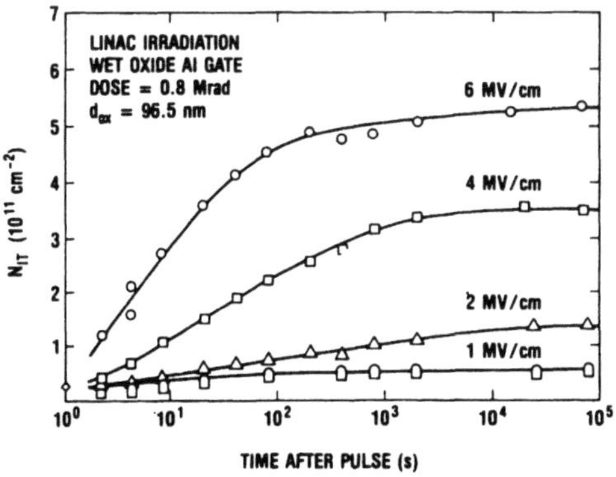

Figure 11. Integrated interface trap density between midgap and inversion surface potentials as function of time following pulsed electron-beam exposure for several values of oxide field [64]

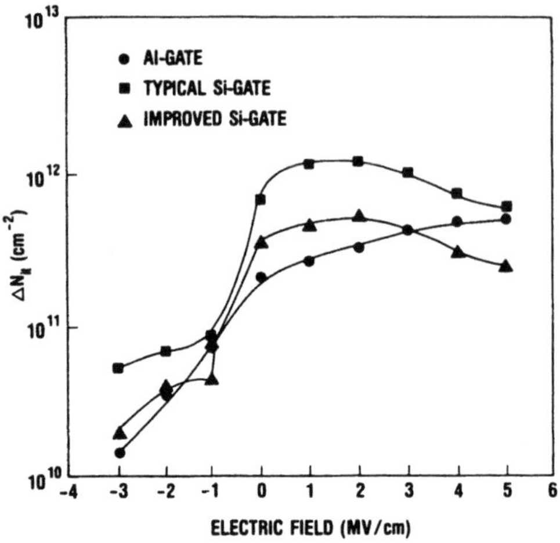

Figure 12. Oxide field dependence of radiation-induced interface trap buildup following 1 Mrad(SiO_2) irradiation in three MOS capapcitors having different gate structures [73]

increase in interface trap density between midgap and inversion following 1 Mrad(SiO_2) irradiation as a function of oxide electric field for both positive and negative polarities. The ΔN_{it} for the Si-gate samples contained both prompt and delayed time-dependent components. In all three samples, the ΔN_{it} is much less under negative bias polarity than positive polarity (note the log scale for ΔN_{it}). The N_{it} buildup for the Al-gate capacitors increases with increasing positive field in general agreement with the earlier data of figure 11. However, the Si-gate capacitors exhibit a rapid increase in N_{it} production for positive fields up to 1 MV/cm, but peaks in the range from 1 to 2 MV/cm and then drops off somewhat at higher fields. It is clear that there are some qualitative differences in the mechanisms of N_{it} buildup in Al-gate and Si-gate systems. These differences are not understood at the present time although they are likely to be associated with differences in the nature of the generation mechanisms for the prompt and delayed time-dependent components. Note also in figure 12 that for normal operating fields of 1 to 2 MV/cm, the ΔN_{it} for the typical commercial Si-gate process is about five times greater than that for the Al-gate samples. This increased ΔN_{it} is commonly observed for Si-gate technologies. The magnitude of the interface trap production in the "improved" Si-gate process in the figure is comparable to that in the Al-gate process.

We have attempted with the discussion thus far to illustrate some of the major observations that have been made concerning the buildup of radiation-induced interface traps, such as the existence of prompt and delayed time-dependent components, the strong bias polarity effect, the dependence on the magnitude for positive fields, and the strong dependence on oxide processing including the differences between aluminum gate and polysilicon gate structures. At the present time, the precise mechanisms responsible for the interface trap buildup are not well understood, and in fact the general area of SiO_2/Si interface traps and their generation under various stresses (radiation, high electric fields, hot carrier injection, etc) is a complex subject of intense current interest and debate. It is beyond the scope of the present discussion to attempt a review of this area and particularly to sort it out. We conclude this section simply by noting without discussion some other observations that have been made concerning the characteristics of radiation-induced interface traps and factors that may be contributing to their buildup.

(1) There is much evidence that N_{it} buildup is associated with the hole transport, trapping and annealing processes [63-69,74,75].

(2) There may also be a strong correlation of the buildup with hydrogen or water content of the oxide, and in fact arguments have been made that the time-dependent buildup is associated with the release (during the hole transport phase) and diffusion of hydrogen, or a water-related species, from the oxide bulk to the interface [76-81].

(3) The amount of interfacial strain--which is highly variable with processing--seems to be an important factor [56,69,82-84].

(4) The rate of buildup of the time-dependent N component increases with temperature; however, the final density at long times is essentially independent of temperature [64,66,79].

(5) Annealing of interface traps has not been observed at normal operating temperatures (64,66); significant annealing occurs only for T >150°C [105].

(6) In many cases, a sublinear dependence of ΔN_{it} on dose $\sim D^{2/3}$ has been observed, particularly for the delayed time-dependent component [64,67,86,87].

(7) The magnitude of N_{it} buildup is generally observed to decrease with decreasing oxide thickness [23,88-91].

(8) The energy distributions of radiation-induced interface traps are commonly observed to be U-shaped, with a minimum near midgap and rising toward both the conduction and valence band edges [63,64,92]. However, there is debate on the existence of intermediate structures, such as peaks [93].

(9) The interface traps seem to be amphoteric in nature, that is, they have a net negative charge (acceptor-like) when the Fermi level (E_F) at the surface is in the top half of the Si bandgap; they have net positive charge (donor-like) when E_F is in the bottom half of the bandgap; and they are charge neutral when E_F is near midgap [93-96]. This latter statement is important for charge separation (between ΔV_{ot} and ΔV_{it}) techniques.

(10) The evidence originating from ESR spectroscopy indicates that (at least for a large fraction) the microscopic structure of the interface trap is that of an Si atom at the interface back-bonded to three Si atoms on the silicon side and having a single broken dangling bond extending into the SiO_2). This is called the P_b center by the ESR spectroscopists [97]. The amphoteric nature of the states is attributed to the broken dangling bond containing 0, 1, or 2 electrons, with the charge state of the P_b center as a whole therefore being positive, neutral, or negative, respectively [93,95,98,99]. We should qualify these statements, however, to the extent that most of the ESR work to date has been done on [111] Si interfaces to obtain the maximum ESR signals; not much has been done on the technologically more important [100] Si interface. In fact, there is much

active discussion at present as to whether the midgap charge neutrality condition actually holds for the [100] Si interface.

(11) Since the radiation-induced interface traps lie at or very near the SiO_2/Si interface, they--especially when charged--introduce additional scattering centers for the conducting channel carriers in MOSFET's; these are in addition to the normal scattering from surface roughness, ionized dopants or other impurities in the surface region, and lattice vibrations (phonons). As a consequence, ΔN_{it} can lead to reduction of the carrier mobilities and hence to degradation in the transconductance [100]. This is normally a secondary effect in the low to moderate dose range (< 1 Mrad), but can become a serious factor in circuit degradation or failure at high dose levels by causing a significant decrease in circuit speed.

Superrecovery (Rebound Effect)

We have noted that deeply trapped holes (ΔN_{ot}) can anneal out over very long periods of time. We also mentioned that interface traps ΔN_{it} have not been observed to anneal out at normal operating temperatures. Furthermore, ΔN_{it} may continue to build up over time. Since for an n-channel device under positive bias the interface traps contribute net negative charge and therefore a positive contribution (ΔV_{it}) to the threshold voltage, these factors taken together may lead to the phenomenon called superrecovery [101] or rebound [102]. Namely, at the end of an irradiation, the threshold voltage for an n-channel MOSFET is generally shifted negatively, since the (negative) voltage contribution ΔV_{ot} from the trapped holes usually dominates ΔV_{it}, at least for doses up to 1 Mrad or so. However, as annealing of the trapped holes occurs and/or further buildup of the interface trapped charge occurs, the threshold voltage may actually anneal back to and past its initial preirradiation value; i.e., ΔV_T may go positive at some late time after cessation of the radiation. If the positive voltage excursion is sufficiently large, circuit failure may result from this effect as well as for a negative excursion.

The rebound effect is clearly demonstrated by the data shown in Figure 13 [102]. Here, the threshold voltage of a polysilicon gate n-channel MOSFET, as well as the separate components ΔV_{it} and ΔV_{ot}, is plotted against time after exposure to 1-Mrad Co^{60} irradiation, which is completed in 1 hour. A gate bias of + 10 V was maintained during both irradiation and the anneal time. Data for two temperatures are shown, for 25°C and for 125°C (which accelerates the rate of hole annealing). At neither temperature is there observed any annealing of ΔN_{it}, but rather some slight increase in ΔV_{it} after cessation of the irradiation is discernible. The important point, of course, is that V_T, which initially at the end of the irradiation is shifted negatively by about a volt, then returns in the positive direction far past its preirradiation value. In fact, after 100 hours of anneal at 125°C, ΔV_{ot} is completely annealed, leaving a final positive shift in V_T of +2.5 V, which is due entirely to interface trapped charge (most of which was already present at the end of irradiation).

There are some troublesome possibilities associated with this effect. For example, a device may initially be in failure at the end of a test irradiation because of a sufficiently large negative voltage shift, and then as the trapped holes anneal, the device may begin operating normally at some time and continue operating until the device fails from too great a positive threshold voltage excursion. We may also note in this regard that some hardening schemes in the past have attempted to rely on the

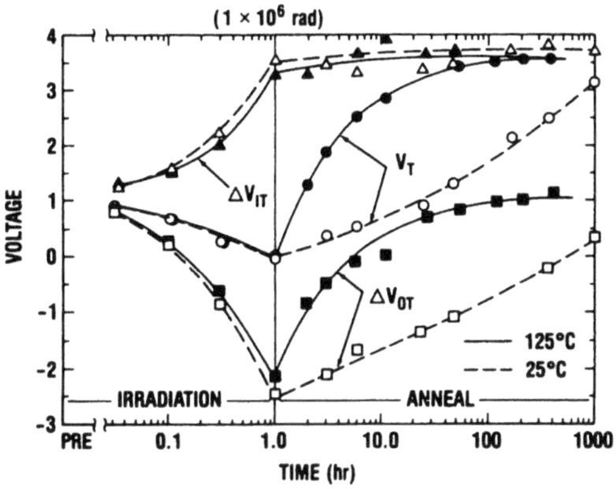

Figure 13. Threshold voltage shift of n-channel MOS transistor during irradiation and anneal with V_T separated into shifts due to interface trapped charge (ΔV_{it}) and oxide trapped charge (ΔV_{ot}). Long term response illustrates rebound effect [102]

compensation of trapped positive (hole) charge by net negative interface trapped charge. However, this type of hardening fix is risky at best, because utmost control must be maintained over the oxide processing to avoid too much radiation-induced N_{it} and consequently failure by rebound at long times. Such careful control of the processing is usually not possible in the long run for production lines. The optimum hardening approach is to minimize both ΔN_{ot} and ΔN_{it} [73]. This entire effect of "rebound" should be carefully evaluated when selecting ICs for long term (years) exposure in space using short term exposures to radiation such as typical Co^{60} irradiations.

Conclusion:

In this paper we have presented a general picture of the radiation response of MOS devices in an ionizing radiation environment such as the natural radiation environment of space. The focus of this paper has been on threshold voltage shifts from effects in the gate oxide. Even in this area there is more work needed to understand the physics and chemistry of interface state creation and the anneal of trapped positive charge. In addition, with the advancement of technology there is much research to be done to understand the leakage current effects in field oxides, trenches, sidewalls and back channels.

References

1. See, for example, the December issues of the IEEE Transaction on Nuclear Science (1965-1988).

2. G.A. Ausman, Jr., and F.B. McLean, Appl. Phys. Lett., 26, (1975), 173.

3. O.L. Curtis, J.R. Srour, and K.Y. Chiu, J. Appl. Phys., 45, (1974), 406.

4. H. H. Sander and B. L. Gregory, IEEE Trans. Nucl. Sci. NS-22 (1975), 2157.

5. H. E. Boesch, Jr., and J. M. McGarrity, IEEE Trans. Nucl. Sci. NS-23, (1976), 1520.

6. J. M. Benedetto and H. E. Boesch, Jr., IEEE Trans. Nucl. Scie. NS-33 (1986), 1318.

7. T.R. Oldham, J. Appl. Phys. (1985), 2695.j

8. M. von Smoluchowski, Ann. Phys. (Leipzig) 44 (1915), 1103.

9. L. Onsager, Phys. Rev. (1938), 554.

10. G. Jaffe, Ann. Phys. (Leipzig) 42 (1913), 303.

11. P. Langevin, Ann. Chim. Phys. 28 (1903), 289; 28 (1903), 433.

12. T. R. Oldham and J. M. McGarrity, IEEE Trans. Nucl. Sci. NS-28 (1981), 3975; NS-30 (1983), 4377.

13. H. E. Boesch, Jr., F. B. McLean, J. M. McGarrity, and G. A. Ausman, Jr., IEEE Trans. Nucl. Sci. NS-22 (1975), 2163.

14. R. C. Hughes, Appl. Phys. Lett. 26 (1975), 436.

15. 15. R. C. Hughes, E. P. ErNiss, and H. J Stein, IEEE Trans. Nucl. Sci. NS-22 (1975), 227.

16. F. B. McLean, G. A. Ausman, Jr., H. E. Boesch, Jr., and J. M. McGarrity, J. Appl. Phys. 47 (1976), 1529.

17. F. B. McLean, H. E. Boesch, Jr., and J. M. McGarrity, IEEE Trans. Nucl. Sci. NS-23 (1976), 1506.

18. J. R. Srour, S. Othmer, O. L. Curtis, Jr., and K. Y. Chiu, IEEE Trans. Nucl. Sci. NS-23 (1976), 1513.

19. O. L. Curtis, Jr. and J. R. Srour, J. Appl. Phys. 48 (1977), 3819.

20. R. C. Hughes, Phys. Rev. B 15 (1977), 2012.

21. H. E. Boesch, Jr., J. M. McGarrity, and F. B. McLean, IEEE Trans. Nucl. Sci. NS-25 (1978), 1012.

22. F. B. McLean, H. E. Boesch, Jr., and J. M. McGarrity, in The Physics of SiO_2 and Its Interfaces, S. T. Pantelides, ed., Pergamon Press, New York (1978), p.19.

23. H. E. Boesch, Jr., F. B. McLean, J. M. McGarrity, and P.S. Winokur IEEE Trans. Nucl. Sci. NS-25 (1978), 1239.

24. H. E. Boesch, Jr. and F. B. McLean, IEEE Trans. Nucl. Sci. NS-32 (1985), 3940.

25. E. W. Montroll and G. H. Weiss, J. Math. Phys. (1965), 167.

26. H. Scher and M. Lax, Phys. Rev. B 7 (1973), 4491.

27. H. Scher and E. W. Montroll, Phys. Rev. B 12 (1975), 2455.

28. G. Pfister and H. Scher, Adv. Phys. 27 (1978), 744.

29. F. B. McLean and G. A. Ausman, Jr., Phys. Rev. B 15 (1977), 1052.

30. G. F. Derbenwick and H. H. Sander, IEEE Trans. Nucl. Sci. NS-24 (1977), 2244.

31. G. F. Derbenwick and B. L. Gregory IEEE Trans. Sci. NS-22 (1975), 2151.

32. K. H. Zaininger, Appl. Phys. Lett. 8 (1966), 140.

33. A. S. Grove and E. H. Snow, Proc. IEEE 54 (1966), 894.

34. K. H. Zaininger, IEEE Trans. Nucl. Sci. NS-13 (1966), 237.j

35. R. J. Powell and G. F. Derbenwick, IEEE Trans, Nucl. Sci. NS-18 (1971), 99.

36. J. P. Mitchell and D. G. Denure, Solid State Elect. 16 (1973). 825.

37. J. M. Aitken, D. J. DiMaria, and D. R. Young, IEEE Trans. Nucl. Sci. NS-23 (1976), 1526.

38. J. L. Peel and R. C. Eden, IEEE Trans. Nucl. Sci. NS-18 (1971), 84.

39. J. P. Mitchell, IEEE Trans. Elec. Dev. ED-14 (1967), 764.

40. J. N. Churchill, T. W. Collins, and F. E. Holmstrom, IEEE Trans. Elec. Dev. Ed-21 (1974), 768.

41. T. W. Collins, F. E. Holmstrom, and J. N. Churchill, IEEE Trans. Nucl. Sci. NS-26 (1979), 5176.

42. T. R. Oldham, A. J. Lelis and F. B. McLean, IEEE Trans. Nucl. Sci. NS-33 (1986), 1203.

43. R. C. Hughes and C. H. Seager, IEEE Trans. Nucl. Sci. NS-30 (1983), 4049.

44. J. M. Aitken and D. R. Young, IEEE Trans. Nucl. Sci. NS-24 (1977), 2128.

45. H. E. Boesch, Jr., F. B. McLean, J. M. Benedetto, J. M. McGarrity, and W. E. Bailey, IEEE Trans. Nucl. Sci. NS-33 (1986), 1191.

46. C. M. Dozier and D. B. Brown, IEEE Trans. Nucl. Sci. NS-27 (1980), 1694.

47. H. E. Boesch, Jr., and F. B. McLean, IEEE Trans. Nucl. Sci. NS-32 (1985), 3940.

48. J. J. Tzou, J. Y. C. Sun, and C. T. Sah, Appl, Phys. Lett. 43, (1983), 861.

49. G. H. Sigel, Jr., E. J. Friebele, R. J. Ginther,a nd D. L. Griscom, IEEE Trans. Nucl. Sci. NS-21 (1974), 56.

50. S. Manzini and A. Modelli, in Insulating Films on Semiconductors J. F. Verwis and D. R. Wolters, eds., North-Holland, New York (1983), p. 112.

51. C. L. Marquardt and jG. H. Sigel, Jr., IEEE Trans. Nucl. Sci. NS-22 (1975), 2234.

52. P. M. Lenahan and P. V. Dressendorfer, IEEE Trans. Nucl. Sci. NS-30 (1983), 4602.

53. R. H. Silsbee, J. Appl. Phys. (1961), 1459.

54. F. J. Feigl, W. B. Fowler, and K. L. Yip, solid State Comm. 14 (1974), 225.

55. D. L. Griscom, Nucl. Instr. Methods Phys. Res. 131 (1984), 481.

56. F. J. Grunthaner, P. J. Grunthaner, and J. Maserjian, IEEE Trans. Nucl. Sci. NS-29 (1982). 1462.

57. P. S. Winokur, IEEE Trans. Nucl. Sci. NS-29 (1982), 2102.

58. P. S. Winokur, K. G. Kerris, and L. Harper, IEEE Trans. Nucl. Sci. NS-30 (1983), 4326.

59. J. M Benedetto, H. E. Boesch, Jr., F. B. McLean, and J. P. Mize, IEEE Trans. Nucl. Sci. NS-32 (1985), 3916.

60. E. C. Ross and J. T. Wallmark, RCA REview 30 (1969), 366.

61. F. B. McLean, A Direct Tunneling Model of Charge Transfer at the Insulator-Semiconductor Interface in MIS Devices, Harry diamond Laboratories, HDL-TR-1765 (October 1976).

62. A. H. Johnston and S. B. Roeske, IEEE Trans. Nucl. Sci. NS-33 (1986), 1487.

63. p. S. Winokur, J. M. McGarrity, and H. E. Boesch, Jr., IEEE Trans. Nucl. Sci. NS-23 (1976), 1580.

64. P. S. Winokur, H. E. Boesch,Jr., J. M. McGarrity, and F. B. McLean, IEEE Trans. Nucl. Sci. NS-24 (1977), 2113.

65. J. M. McGarrity, P. S. Winokur, H. E. Boesch, Jr., and F. B. McLean, in The Physics of SiO_2 and Its Interfaces, S. T. Pantelides, ed., Pergamon Press, New York (1978), p. 428.

66. P. S. Winokur, H. E. Boesch, Jr., J. M. McGarrity, and F. B. McLean, J. Appl. Phys. (1979), 3492.

67. P. S. Winokur and H. E. Boesch, Jr., IEEE Trans. Nucl. Sci. NS-27 (1980), 1647.

68. H. E. Boesch, Jr., and F. B. McLean, J. Appl. Phys. 60 (1986), 448.

69. P. S. Winokur, F. B. McLean, and H. E. Boesch, Jr., Physical Processes Associated with Radiation-Induced Interface States, Harry Diamond Laboratories, HDL-TR-2081 (April 1986).

70. H. E. Boesch, Jr., IEEE Trans. Nucl. Sci. NS-29 (1982), 1446.

71. H. E. Boesch, JR. and T. L. Taylor, IEEE Trans. Nucl. Sci. NS-31 (1983), 1178.

72. J. R. Schwank, P. S. Winokur, F. W. Sexton, D. M. Fleetwood, J. H. Perry, P.V. Dressendorfer, D. T. Sanders, and D. C. Turpin, IEEE Trans. Nucl. Sci. NS-33 (1986), 1178.

73. P. S. Winokur and M. M. Sokoloski, Appl. Phys. Lett. 28 (1976), 627.

75. S. K. Lai, J. Appl. Phys. 54 (1983), 2540.

76. C. T. Sah, IEEE Trans. Nucl. Sci. (1976), 1563.

77. A. G. Revesz, IEEE Trans. Nucl. Sci. NS-24 (1977), 2102.

78. C. M. Svensson, in The Physics of SiO_2 and Its Interfaces S. T. Pantelides, ed., Pergamon Press, NEw York (1978), p 328.

79. F. B. McLean, IEEE Trans. Nucl. Sci. NS-27 (1980), 1651.

80. D. L. Griscom, J. Appl. Phys. 58 (1985), 2524.

81. D. B. Brown, IEEE Trans. Nucl. Sci. NS-32 (1985), 3900.

82. E. P. EerNisse and G. F. Derbenwick, IEEE Trans. Nucl. Sci. NS-23 (1976), 1534.

83. V. Zekeriya and T. P. Ma, IEEE Trans. Nucl. Sci. NS-3 (1984), 1261.

84. F. Toyokawa, M. Tsukija, M. Sakamoto, and K. Koboyashi, IEEE Trans. Nucl. Sci. NS-33, (1986), 1210.

85. M. L. Reed and J. D. Plummer, IEEE Trans. Nucl. Sci. NS-30 (1983), 4054.

87. C. M. Dozier and D. B. Brown, IEEE Trans. Nucl. Sci. NS-30 (1983), 4382.

88. T. P. Ma, Appl. Phys. Lett. 27 (1975), 615.

89. C. R. Viswanathan and J. Maserjian, IEEE Trans. Nucl. Sci. NS-23 (1976), 1540.

90. N. Naruka, M. Yoshida, K. Maeghuchi, and H. Tango, IEEE Trans. Nucl. Sci. NS-27 (1980), 1647.

91. N. S. Saks, M. G. Ancona, and J. A. Modolo, IEEE Trans. Nucl. Sci. NS-33 (1986), 1185.

92. C. T. Sah, J. Y. C. Sun, and J. J. T. Tzou, J. Appl. Phys. 53 (1982), 8886.

93. E. H. Poindexter, G. J. Gerardi, M. E. Rueckel, P. J. Caplan, N. M. Johnson, and D. K. Biegelsen, J. Appl. Phys. 56 (1984), 2844.

94. T. P. Ma, G. Scoggan, and R. Leone, Appl. Phys. Lett. 27 (1975), 61.

95. P. M. Lenahan and P. V. Dressendorfer, J. Appl. Phys. 55 (1984), 3495.

96. M. Knoll, D. Braunig, and W. R. Fahrner, IEEE Trans. Nucl. Sci. NS-29 (1982), 1471.

97. Y. Nishi, Jpn. J. Appl. Phys. 5 (1966), 333.

98. E. H. Poindexter and P. J. Caplan, Prog. Surf. Sci. 14 (1983), 201.

99. P. M. Lenahan and P. V. Dressendorfer, Appl. Phys. Lett. 41 (1982), 542.

100. S. C. Sun and J. D. Plummer, IEEE Trans. Elec. Devices ED-27 (1980), 1497.

101. A. H. Johnston, IEEE Trans. Nucl. Sci. NS-31 (1984), 1427.

102. J. R. Schwank, P. S. Winokur, P. J. McWhorter, F. W. Sexton, P. V. Dressendorfer, and D. C. Turpin, IEEE Trans. Nucl. Sci. NS-31 (1984), 1434.

Crystalline Silicon Solar Cells

R. Mertens

IMEC, Kapeldreef 75, B-3030 Leuven, Belgium

1. INTRODUCTION

Photovoltaic cells can be divided in crystalline and amorphous types. Amorphous materials, such as amorphous silicon, offer several advantages to make solar cells, such as the low deposition temperature, the high absorption coefficient and the compatibility with foreign substrates, making them attractive materials for thin film cells. On the other hand, crystalline silicon yields higher potential and practical cell efficiencies and a better cell stability.
Today the majority of solar cells are of the crystalline type; only in the field of consumer electronics (e.g. calculators) amorphous silicon dominates. Most scientists believe that it will take a considerable number of years before crystalline solar cells will gradually be substituted by amorphous cells in power applications (> 100 W).

This paper is organized as follows: first the efficiency of a solar cell is discussed in terms of the thermodynamic limit and in terms of the achievable technological limits. This is followed by a section dealing with high efficiency silicon solar cells. Section 4 is concerned with semicrystalline silicon solar cells. Finally some conclusions are given in section 5.

2. CRYSTALLINE SILICON SOLAR CELL EFFICIENCY

The key parameter that determines the economics of a photovoltaic system is the module efficiency. The break-even module cost, for which electricity can be produced at a given price per kWh, is a strong function of the module efficiency. Module efficiencies in the 15 to 20 % range will be required in most applications to reach cost levels comparable to those of conventional energy sources.

Crystalline semiconductors are characterized by a well defined bandgap E_g that separates the conduction band in which the electrons are free from the valence band in which the electrons are immobile. The bandgap corresponds to the binding energy of the outer electrons.

It has been proven by Shockley (1) that the laws of thermodynamics allow us to calculate a one-to-one relation between the theoretical cell efficiency and the bandgap of the material, if the light absorbing part of the cell is entirely made of one material only. The efficiency is defined as the ratio between the generated electrical power and the incident solar power. In reality the conversion efficiencies may be considerably smaller since the

Springer Series in Materials Science, Vol. 13
Semiconductor Silicon Editors: G. C. Harbeke and M. J. Schulz

theoretical calculation only takes unavoidable losses into account. Practical solar cells are made with non-ideal materials and technologies. Table I compares, for the case of silicon, the maximum efficiency obtained to date with the thermodynamic limit and with the maximum semi-empirical efficiency, corresponding to a reasonable estimate of the practical technological limits.

Table I: Maximum conversion efficiencies of crystalline silicon solar cells

	efficiency (%)
thermodynamic limit	32
semi-empirical limit	24
experimental result	22

In the case of silicon the semi-empirical limit is estimated to be 75% of the thermodynamic limit, the difference being caused by the non-ideality of the practical materials and technology. If more than one semiconductor material is used to make the active part of the cell, as is the case for multijunction structures, the maximum efficiency can be higher.

3. HIGH EFFICIENCY SILICON SOLAR CELLS

In recent years a breakthrough has been achieved in the field of high efficiency crystalline silicon solar cells. Two groups, respectively at Stanford University and at the University of New South Wales, did pioneering work to break through the 20% barrier of energy conversion. These achievements were possible because of the progress made in the device physics of high efficiency silicon solar cells, particularly because of the better understanding of heavy doping effects in silicon and of surface recombination. An even more important step was the progress in technology related to the material (large diffusion lengths), surface passivation techniques and the use of double layer antireflection coatings.

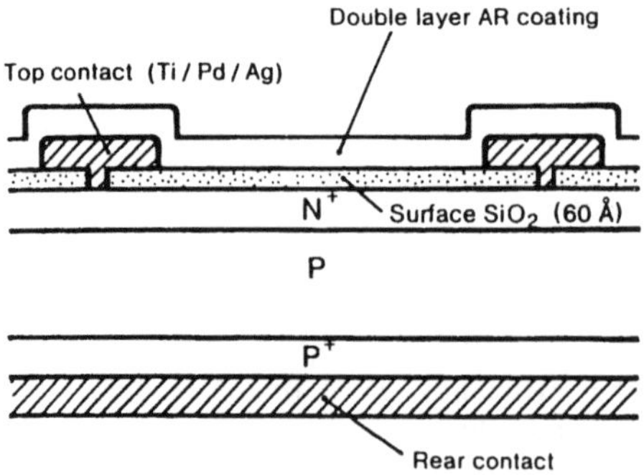

Fig. 1 Diagram of the passivated emitter solar cell (after (2))

Green (2) obtained very large open circuit voltages with the PESC cell (Passivated Emitter Solar Cell), schematically represented in figure 1. In this cell the surface is passivated by a 60 Angström SiO_2 layer. Recombination under the metal fingers is limited by using contact holes that are considerably smaller than the finger width.

A further improvement, also introduced by Green, was the use of micro-grooving of the cell surface to lower the reflection loss and to increase the collection efficiency. Such a cell is schematically represented in figure 2; this cell was the first non-concentrating silicon cell to exceed 20% energy conversion efficiency. (AM 1.5 solar spectrum).

For concentrating silicon solar cells Stanford has been the leading group. Sinton et al. (3) developed the Point Contact Solar Cell (Fig. 3).

The structure is of the interdigitated back contact type. All junctions and contacts are at the backside. The junction and metal contact areas are kept very small and a textured light-trapping scheme is used to obtain optical confinement. This type of cell has yielded a conversion effiency of more than 27% at 100 suns concentration. Even higher efficiencies can be expected from thin solar cells with optical confinement. The potentially higher efficiencies in such cells result from the smaller recombination volume; the thermodynamically limiting efficiencies for such thin cells are larger than for thick cells. Fig. 4 shows a possible practical implementation

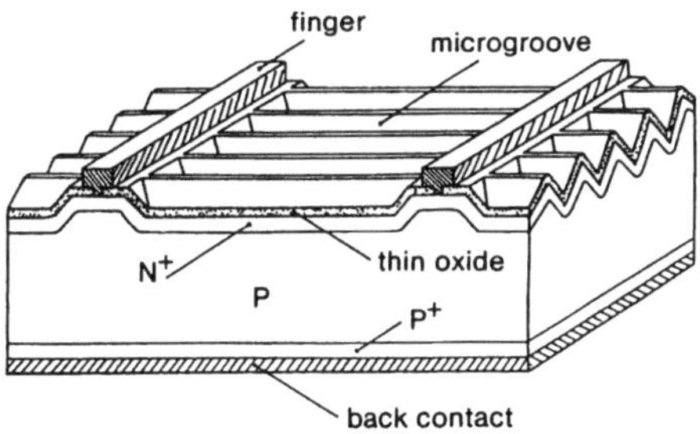

Fig. 2 The microgrooved PESC cell (after (2))

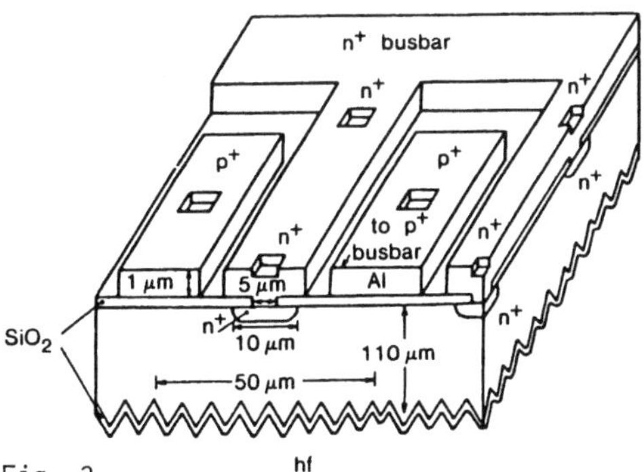

Fig. 3

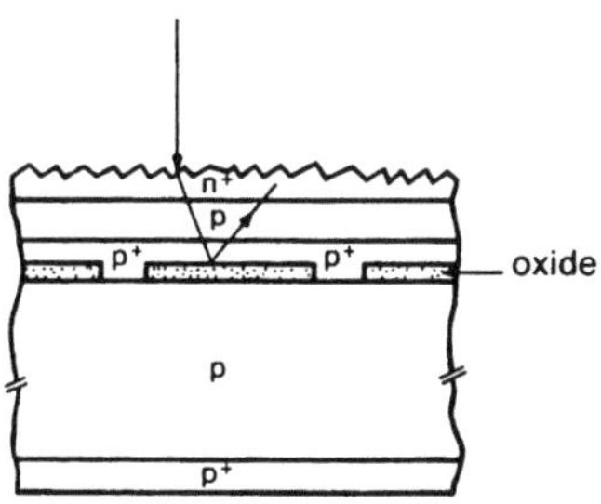

Fig. 4

Fig. 3 Diagram of the Point Contact Solar Cell (after (3))

Fig. 4: Principle of a thin film crystalline silicon solar cell with optical confinement. Such a structure potentially yields very high efficiencies

of such a cell. A thin film of silicon (20 to 50 μm thick) is epitaxially grown on an oxidized wafer.

In order to obtain epitaxial growth the oxide is selectively etched such that lateral epitaxial overgrowth is possible.

4. SEMICRYSTALLINE SILICON SOLAR CELLS

The previous section dealt with maximum efficiencies that have been obtained in different laboratories. This section discusses technologies to lower the cost of crystalline silicon solar cells.

Although silicon is after oxygen the most abundant material in the earth's crust monocrystalline 'electronic-grade' wafers are expensive. They necessitate energy inefficient production steps characterized by a low throughput, involving capital intensive equipment. Monocrystalline electronic-grade silicon wafers are at least an order of magnitude too expensive to be used in large scale photovoltaic applications. Fortunately experiments have indicated that solar cells do not require ultra-high purity and perfectly crystalline wafers to yield reasonable efficiencies. Therefore cheaper silicon sheets are being developed using less pure polysilicon feedstock and new crystallization techniques.

Less pure, cheaper polysilicon may be produced by a number of techniques. One of the more promising methods is the use of ultrapure quartz and carbon as starting materials such that after reduction sufficiently pure silicon is obtained. Another technique uses aluminium rather than carbon as reactive element to reduce quartz. The overall aim of these research programs is to produce sufficiently pure silicon of 'solar grade' quality at a price of 10 - 15 $/kg.

The new crystallization techniques can be divided in three categories: ingot, ribbon and thin film.

New ingot methods are being developed as an alternative to the conventional crystal pulling. The most succesful technique is the casting method that offers a much larger throughput, a lower capital investment and a lower energy consumption than the crystal pulling method. To date a production of cast silicon exists; the market price is around 180$/m^2 compared to the 400 $/m^2 for pulled silicon. To some extent this cost advantage is partly offset by the lower cell efficiency on cast material, which is of semicrystalline and not of monocrystalline nature.

Ribbon technologies directly produce silicon sheets of appropriate thickness (200 microns) eliminating the slicing operation. Many alternatives for ribbon growth have been proposed; however most of them are characterized by a low throughput such that they are not industrially viable. New technologies such as foil casting (4) or S-web ribbons (5) do not have this disadvantage. Todate several ribbon-type silicon solar cells are in production or in pilotproduction stage.

A third alternative (6) to produce silicon sheet for solar cells is to deposit a thin film of 10 to 20 microns of high quality silicon on a conductive substrat, which has to be cheap (figure 5).

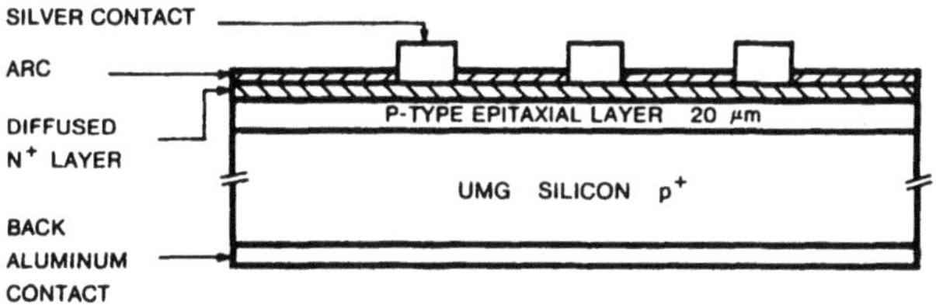

Fig. 5 Epitaxial silicon solar cell on an upgraded metallurgical grade substrate (6)

This is possible because 85% of the useful solar radiation is absorbed within the first 20 microns of a silicon wafer, which is normally several hundreds microns thick only for mechanical reasons. The silicon film is deposited from the vapour phase. As substrates one normally uses silicon wafers of metallurgical grade quality which ar potentially cheap but too impure for direct cell fabrication.

An important step in the manufacturing of silicon solar cells is the fabrication of the cell in the silicon sheet. In recent years major cost reductions have been possible by using high-throughput processes, without using vacuum steps, with high yield and by introducing automation. Today most cell manufacturers use screenprinting technology for metallization (6). A variety of techniques is being used for the junction formation and for the deposition of the antireflective coating.

After cell fabrication and testing the cells must be assembled and encapsulated in module form. Here too, high throughput processes, amenable to automation, have been introduced.

To date crystalline silicon modules cost, if purchased in large quantities, 5 to 6 $ per peak watt; 40 to 50% is attributed to the silicon sheet whereas cell manufacturing and module encapsulation are responsible for 25 to 30% each. As a result of ongoing development programs one can expect that this price will decrease in the coming years. Most experts believe that module prices between 2 and 2.5 $ per peak Watt will be reached with crystalline silicon technology; the ultimate price level will be determined by the outcome of the research and development programmes, especially those on silicon materials.

5. CONCLUSIONS

The economics of a photovoltaic system are very sensitive to the module efficiency. Module efficiencies in the 15 to 20% range will be required in most applications to reach cost levels comparable to those of conventional energy sources.
Considerable progress has been made in the field of cell conversion efficiency since efficiencies higher than 20% have been reported by several laboratories.
In a parallel effort, cheaper silicon wafer production techniques and cell manufacturing methods have been introduced making crystalline silicon solar cells still the most promising option to reach long term low cost goals.

REFERENCES

1. W. Shockley and H. Queisser: J. Appl. Phys., 32, p. 510 (1961)
2. M.A. Green, "High Efficiency Silicon Solar Cells": Trans. Tech. Publications (1987)
3. R.A. Sinton, Y. Kwark and R.M. Swanson: 14th Project Integration Meeting, Photovoltaic Concentrator Technology project, p. 117 (1986)
4. D. Helmreich and J. Geissler: Proceedings of the 5th E.C. Photovoltaic Solar Energy Conference, Athens, Greece, p. 955 (1983).
5. J. Grabmaier, G. Greiger and R. Falckenberg: Proceedings of the First International Photovoltaic Solar Energy Conference, Kobe, Japan, p. 63 (1984)
6. Van Overstraeten and R.P. Mertens: Physics, Technology and Use of Photovoltaics, Adam Hilger Ltd (1986)

Index of Contributors

MIX
Papier aus verantwortungsvollen Quellen
Paper from responsible sources
FSC® C105338

If you have any concerns about our products,
you can contact us on
ProductSafety@springernature.com

In case Publisher is established outside the EU,
the EU authorized representative is:
Springer Nature Customer Service Center GmbH
Europaplatz 3, 69115 Heidelberg, Germany

Printed by Libri Plureos GmbH
in Hamburg, Germany